PENNSYLVANIA COLLEGE OF TECHNOLOGY LIBRARY

5 0608 01170953 1

DISCARDED

O9-ABF-086

SIGNIFICANT CHANGES TO THE

INTERNATIONAL BUILDING CODE®

2012 EDITION

DOUGLAS W. THORNBURG, AIA

JOHN R. HENRY, P.E.

& JAY WOODWARD

DELMAR
CENGAGE Learning™

Australia • Brazil • Japan • Korea • Mexico • Singapore • Spain • United Kingdom • United States

Madigan Library
Pennsylvania College
of Technology

One College Avenue
Williamsport, PA 17701

NOV 0 6 2013

Significant Changes to the International Building Code® 2012 Edition
International Code Council

Douglas W. Thornburg, John R. Henry, and Jay Woodward

Delmar Cengage Learning Staff :

Vice President, Technology and Trades
Professional Business Unit: Gregory L. Clayton

Director of Building Trades:
Taryn Zlatin McKenzie

Executive Editor: Robert Person

Development: Nobina Preston

Director of Marketing: Beth A. Lutz

Marketing Manager: Marissa Maiella

Marketing Communications Manager:
Nicole McKasty Stagg

Production Director: Wendy Troeger

Senior Content Project Manager: Stacey Lamodi

Art Director: Benjamin Gleeksman

ICC Staff :

Senior Vice President, Business and Product
Development: Mark A. Johnson

Deputy Senior Vice President, Business and
Product Development: Hamid Naderi

Technical Director, Product Development:
Doug Thornburg

Director, Project and Special Sales:
Suzane Nunes Holten

Senior Marketing Specialist: Dianna Hallmark

© 2012 International Code Council

Line illustrations copyright © 2012 by International Code Council

ALL RIGHTS RESERVED. No part of this work covered by the copyright herein may be reproduced, transmitted, stored or used in any form or by any means graphic, electronic, or mechanical, including but not limited to photocopying, recording, scanning, digitizing, taping, Web distribution, information networks, or information storage and retrieval systems, except as permitted under Section 107 or 108 of the 1976 United States Copyright Act, without the prior written permission of the publisher.

For product information and technology assistance, contact us at
Cengage Learning Customer & Sales Support, 1-800-354-9706

For permission to use material from this text or product,
submit all requests online at **www.cengage.com/permissions**
Further permissions questions can be emailed to
permissionrequest@cengage.com

Library of Congress Control Number: 2011921377

ISBN-13: 978-1-111-54246-7

ISBN-10: 1-111-54246-5

ICC World Headquarters
500 New Jersey Avenue, NW
6th Floor
Washington, D.C. 20001-2070
Telephone: 1-888-ICC-SAFE (422-7233)
Website: http://www.iccsafe.org

Delmar
Executive Woods
5 Maxwell Drive
Clifton Park, NY 12065
USA

Cengage Learning is a leading provider of customized learning solutions with office locations around the globe, including Singapore, the United Kingdom, Australia, Mexico, Brazil, and Japan. Locate your local office at
www.cengage.com/global

Cengage Learning products are represented in Canada by Nelson Education, Ltd.

Visit us at **www.InformationDestination.com**

For more learning solutions, please visit our corporate website at
www.cengage.com

Notice to the Reader

Publisher does not warrant or guarantee any of the products described herein or perform any independent analysis in connection with any of the product information contained herein. Publisher does not assume, and expressly disclaims, any obligation to obtain and include information other than that provided to it by the manufacturer. The reader is expressly warned to consider and adopt all safety precautions that might be indicated by the activities described herein and to avoid all potential hazards. By following the instructions contained herein, the reader willingly assumes all risks in connection with such instructions. The publisher makes no representations or warranties of any kind, including but not limited to, the warranties of fitness for particular purpose or merchantability, nor are any such representations implied with respect to the material set forth herein, and the publisher takes no responsibility with respect to such material. The publisher shall not be liable for any special, consequential, or exemplary damages resulting, in whole or part, from the readers' use of, or reliance upon, this material.

Printed in China
3 4 5 6 7 17 16 15 14 13

Contents

Preface

The purpose of *Significant Changes to the International Building Code® 2012 Edition* is to familiarize building officials, fire officials, plans examiners, inspectors, design professionals, contractors, and others in the construction industry with many of the important changes in the 2012 International Building Code® (IBC®). This publication is designed to assist those code users in identifying the specific code changes that have occurred and, more important, understanding the reason behind the change. It is also a valuable resource for jurisdictions in their codeadoption process.

Only a portion of the total number of code changes to the IBC are discussed in this book. The changes selected were identified for a number of reasons, including their frequency of application, special significance, or change in application. However, the importance of those changes not included is not to be diminished. Further information on all code changes can be found in the *Code Changes Resource Collection,* available from the International Code Council® (ICC®). The resource collection provides the published documentation for each successful code change contained in the 2012 IBC since the 2009 edition.

This book is organized into seven general categories, each representing a distinct grouping of code topics. It is arranged to follow the general layout of the IBC, including code sections and section number format. The table of contents, in addition to providing guidance in use of this publication, allows for quick identification of those significant code changes that occur in the 2012 IBC.

Throughout the book, each change is accompanied by a photograph, an application example, or an illustration to assist and enhance the reader's understanding of the specific change. A summary and a discussion of the significance of the changes are also provided. Each code change is identified by type, be it an addition, modification, clarification, or deletion.

The code change itself is presented in a format similar to the style utilized for code-change proposals. Deleted code language is shown with a strike-through, whereas new code text is indicated by underlining. As a result, the actual 2012 code language is provided, as well as a comparison with the 2009 language, so the user can easily determine changes to the specific code text.

As with any code-change text, *Significant Changes to the International Building Code 2012 Edition* is best used as a study companion to the 2012 IBC. Because only a limited discussion of each change is provided, the code itself should always be referenced in order to gain a more comprehensive understanding of the code change and its application.

The commentary and opinions set forth in this text are those of the authors and do not necessarily represent the official position of the ICC. In addition, they may not represent the views of any enforcing agency, as such agencies have the sole authority to render interpretations of the IBC. In many cases, the explanatory material is derived from the reasoning expressed by the code-change proponent.

Comments concerning this publication are encouraged and may be directed to the ICC at *significantchanges@iccsafe.org*.

About the International Building Code®

Building officials, design professionals, and others involved in the building construction industry recognize the need for a modern, up-to-date building code addressing the design and installation of building systems through requirements emphasizing performance. The *International Building Code* (IBC), in the 2012 edition, is intended to meet these needs through model code regulations that safeguard the public health and safety in all communities, large and small. The IBC is kept up to date through the open code-development process of the International Code Council (ICC). The provisions of the 2009 edition, along with those code changes approved through 2010, make up the 2012 edition.

The ICC, publisher of the IBC, was established in 1994 as a nonprofit organization dedicated to developing, maintaining, and supporting a single set of comprehensive and coordinated national model building construction codes. Its mission is to provide the highest quality codes, standards, products, and services for all concerned with the safety and performance of the built environment.

The IBC is 1 of 14 International Codes® published by the ICC. This comprehensive building code establishes minimum regulations for buildings systems by means of prescriptive and performance-related provisions. It is founded on broad-based principles that make possible the use of new materials and new building designs. The IBC is available for adoption and use by jurisdictions internationally. Its use within a governmental jurisdiction is intended to be accomplished through adoption by reference, in accordance with proceedings establishing the jurisdiction's laws.

Acknowledgments

A special thank you is extended to Scott Stookey, Senior Technical Staff with ICC for his assistance with the fire protection portions of this text. Thanks also to ICC staff members Alan Carr, Kim Paarlberg, Bill Rehr, and Kermit Robinson for their valued review and input.

About the Authors

Douglas W. Thornburg, AIA, CBO
International Code Council
Technical Director of Product Development

Douglas W. Thornburg is the Technical Director of Product Development for the International Code Council (ICC), where he provides leadership in technical development and positioning of support products for the council. In addition, Doug develops and reviews technical products, reference books, and resource materials relating to construction codes and their supporting documents. Prior to employment with the ICC in 2004, he spent nine years as a code consultant and educator on building codes. Formerly Vice-President/Education for the International Conference of Building Officials (ICBO), Doug continues to present building code seminars nationally and has developed numerous educational texts and resource materials, including *the IBC Handbook—Fire- and Life-Safety Provisions*. He was presented with ICC's inaugural Educator of the Year Award in 2008, in recognition of his outstanding contributions to education and professional development. A graduate of Kansas State University and a registered architect, Doug has more than 30 years of experience in building code training and administration, including 10 years with the ICBO and 5 years with the City of Wichita, Kansas. He is certified as a building official, building inspector, and plans examiner, as well as in seven other code enforcement categories.

John R. Henry, P.E.
International Code Council
Principal Staff Engineer

John R. Henry is a Principal Staff Engineer with the International Code Council (ICC) Business and Product Development Department, where he is responsible for the research and development of technical resources pertaining to the structural engineering provisions of the *International Building Code* (IBC). John also develops and presents technical seminars on the structural provisions of the IBC. He has a broad range of experience that includes structural design in private practice, plan-check engineering with consulting firms and building department jurisdictions, and 14 years as an International Conference of Building Officials (ICBO)/ICC Staff Engineer. John graduated with honors from California State University in Sacramento with a Bachelor of Science Degree in Civil Engineering and is a Registered Civil Engineer in the State of California. He is a member of the American Society of Civil Engineers (ASCE) and the Structural Engineers Association of California (SEAOC) and is an ICC Certified Plans Examiner. John has written several articles on the structural provisions of the IBC that have appeared in *Structure Magazine* and *Structural Engineering and Design* magazine's Code Series. He is also the coauthor with S. K. Ghosh, PhD, of the IBC Handbook—Structural Provisions.

Jay Woodward
International Code Council
Senior Staff Architect

Jay is a senior staff architect with the ICC's Business and Product Development department and works out of the Lenexa, Kansas, Distribution Center. His current responsibilities include serving as the Secretariat for the ICC A117.1 standard committee and assisting in the development of new ICC publications.

With more than 28 years of experience in building design, construction, code enforcement, and instruction, Jay's experience provides him with the ability to address issues of code application and design for code enforcement personnel as well as architects, designers, and contractors. Jay has previously served as the Secretariat for the ICC's *International Energy Conservation Code* and the *International Building Code's* Fire Safety Code Development committee.

A graduate of the University of Kansas and a registered architect, Jay has also worked as an architect for the Leo A. Daly Company in Omaha, Nebraska; as a building Plans Examiner for the City of Wichita, Kansas; and as a Senior Staff Architect for the International Conference of Building Officials (ICBO) prior to working for the ICC. He is also author of *Significant Changes to the A117.1 Accessibility Standard 2009 Edition.*

About the ICC

The ICC is a nonprofit membership association dedicated to protecting the health, safety, and welfare of people by creating better buildings and safer communities. The mission of the ICC is to provide the highest-quality codes, standards, products, and services for all concerned with the safety and performance of the built environment. The ICC is the publisher of the family of the International Codes® (I-Codes®), a single set of comprehensive and coordinated model codes. This unified approach to building codes enhances safety, efficiency, and affordability in the construction of buildings. The ICC is also dedicated to innovation, sustainability, and energy efficiency. In addition, the ICC Evaluation Service, an ICC subsidiary, issues Evaluation Reports for innovative products and Reports of Sustainable Attributes Verification and Evaluation (SAVE).

Headquarters:
500 New Jersey Avenue, NW, 6th Floor
Washington, DC 20001-2070

District Offices:
Birmingham, AL; Chicago. IL; Los Angeles, CA

1-888-422-7233
www.iccsafe.org

PART 1

Administration

Chapters 1 and 2

- ■ **Chapter 1** Administration
- ■ **Chapter 2** Definitions

The provisions of Chapter 1 address the application, enforcement, and administration of subsequent requirements of the code. In addition to establishing the scope of the International Building Code (IBC), the chapter identifies which buildings and structures come under its purview. A building code, as with any other code, is intended to be adopted as a legally enforceable document to safeguard health, safety, property, and public welfare. A building code cannot be effective without adequate provisions for its administration and enforcement. Chapter 2 provides definitions for terms used throughout the IBC. Codes, by their very nature, are technical documents, and as such, literally every word, term, and punctuation mark can add to or change the meaning of the intended result. ■

102.4

Conflicting Provisions between Codes and Standards

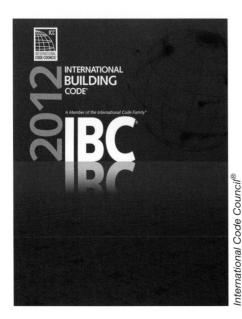

Codes vs. Standards

CHANGE TYPE: Clarification

CHANGE SUMMARY: The hierarchy between the IBC and its referenced standards has been further described to clarify the extent of a referenced standard's application.

2012 CODE: 102.4 Referenced Codes and Standards. The codes and standards referenced in this code shall be considered part of the requirements of this code to the prescribed extent of each such reference and as further regulated in Sections 102.4.1 and 102.4.2.

102.4.1 Conflicts. Where ~~differences~~ conflicts occur between provisions of this code and referenced codes and standards, the provisions of this code shall apply.

102.4.2 Provisions in Referenced Codes and Standards. Where the extent of the reference to a referenced code or standard includes subject matter that is within the scope of this code or the International Codes listed in Section 101.4, the provisions of this code or the International Codes listed in Section 101.4, as applicable, shall take precedence over the provisions in the referenced code or standard.

CHANGE SIGNIFICANCE: The IBC is, for the most part, a performance-based code, relying on numerous referenced standards to assist in its application. Where standards are referenced in the body of the IBC, the applicable portions of the standard relating to the specific code provision under consideration are considered a part of the code. Conflicts between the IBC and the various standards it references are to be expected, as there is not necessarily a conscious effort to see that the publications are completely compatible with each other. As a result, it is critical that the code indicate that its provisions are to be applied over those of a referenced standard where such conflicts exist. Additional language has been provided to address the hierarchy between the IBC and its referenced standards and to clarify the extent of a referenced standard's application.

New Section 102.4.2 expands upon the provisions by making it clear that, even if a referenced standard contains requirements that parallel the IBC in the standard's own duly referenced section(s), the provisions of the IBC will always take precedence. It is intended that the requirements of a referenced standard supplement the IBC provisions in those areas not already addressed by the code. In those cases where parallel or conflicting requirements occur, the IBC provisions are always to be applied.

As an example, IBC Section 415.8.4 mandates that "the construction and installation of dry cleaning plants shall be in accordance with the requirements of the IBC; the *International Mechanical Code* (IMC); the *International Plumbing Code* (IPC); and NFPA 32, *Standard for Dry Cleaning Plants*. Although NFPA 32 addresses construction and installation criteria for dry cleaning plants, only those portions of the standard that are not addressed within the IBC, IMC, and IPC are applicable. The requirements in NFPA 32 are intended to simply supplement the construction and installation provisions established in the International codes.

CHANGE TYPE: Addition

CHANGE SUMMARY: Mandatory conditions regarding the evaluation of modifications to flood-resistant construction provisions are now specifically identified and the building official has been given the authority to make such decisions.

2012 CODE: IBC 104.10 Modifications. Wherever there are practical difficulties involved in carrying out the provisions of this code, the building official shall have the authority to grant modifications for individual cases, upon application of the owner or owner's representative, provided the building official shall first find that special individual reason makes the strict letter of this code impractical and the modification is in compliance with the intent and purpose of this code and that such modification does not lessen health, accessibility, life and fire safety, or structural requirements. The details of action granting modifications shall be recorded and entered in the files of the department of building safety.

104.10.1 Flood Hazard Areas. The building official shall not grant modifications to any provision required in flood hazard areas as established by Section 1612.3 unless a determination has been made that:

1. A showing of good and sufficient cause that the unique characteristics of the size, configuration, or topography of the site render the elevation standards of Section 1612 inappropriate.

2. A determination that failure to grant the variance would result in exceptional hardship by rendering the lot undevelopable.

3. A determination that the granting of a variance will not result in increased flood heights, additional threats to public safety, extraordinary public expense, cause fraud on or victimization of the public, or conflict with existing laws or ordinances.

4. A determination that the variance is the minimum necessary to afford relief, considering the flood hazard.

5. Submission to the applicant of written notice specifying the difference between the design flood elevation and the elevation to which the building is to be built, stating that the cost of flood insurance will be commensurate with the increased risk resulting from the reduced floor elevation and stating that construction below the design flood elevation increases risks to life and property.

CHANGE SIGNIFICANCE: The building official is granted authority to permit modifications to the requirements of the code under certain specified circumstances. Although the provisions of Section 104.10 have never limited the extent of such authority, Appendix Section G105 has previously indicated that variances regarding flood-resistant construction shall only be issued by the board of appeals based upon compliance with five stated conditions of issuance. These conditions have been relocated to the body of the IBC regarding the evaluation of modifications to flood-resistant construction provisions and the building official has properly been given the authority to make such decisions.

104.10.1
Code Modifications for Flood Hazard Areas

Flood hazard area

104.10.1 continues

104.10.1 continued   In order to be consistent with the requirements of the National Flood Insurance Program (NFIP) for the allowance of variances, Section 104.10.1 now identifies the specific issues the building official must consider prior to granting any modification to a flood hazard area provision. The NFIP requires that the authority having jurisdiction carefully consider a number of issues and make an informed determination as to the worthiness of the requested modification. Specifically assigning this responsibility to the building official rather than the board of appeals is consistent with the long-standing application of the IBC.

105.2, #B2
Fences Exempt from Permits

CHANGE TYPE: Modification

CHANGE SUMMARY: The allowance for fences to be exempt from permit requirements based on height was revised in a manner that maintains the spirit of the provision while at the same time allowing for minor variances that may exceed the previously-established fence height limit of 6 feet.

2012 CODE: 105.2 Work Exempt from Permit. Exemptions from permit requirements of this code shall not be deemed to grant authorization for any work to be done in any manner in violation of the provisions of this code or any other laws or ordinances of this jurisdiction. Permits shall not be required for the following:

Building:

1. (no changes to text)

2. Fences not over ~~6 feet (1829 mm)~~ 7 feet (2134 mm) high.

3.-13. (no changes to text)

CHANGE SIGNIFICANCE: A permit is typically required where any work regulated by the IBC takes place. The purpose of a permit is to cause the work to be reviewed, inspected, and approved to determine compliance with the code. There is some work that, although regulated by the code, is viewed to be so minor in nature that the issuance of a permit is deemed to be unnecessary. Fences are classified as Group U structures and are regulated to a limited degree. As a result, a building permit is required for the installation or construction of a fence unless it does not exceed an established height.

Traditionally, those fences not exceeding 6 feet in height were exempt from permit requirements due to the limited hazard or concern posed by such structures. However, the construction or installation of fences greater than 6 feet in height has required that the owner or authorized agent obtain a building permit in order to evaluate and approve the fence for potential issues such as structural integrity, effect on required light and ventilation, and fire department access. The 6-foot threshold has allowed the vast majority of fences to fall under the exemption from permits, including most fences in residential applications. However, with the potential variations in adjacent ground level and variable methods of construction, it is not uncommon for 6-foot-high fences to have portions above the 6-foot limit. Decorative elements at support posts also often exceed the 6-foot limitation. Therefore, a modification was made that maintains the spirit of the provision while at the same time allowing for minor variances that may exceed the established fence height limit.

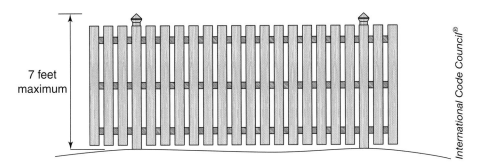

7 feet maximum

Fence not requiring a permit

202

Definitions

CHANGE TYPE: Clarification

CHANGE SUMMARY: For consistency and usability purposes, all definitions of terms specifically defined in the IBC have now been moved to a single location in Chapter 2.

2012 CODE: 310.2 Definitions. ~~The following words and terms shall, for the purposes of this section and as used elsewhere in this code, have the meanings shown herein.~~ The following terms are defined in Chapter 2:

CONGREGATE LIVING FACILITIES. ~~A building or part thereof that contains sleeping units where residents share bathroom and/or kitchen facilities.~~

202 Definitions.

CONGREGATE LIVING FACILITIES. ~~See Section 310.2.~~ A building or part thereof that contains sleeping units where residents share bathroom and/or kitchen facilities.

(The preceding example reflects the format change for all definitions in the IBC.)

CHANGE SIGNIFICANCE: Throughout the IBC, specific terms are used in a manner that differs from their ordinarily accepted meaning. Such terms are necessarily defined in order to clarify their meaning within the context of the code. In the past, these definitions have been found in various locations throughout the IBC. For consistency and usability purposes, all definitions have now been moved to a single location in Chapter 2.

Congregate living facility

International Code Council®

There are more than 700 definitions in the IBC. Historically, approximately 10 percent of the definitions have been located in Chapter 2. The other 90 percent have been scattered in more than 40 locations throughout the remainder of the code. With all defined terms now italicized, it is likely that more code users will research the definitions rather than rely on their assumption of the definition of a term. By relocating all of the definitions to a single location in Chapter 2, the IBC will become more user-friendly. It should be noted that only the definitions themselves have been relocated. The specifically defined terms continue to be listed in their previous locations to remind the code user that a definition of the term can be found in Chapter 2.

The application of the International Building Code to a structure is typically initiated through the provisions of Chapters 3, 5, and 6. Chapter 3 establishes one or more occupancy classifications based upon the anticipated uses of a building. The appropriate classifications are necessary to properly apply many of the code's nonstructural provisions. The requirements of Chapter 6 deal with classification as to construction type, based on a building's materials of construction and the level of fire resistance provided by such materials. Limitations on a building's height and area, set forth in Chapter 5, are directly related to the occupancies it houses and its type of construction. Chapter 5 also provides the various methods available to address conditions in which multiple uses or occupancies occur within the same building. Chapter 4 contains special detailed requirements based on unique conditions or uses that are found in some buildings. ■

303.1.3

Assembly Rooms Associated with Group E Occupancies

CHANGE TYPE: Clarification

CHANGE SUMMARY: The allowance for a Group E classification of accessory assembly spaces in school buildings has been clarified so as to not confuse the provision with the mixed-occupancies requirements dealing with accessory occupancies as regulated by Section 508.2.

2012 CODE: 303.1 Assembly Group A.   Assembly Group A occupancy includes, among others, the use of a building or structure, or a portion thereof, for the gathering of persons for purposes such as civic, social, or religious functions; recreation; food or drink consumption; or awaiting transportation.

Exceptions:

~~1.~~ **303.1.1 Small Buildings and Tenant Spaces.**   A building or tenant space used for assembly purposes with an *occupant load* of less than 50 persons shall be classified as a Group B occupancy.

~~2.~~ **303.1.2 Small Assembly Spaces.**   The following rooms and spaces shall not be classified as Assembly occupancies:

    <u>1.</u>   A room or space used for assembly purposes with an *occupant load* of less than 50 persons and accessory to another occupancy shall be classified as a Group B occupancy or as part of that occupancy.

    ~~3.~~<u>2.</u>   A room or space used for assembly purposes that is less than 750 square feet (70 m^2) in area and accessory to another occupancy shall be classified as a Group B occupancy or as part of that occupancy.

High school gymnasium/auditorium

International Code Council®

4. 303.1.3 Associated with Group E Occupancies. ~~Assembly areas that are accessory to Group E occupancies are not considered separate occupancies except when applying the assembly occupancy requirements of Chapter 11.~~ A room or space used for assembly purposes that is associated with a Group E occupancy is not considered a separate occupancy.

5. 303.1.4 Accessory to Places or Religious Worship. Accessory religious educational rooms and religious auditoriums with occupant loads of less than 100 are not considered separate occupancies.

CHANGE SIGNIFICANCE: Where persons gather for civic, social, or religious functions; recreation; food or drink consumption; and similar activities, the function is considered "assembly" in nature. Classification as a Group A occupancy is typically warranted, unless the space is relatively small or the occupant load is relatively low. In addition, assembly spaces—such as gymnasiums and auditoriums—directly related to Group E educational occupancies are not generally classified as Group A occupancies but rather as simply portions of the Group E building. The allowance for the Group E classification of "accessory" assembly spaces in school buildings has been clarified by modifying the code to address "associated" assembly spaces so as to not confuse the provision with the mixed-occupancies requirements dealing with accessory occupancies as regulated by Section 508.2. The application of the provision continues to be appropriate to those assembly areas of school buildings—such as gymnasiums and auditoriums—that are primarily an extension of the educational function.

The reference to Chapter 11 was also removed as it was deemed unnecessary in the application of accessibility provisions as they apply to assembly areas. The accessibility requirements for fixed-seating facilities, dining areas, and other assembly seating areas are based on the general function of assembly activities and not tied to an occupancy classification. In addition, the assembly means of egress provisions of Section 1028 are also identified as applicable to assembly spaces within Group E occupancies. A number of other text changes were made throughout the code to focus on the use of the space for assembly purposes, rather than the occupancy classification.

303.3

Occupancy Classification of Casino Gaming Floors

CHANGE TYPE: Addition

CHANGE SUMMARY: The classification of a casino gaming floor is now specifically identified as a Group A-2 occupancy.

2012 CODE: <u>**303.3 Assembly Group A-2.**</u> Assembly uses intended for food and/or drink consumption including, but not limited to:

Banquet halls

<u>Casinos (gaming areas)</u>

Night clubs

Restaurants, cafeterias, and similar dining facilities (including associated commercial kitchens)

Taverns and bars

CHANGE SIGNIFICANCE: Assembly uses classified as Group A occupancies are further subclassified into one of five occupancy groups. Many assembly uses are specifically identified as to which classification they most typically belong through the listing of various uses found within each subclassification. Casino gaming floors have traditionally been considered as Group A occupancies where the occupant load is 50 or more persons; however, there has been disagreement over the specific classification of such uses as they previously have not been listed in the code. The classification of a casino gaming floor is now specifically identified as a Group A-2 occupancy.

Assigning an occupancy group to a casino gaming floor has varied due to the lack of any specific mention as to its proper classification. Although the degree of hazard has caused some to historically classify the use as a Group A-2 occupancy, the lack of a specific mention often resulted in applying the default provisions associated with Group A-3 occupancies. And although a casino gaming floor does not seem to fit into a classification reserved for food and/or drink consumption, it has been determined that there are similar hazard characteristics with other uses classified as Group A-2. There are distracting lights, sounds, decorations, and, in many

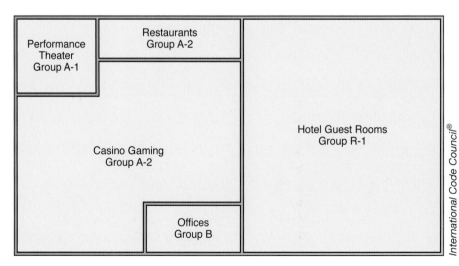

Classification of casino and related uses.

International Code Council®

cases, alcoholic beverages are being consumed. Due to the various distractions, it is possible that the occupants will become disoriented and confused in an emergency situation and have difficulty locating the means of egress.

Classification of a casino gaming floor as a Group A-2 occupancy allows for application of the necessary provisions to address the expected hazards. It should be noted that small casino gaming facilities may be classified as Group B where they meet the conditions of Section 303.1.1 or 303.1.2. It is also important to note that the Group A-2 classification is limited to the gaming areas only. Other areas in a casino that may be associated with the gaming activities—such as restaurants, theaters, guest rooms, and administrative areas—are to be classified based upon their own individual function.

303.3, 306.2

Occupancy Classification of Commercial Kitchens

Commercial kitchen

International Code Council®

CHANGE TYPE: Clarification

CHANGE SUMMARY: The appropriate occupancy classification of a commercial kitchen has been clarified based upon the kitchen's relationship, or lack of a relationship, to dining facilities.

2012 CODE: <u>303.3 Assembly Group A-2.</u> Assembly uses intended for food and/or drink consumption including, but not limited to:

Banquet halls

Casinos (gaming areas)

Night clubs

Restaurants<u>, cafeterias, and similar dining facilities (including associated commercial kitchens)</u>

Taverns and bars

306.2 Moderate-hazard Factory Industrial, Group F-1. Factory industrial uses which are not classified as Factory Industrial F-2 Low Hazard shall be classified as F-1 Moderate Hazard and shall include, but not be limited to, the following:

Food processing <u>and commercial kitchens not associated with restaurants, cafeterias, and similar dining facilities.</u>

(no changes to other uses on the list)

CHANGE SIGNIFICANCE: Commercial kitchens have historically been characterized as two different types, those that are directly associated with a restaurant or similar dining facility and those that are independent of any related dining area, such as a catering business. The appropriate occupancy classification of commercial kitchens has been clarified through text changes in three different areas of the code.

In Table 508.4 regulating separated occupancies, footnote d has been eliminated to help provide clarity to the classification of a commercial kitchen. The past presence of the footnote eliminating any required fire separation between a commercial kitchen and the restaurant seating area that it serves often led to a conclusion that the commercial kitchen needed to be classified differently than the associated dining area. It was occasionally assumed that if they were intended to both be classified as the same occupancy, that of the restaurant seating area, then there was no relevance to the footnote. However, common practice has always been to include the kitchen area as an extension of the restaurant seating area, causing both spaces to be considered as Group A-2, or Group B for smaller restaurants. In order to clarify the appropriate occupancy classification of the associated kitchen, the footnote has been deleted.

To further identify the classification of the two types of commercial kitchens, additional language has been added to the code listings of those uses classified as Group A-2 and Group F-1 occupancies. Commercial kitchens associated with restaurants, cafeterias, and similar dining facilities, are now considered as a portion of the Group A-2 occupancies classification. Extending this concept, a kitchen associated with a small

Group B restaurant would simply be classified as a portion of the Group B occupancy. Although a commercial kitchen does not pose the same types of hazards as an assembly use, the allowance for a similar classification has generally been considered as an appropriate decision. Where the commercial kitchen is not associated with a dining facility, such as a catering business, the kitchen is to be classified as a Group F-1 occupancy in the same manner as any other food processing operations.

Table 307.1(1), Section 307.4

Facilities Generating Combustible Dusts

CHANGE TYPE: Modification

CHANGE SUMMARY: In the determination of occupancy classification for a facility where combustible dusts are anticipated, a technical report and opinion must now be provided to the building official that provides all necessary information for a qualified decision as to the potential combustible dusts hazard.

2012 CODE:

TABLE 307.1(1) Maximum Allowable Quantity Per Control Area of Hazardous Materials Posing a Physical Hazard

Material	Class	Group When the Maximum Allowable Quantity is Exceeded	Storage[b]			Use-Closed Systems[b]			Use-Open Systems[b]	
			Solid pounds (cubic feet)	Liquid gallons (pounds)	Gas cubic feet at NTP	Solid pounds (cubic feet)	Liquid gallons (pounds)	Gas cubic feet at NTP	Solid pounds (cubic feet)	Liquid gallons (pounds)
Combustible Dust	N/A	H-2	Note q	N/A	N/A	Note q	N/A	N/A	Note q	N/A

q. Where manufactured, generated or used in such a manner that the concentration and conditions create a fire or explosion hazard based on information prepared in accordance with Section 414.1.3.

(no changes to remainder of table and footnotes)

307.4 High-hazard Group H-2. Buildings and structures containing materials that pose a deflagration hazard or a hazard from accelerated burning shall be classified as Group H-2. Such materials shall include, but not be limited to, the following:

> Combustible dusts where manufactured, generated, or used in such a manner that the concentration and conditions create a fire or explosion hazard based on information prepared in accordance with Section 414.1.3.

(no changes to other materials on list)

CHANGE SIGNIFICANCE: Combustible dusts are considered as finely divided solid material that is less than 420 microns in diameter which, when dispersed in air in the proper proportions, could be ignited by a flame, spark, or other source of ignition. Examples include organic materials such as wheat flour or corn meal in a food manufacturing plant, pharmaceuticals, wood flour produced during sanding operations in a furniture manufacturing plant, or powdered plastics in a manufacturing environment. The hazard presented by uncontrolled combustible dusts is so great that classification as a Group H-2 occupancy occurs where the concentration and conditions under which the dusts are manufactured, generated, or used are such that a fire or explosion hazard is created. Reference is now made to Section 414.1.3, which requires a technical report and opinion be provided to the building official that provides all necessary information for a qualified decision as to the potential combustible dusts hazard.

A comprehensive discussion on the evaluation of combustible dusts hazards can be found in the *Significant Changes to the International Fire Code,* 2012 Edition, authored by Scott Stookey.

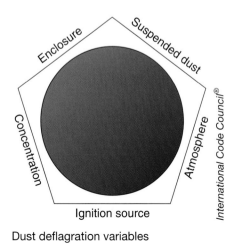

Dust deflagration variables

CHANGE TYPE: Clarification

CHANGE SUMMARY: A number of new definitions related to care facilities have been added and some existing definitions have been revised to provide clarity and consistency in application.

2012 CODE: ~~308.3.1~~ **308.2 Definitions.**   ~~The following words and terms shall, for the purposes of this section and as used elsewhere in this code, have the meanings shown herein.~~ The following terms are defined in Chapter 2:

> 24-HOUR CARE
>
> CUSTODIAL CARE
>
> DETOXIFICATION FACILITIES
>
> ~~CHILD~~ FOSTER CARE FACILITIES
>
> HOSPITALS AND ~~MENTAL~~ PSYCHIATRIC HOSPITALS
>
> INCAPABLE OF SELF PRESERVATION
>
> MEDICAL CARE
>
> NURSING HOMES

202 Definitions.

24 HOUR CARE.   The actual time that a person is an occupant within a facility for the purpose of receiving care. It shall not include a facility that is open for 24 hours and is capable of providing care to someone visiting the facility during any segment of the 24 hours.

CUSTODIAL CARE.   Assistance with day-to-day living tasks; such as assistance with cooking, taking medication, bathing, using toilet facilities, and other tasks of daily living. Custodial care include occupants who evacuate at a slower rate and/or who have mental and psychiatric complications.

DETOXIFICATION FACILITIES.   Facilities that ~~serve patients who are~~ provide~~d~~ treatment for substance abuse ~~on a 24-hour basis and~~ serving care recipients who are incapable of self-preservation or who are harmful to themselves or others.

~~CHILD~~ FOSTER CARE FACILITIES.   Facilities that provide care ~~on a 24-hour basis~~ to more than five children, 2½ years of age or less.

HOSPITALS AND ~~MENTAL~~ PSYCHIATRIC HOSPITALS.   Facilities ~~buildings, or portions thereof used on a 24-hour basis~~ that provide care or treatment for the medical, psychiatric, obstetrical, or surgical treatment of ~~inpatients who~~ care recipients that are incapable of self-preservation.

INCAPABLE OF SELF PRESERVATION.   Persons because of age; physical limitations; mental limitations; chemical dependency, or medical treatment cannot respond as an individual to an emergency situation.

MEDICAL CARE.   Care involving medical or surgical procedures, nursing, or for psychiatric purposes.

308.2, 202 continues

308.2, 202 continued

Group I-2 nursing home

NURSING HOMES. ~~Nursing homes are long-term care~~ Facilities <u>that provide care</u> ~~on a 24-hour basis~~, including both intermediate care facilities and skilled nursing facilities~~, serving more than five persons and~~ <u>where</u> any of the persons are incapable of self-preservation.

CHANGE SIGNIFICANCE: The special provisions of Section 308 addressing Group I occupancies can vary significantly based upon the specific type of institutional use involved. Requirements are in part based upon the number, capabilities, and condition of occupants, as well as the services rendered. In order to correctly assign code requirements based upon the perceived hazards, it is important that the various types of uses and terms be clearly defined. A number of new definitions have been added, and some existing definitions have been revised to provide clarity and consistency in terminology. The new definitions specifically describe each type of care or facility and identify the distinct differences between them. Some terms were consolidated to be more descriptive of a group of occupants, yet generic enough to be used interchangeably.

The term "24-hour care" has been introduced to clarify that it is applicable to the length of stay within the facility by the person receiving care, rather than the hours of operation of the care facility. "Custodial care" includes those conditions where the care recipient needs assistance with daily living tasks, typically for extended periods of time. It is expected that individuals under custodial care are generally capable of self-preservation under emergency conditions; however, their evacuation time is more lengthy than that expected of the general population. The term "incapable of self-preservation" has been utilized in the IBC since its first edition to describe individuals who need physical assistance from others during evacuation procedures; however, the term has never before been specifically defined. The general term "medical care" is also defined to include the major aspects of health care, including medical and surgical procedures.

Infant and toddler care, previously defined as "child care," is now considered "foster care" to differentiate it from the child care activities that occur in a day care environment. The term "mental hospital" has been revised to "psychiatric hospital" to reflect the current thinking within the industry. In both the foster care and psychiatric hospital definitions, the reference to 24-hour care has been deleted as it is redundant to the general description of a Group I-2 occupancy. Other minor modifications include the change from the term "patient" to "care recipient" and the reference to "nurse" is now "care provider."

International Code Council®

CHANGE TYPE: Modification

CHANGE SUMMARY: A Group I-2 occupancy classification is now only applicable to those medical facilities where six or more individuals incapable of self-preservation are receiving care.

2012 CODE: 308.3 308.4 Institutional Group I-2.   This occupancy shall include buildings and structures used for medical, ~~surgical, psychiatric, nursing or custodial~~ care on a 24-hour basis for more than five persons who are incapable of self-preservation. This group shall include, but not be limited to, the following:

Foster ~~Child~~ care facilities

Detoxification facilities

Hospitals

Nursing homes

~~Mental~~ Psychiatric hospitals

308.4.1 Five or Fewer Persons Receiving Care.   A facility such as the above with five or fewer persons receiving such care shall be classified as Group R-3 or shall comply with the *International Residential Code* provided an automatic sprinkler system is installed in accordance with Section 903.3.1.3 or Section P2904 of the *International Residential Code.*

CHANGE SIGNIFICANCE: Group I-2 occupancies include those medical care functions where the recipients receive care on a 24-hour basis, such as nursing homes and hospitals. It is anticipated that most of the care recipients are incapable of self-preservation and require the assistance of others under fire or other emergency conditions. A care-recipient threshold

308.4 continues

308.4

Occupancy Classification for Medical Care Facilities

Group I-2 hospital

308.4 continued

has now been established to limit the Group I-2 classification only to those facilities where six or more individuals are receiving care. Where the number of care recipients does not exceed five, a classification of Group R-3 is most appropriate. As an alternative, such care facilities with five or fewer care recipients may also be regulated under the provisions of the *International Residential Code* (IRC), rather than the IBC, if the building is provided with a fire sprinkler system. The sprinkler system must be installed in accordance with the requirements of NFPA 13D, *Installation of Sprinkler Systems in One- and Two-family Dwellings and Manufactured Homes*, or those set forth in IRC Section 2904, *Dwelling Unit Fire Sprinkler Systems*. The reduction in requirements provided through classification as a Group R-3 occupancy or IRC-regulated building is consistent with that provided for other institutional uses. The code typically recognizes that such small occupant loads in institutional or educational environments can be adequately addressed for fire and life safety through the provisions for dwelling units.

CHANGE TYPE: Modification

CHANGE SUMMARY: The allowance for constructing Group R-4 supervised residential facilities under the *International Residential Code* has been eliminated.

2012 CODE: 310.6 Residential Group R-4. ~~Residential occupancies shall include buildings arranged for occupancy as residential care/assisted living facilities including more than five but not more than 16 occupants, excluding staff.~~

This occupancy shall include buildings, structures, or portions thereof for more than five but not more than 16 persons, excluding staff, who reside on a 24-hour basis in a supervised residential environment and receive custodial care. The persons receiving care are capable of self-preservation. This group shall include, but not be limited to, the following:

Alcohol and drug centers

Assisted living facilities

Congregate care facilities

Convalescent facilities

Group homes

Halfway houses

Residential board and custodial care facilities

Social rehabilitation facilities

Group R-4 occupancies shall meet the requirements for construction as defined for Group R-3, except as otherwise provided for in this code. ~~or shall comply with the *International Residential Code* provided the building~~

310.6 continues

310.6

Uses Classified as Group R-4 Occupancies

International Code Council®

Group R-4 halfway house

310.6 continued ~~is protected by an~~ *~~automatic sprinkler system~~* ~~installed in accordance with Section 903.2.8.~~

CHANGE SIGNIFICANCE. Facilities where occupants are ambulatory but live in a residential environment where supervised custodial and/or personal care services are provided are classified as Group I-1 occupancies unless the number of persons receiving care does not exceed 16. The Group R-4 classification is applicable where more than 5, but no more than 16, persons are housed. The direct relationship between Groups I-1 and R-4 is now more obvious because the laundry list of such types of uses is consistent between both occupancy groups. The only difference between the two classifications is the number of care recipients, as the expectation for both occupancy groups is that the individuals, although supervised, are individually capable of responding to an emergency without physical assistance from others.

Although the new listing of common Group R-4 occupancies simply provides clarification to the code, a second change modifies the provisions by removing the allowance for constructing the described type of supervised care facilities under the *International Residential Code*. Only the provisions of the IBC for Group R-3 or R-4 occupancies are now applicable for custodial care and similar facilities. This is consistent with the change affecting Group R-3 care facilities that limits the application of the IRC to such facilities with five or fewer individuals receiving care. A companion code change in Section 903.2.8 now allows for the use of an NFPA 13D fire sprinkler system in Group R-4 occupancies. Previously, the installation of an NFPA 13R system was the minimum requirement for such occupancies. The allowance for use of an NFPA 13D system is consistent with the requirements of the standard and with court findings regarding nondiscrimination issues involving group homes.

402

Open Mall Buildings

CHANGE TYPE: Clarification

CHANGE SUMMARY: A variety of changes have now been made to clarify the open mall building provisions that were originally developed for covered mall conditions.

2012 CODE: 402.1 ~~Scope.~~ Applicability. The provisions of this section shall apply to buildings or structures defined herein as covered <u>or open</u> mall buildings not exceeding three floor levels at any point nor more than three stories above grade plane. Except as specifically required by this section, covered <u>and open</u> mall buildings shall meet applicable provisions of this code.

Exceptions:

1. Foyers and lobbies of Groups B, R-1, and R-2 are not required to comply with this section.

2. Buildings need not comply with the provisions of this section when they totally comply with other applicable provisions of this code.

CHANGE SIGNIFICANCE: The increasingly popular concept to create large-scale projects resembling covered mall buildings without roofs over the pedestrian circulation areas was newly addressed in the 2009 IBC through the introduction of "open mall building" provisions. It was recognized that designs featuring various "tenant space" buildings and "anchor buildings" situated around unroofed pedestrian ways (open malls) were very similar to those for covered mall buildings, including corresponding code requirements. The new allowance for open mall buildings recognized that the

402 continues

Open mall building

International Code Council®

402 continued   same benefits should be available as for enclosed structures, provided the appropriate measures are taken. Where the mall area is open to the sky, equivalent or better life safety and property protection is provided. A variety of changes have now been made to clarify those provisions that were originally developed only for covered mall conditions. Although the general provisions were intended to be applied equally to both open mall buildings and covered mall buildings, a number of the previous requirements did not fully address open mall conditions. It was deemed necessary to modify some of the past requirements for covered mall buildings in order to make them applicable to open mall conditions. The only new concept is the establishment of an "open mall building perimeter line" that is to be used to identify the boundary between what is considered to be part of the open mall building and what is outside of the building. This allows for the proper application of a variety of provisions, including those dealing with floor area and means of egress. By definition, the perimeter line encircles all buildings which comprise the open mall building, including the open-air walkways and courtyards. Anchor buildings are considered as outside of the building perimeter line.

CHANGE TYPE: Modification

CHANGE SUMMARY: The minimum number of fire service access elevators required in applicable high-rise buildings has been increased from one to two where multiple elevators are provided in the building.

2012 CODE: 403.6.1 Fire Service Access Elevator. In buildings with an occupied floor more than 120 feet (36 576 mm) above the lowest level of fire department vehicle access, no fewer than ~~one~~ two fire service access elevators, or all elevators, whichever is less, shall be provided in accordance with Section 3007. Each fire service access elevator shall have a capacity of not less than 3500 pounds (1588 kg).

CHANGE SIGNIFICANCE: To facilitate the rapid deployment of firefighters, Section 403.6.1 of the 2009 IBC introduced a new requirement for a fire service access elevator in high-rise buildings that have at least one floor level more than 120 feet above the lowest level of fire department

403.6.1 continues

403.6.1
High-Rise Buildings—Fire Service Access Elevators

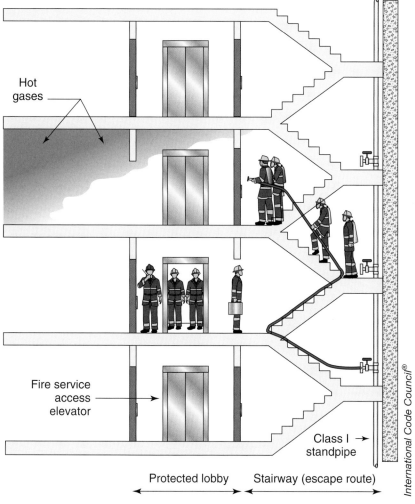

Fire service access elevator.

403.6.1 continued vehicle access. Usable by firefighters and other emergency responders, the specific requirements for this type of elevator are set forth in Section 3007. A fire service access elevator has a number of key features that will allow firefighters to use the elevator for safely accessing an area of a building that may be involved in fire or for facilitating rescue of building occupants. The minimum number of fire service access elevators required in applicable high-rise buildings has been increased from one to two, except for those buildings that are provided with only a single elevator.

The mandate for a second fire service access elevator is based on information that indicates at least two elevators are necessary for firefighting activities in high-rise buildings. In addition, past experience has shown that on many occasions elevators are not available due to shutdowns for various reasons, including problems in operation, routine maintenance, modernization programs, and EMS operations in the building prior to firefighter arrival. A minimum of two fire service elevators provided with all of the benefits afforded to such elevators better ensures that there will be a fire service access elevator available for the firefighters' use in the performance of their duties.

406.4
Public Parking Garages

CHANGE TYPE: Clarification

CHANGE SUMMARY: Those parking structures that fall outside of the scope of Section 406.3 regulating private parking garages are now identified as public parking garages.

2012 CODE: ~~406.2.1 Classification~~ <u>406.4 Public Parking Garages.</u> Parking garages <u>other than private parking garages, shall be classified as public parking garages and shall comply with the provisions of Sections 406.4.2 through 406.4.8 and</u> shall be classified as either <u>an</u> open ~~as defined in Section 406.3,~~ <u>parking garage</u> or <u>an</u> enclosed <u>parking garage</u> ~~and shall meet appropriate criteria of Section 406.4~~. <u>Open parking garages shall also comply with Section 406.5. Enclosed parking garages shall also comply with Section 406.6.</u> ~~Also~~ See Section 510 for special provisions for parking garages.

CHANGE SIGNIFICANCE: Parking garages, as well as other types of structures where motor vehicles are involved, present some unique characteristics that are addressed through the special provisions of Section 406. The sizes and operations of parking garages vary significantly, and such differences are uniquely regulated. The varying requirements have now been reformatted to allow for the provisions to be more clearly applied to the correct situation. In addition, the descriptions of the various types of parking garages have been clarified for consistency purposes.

There are fundamentally two types of parking garages regulated by the IBC: private garages and public garages. Although there is no specific definition for either type of garage, the basis for both classifications is Section 406.3 addressing private garages and carports. Those parking structures that fall outside of the scope of Section 406.3 are now considered as public parking garages. The primary difference between private and public garages is the size of the facility, rather than the use. Strictly

406.4 continues

Public parking garage

International Code Council®

406.4 continued

limited in permissible height and area, private parking garages are typically not commercial in nature. They generally serve only a specific tenant or building and are not open for public use. It is important to note that there is no implication that public parking garages must be open to the public, as they are only considered public in comparison to private garages. A public parking garage is then further characterized as one of two types—either an enclosed parking garage or an open parking garage—and regulated accordingly.

CHANGE TYPE: Addition

CHANGE SUMMARY: A clear horizontal space, whose minimum distance is based on the depth of the open parking garage's exterior wall openings, must now be provided adjacent to any such openings located below grade.

2012 CODE: 406.3.3.1 406.5.2 Openings.   For natural ventilation purposes, the exterior side of the structure shall have uniformly distributed openings on two or more sides. The area of such openings in exterior walls on a tier shall be not less than 20 percent of the total perimeter wall area of each tier. The aggregate length of the openings considered to be providing natural ventilation shall be not less than 40 percent of the perimeter of the tier. Interior walls shall be not less than 20 percent open with uniformly distributed openings.

> **Exception:** Openings are not required to be distributed over 40 percent of the building perimeter where the required openings are uniformly distributed over two opposing sides of the building.

406.5.2.1 Openings below Grade.   Where openings below grade provide required natural ventilation, the outside horizontal clear space shall be one and one-half times the depth of the opening. The width of the horizontal clear space shall be maintained from grade down to the bottom of the lowest required opening.

CHANGE SIGNIFICANCE: The overall fire hazards in a parking garage are relatively low. Because permanently open exterior walls provide sufficient natural ventilation and permit the dissipation of heated gases, open parking garages are viewed as an even lesser hazard because they do not need to rely on mechanical ventilation. As such, a number of code benefits are afforded to open parking garages, including increased allowable heights and areas. There are situations where the required exterior openings of open parking garages are located below the surrounding grade. Previously, no minimum exterior clearance has been mandated between the required openings and adjacent retaining walls or similar enclosures. This has resulted in conditions where the exterior openings were

406.5.2.1 continues

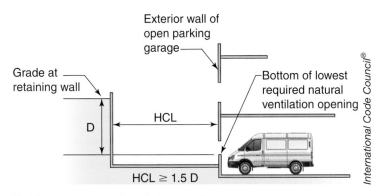

Parking garage openings below grade

406.5.2.1 continued relatively ineffective for ventilation purposes. A new provision mandates that a clear horizontal space be provided adjacent to the garage's exterior openings that allows for adequate air movement through the opening. The dimensional requirements are based upon the provisions of Section 1203.4.1.2, which address openings below grade when such openings are used for the required natural ventilation of a building's occupied spaces.

Where openings in the exterior wall of an open parking garage are located below grade level, some degree of clear space must be provided at the exterior of the openings. As the distance of the openings below the adjoining ground increases, the minimum required exterior clear space also increases proportionately. The horizontal clear space dimension, measured perpendicular to the exterior wall opening, must be at least one and one-half times the distance between the bottom of the opening and the adjoining ground level above. The extent of the required clear space allows for adequate exterior open space to meet the intent and dynamics of natural ventilation requirements for open parking garages.

CHANGE TYPE: Modification

CHANGE SUMMARY: In the determination of permitted area and height increases for open parking garages, the method for determining the amount of openings required to receive such increases has been modified to allow for consistent application.

2012 CODE: ~~406.3.6~~ 406.5.5 Area and Height Increases. The allowable area and height of open parking garages shall be increased in accordance with the provisions of this section. Garages with sides open on three-fourths of the building's perimeter are permitted to be increased by 25 percent in area and one tier in height. Garages with sides open around the entire building's perimeter are permitted to be increased by 50 percent in area and one tier in height. For a side to be considered open under the above provisions, the total area of openings along the side shall not be less than 50 percent of the interior area of the side at each tier and such openings shall be equally distributed along the length of the tier. <u>For purposes of calculating the interior area of the side, the height shall not exceed 7 feet (2134 mm).</u>

Allowable tier areas in Table 406.5.4 shall be increased for open parking garages constructed to heights less than the table maximum. The gross tier area of the garage shall not exceed that permitted for the higher structure. No fewer than three sides of each such larger tier shall have continuous horizontal openings not less than 30 inches (762 mm) in clear height extending not less than 80 percent of the length of the sides and no part of

406.5.5 continues

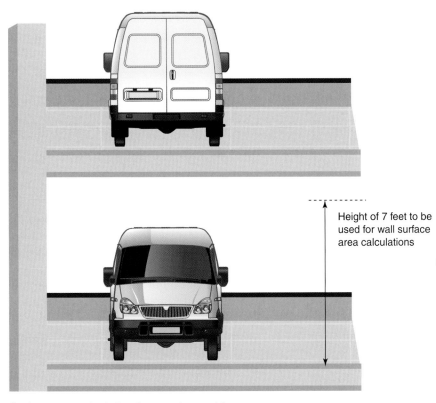

Surface area calculation for opening parking garage

Height of 7 feet to be used for wall surface area calculations

International Code Council®

406.5.5 continued

such larger tier shall be more than 200 feet (60 960 mm) horizontally from such an opening. In addition, each such opening shall face a street or yard accessible to a street with a width not less than 30 feet (9144 mm) for the full length of the opening, and standpipes shall be provided in each such tier.

Open parking garages of Type II construction, with all sides open, shall be unlimited in allowable area where the building height does not exceed 75 feet (22 860 mm). For a side to be considered open, the total area of openings along the side shall not be less than 50 percent of the interior area of the side at each tier, and such openings shall be equally distributed along the length of the tier. For purposes of calculating the interior area of the side, the height shall not exceed 7 feet (2134 mm). All portions of tiers shall be within 200 feet (60 960 mm) horizontally from such openings or other natural ventilation openings as defined in Section 406.5.2. These openings shall be permitted to be provided in courts with a minimum dimension of 20 feet (6096 mm) for the full width of the openings.

CHANGE SIGNIFICANCE: A parking garage with compliant exterior openings allowing for natural ventilation is granted privileges not afforded to parking garages requiring mechanical ventilation due to the lack of such openings. Where the "open" parking garage is used exclusively for parking purposes with no other uses in the building, further benefits are provided by the code regarding the garage's allowable building height and area. In such cases, the allowable height and area is calculated in a different manner than that used for other building types, resulting in an increased allowable height and area over that typically permitted. Table 406.3.5 establishes the height and area limitations for single-occupancy open parking garages of Types I, II, and IV construction. Where the garage is provided with exterior openings in amounts much greater than the minimum required for designation as an open parking garage, increases in area, and possibly height, are available. The method for determining the amount of openings required to receive such increases has been modified to allow for consistent application of the provisions.

Two specific allowances use the percentage of exterior wall that is open as the qualifying criteria. The provisions mandate that for an area increase—and, in one case, a height increase—to apply, some designated portion of the exterior walls of the garage must be at least 50 percent open. In calculating the degree of openness that is required, each tier of the parking garage is to be analyzed individually. Previously, the percentage of openings has been determined simply by dividing the area of the exterior opening by the interior area of the side of each tier. As the tier height increased, the required area of exterior openings was also required to increase in order to qualify for the allowable area and height increases. A change in the method of calculating opening percentage was made that uses 7 feet as the maximum height to be used when determining the interior area of each tier. The dimension of 7 feet is consistent with the minimum height requirement of a parking tier. In the determination of permitted area and height increases, the revised measurement method removes the unnecessary requirement for larger exterior openings based upon a tier height that exceeds the required minimum.

410.6.3, 202
Technical Production Areas

CHANGE TYPE: Clarification

CHANGE SUMMARY: Outdated terminology, such as fly galleries, gridirons, and pinrails, has been replaced by the general and comprehensive term "technical production area," and the special means of egress provisions for such areas have all been relocated to Section 410.

2012 CODE:

SECTION 410
STAGES, ~~AND~~ PLATFORMS, <u>AND TECHNICAL PRODUCTION AREAS</u>

410.2 Definitions. The following ~~words and~~ terms ~~shall, for the purposes of this section and as used elsewhere in this code, have the meanings shown herein~~ <u>are defined in Chapter 2</u>:

~~FLY GALLERY. A raised floor area above a stage from which the movement of scenery and operation of other stage effects are controlled.~~

~~GRIDIRON. The structural framing over a stage supporting equipment for hanging or flying scenery and other stage effects.~~

~~PINRAIL. A rail on or above a stage through which belaying pins are inserted and to which lines are fastened.~~

PLATFORM (definition moved to Chapter 2 with no change)

PROCENIUM WALL (definition moved to Chapter 2 with no change)

STAGE (definition moved to Chapter 2 with no change)

<u>**TECHNICAL PRODUCTION AREA**</u>

<u>**202 Definitions. TECHNICAL PRODUCTION AREA.** Open elevated areas or spaces intended for entertainment technicians to walk on and occupy for servicing and operating entertainment technology systems and equipment. Galleries, including fly and lighting galleries, gridirons, catwalks, and similar areas are designed for these purposes.</u>

<u>**410.6.3 Technical Production Areas.** Technical production areas shall be provided with means of egress and means of escape in accordance with Sections 410.6.3.1 through 410.6.3.5.</u>

<u>**410.6.3.1 Means of Egress.** No fewer than one means of egress shall be provided from technical production areas.</u>

<u>**410.6.3.2 Travel Distance.** The length of exit access travel shall be not greater than 300 feet (91 440 mm) for buildings without a sprinkler system and 400 feet (121 900 mm) for buildings equipped throughout with an automatic sprinkler system in accordance with Section 903.3.1.1.</u>

Technical production area.
(© iStockphoto/hsvrs)

410.6.3, 202 continues

410.6.3, 202 continued

410.6.3.3 Two Means of Egress. Where two means of egress are required the common path of travel shall be not greater than 100 feet (30 480 mm).

> **Exception:** A means of escape to a roof in place of a second means of egress is permitted.

410.6.3.4 Path of Egress Travel. The following exit access components are permitted when serving technical production areas:

1. Stairways
2. Ramps
3. Spiral stairways
4. Catwalks
5. Alternating tread devices
6. Permanent ladders

410.6.3.5 Width. The path of egress travel within and from technical support areas shall be a minimum of 22 inches (559 mm).

CHANGE SIGNIFICANCE: Many auditoriums and performance halls, as well as other types of entertainment and sports venues, are provided with elevated technical support areas used for lighting, sound, scenery, and other performance effects. Such areas may or may not be associated with a stage but are typically an integral part of the production. These spaces are generally limited in floor area, and access is always restricted to authorized personnel. Special means of egress provisions have always been provided that recognize the uniqueness of these areas. In the establishment of these means of egress requirements, outdated terminology—such as fly galleries and gridirons—has now been replaced by the general and comprehensive term "technical production area." In addition, the special means of egress provisions for such areas have all been relocated to Section 410 and revised to eliminate any conflict with Chapter 10.

The new term "technical production areas" is intended to encompass all technical support areas, regardless of their traditional name. These areas are typically used to support entertainment technology from above the performance area. As a result of the new comprehensive definition, the defined terms of "fly gallery" and "gridiron" have been deleted. In addition, the title of Section 410 has been revised to "Stages, Platforms, and Technical Production Areas" in order to recognize that these areas may not necessarily be associated with a stage or platform, such as at sports arenas, and may be considered as stand-alone building elements.

Means of egress provisions relating specifically to technical production areas have previously been located in both Section 410 and in Chapter 10. In addition, the provisions were in conflict with one another and inconsistent in terminology. All of the means of egress requirements relating to technical production areas that modify the general provisions of Chapter 10 are now located in Section 410.6.3. The provisions reflect the special allowances for minimum number of means of egress, maximum travel distance, allowable exit access components, and minimum travel path width.

A related modification was made to the minimum required number of exits or exit access doorways required from a stage. Previously, a minimum of one means of egress was required from each side of the stage. This requirement is now only applicable where two or more exits or exit access doorways are required by Section 1015.1 based upon occupant load and common path of egress travel. It was determined that a stage was no different from other spaces in the building regarding the threshold at which two means of egress are required.

412.4.6.2

Aircraft Hangar Fire Areas

CHANGE TYPE: Modification

CHANGE SUMMARY: Spaces ancillary to the aircraft servicing and storage areas of an aircraft hangar need no longer be included in the fire area size when determining fire suppression requirements.

2012 CODE: 412.4.6.2 Separation of Maximum Single Fire Areas. Maximum single fire areas established in accordance with hangar classification and construction type in Table 412.4.6 shall be separated by 2-hour fire walls constructed in accordance with Section 706. In determining the maximum single fire area as set forth in Table 412.4.6, ancillary uses which are separated from aircraft servicing areas by a fire barrier of not less than one hour, constructed in accordance with Section 707 shall not be included in the area.

CHANGE SIGNIFICANCE: In order to minimize the fire hazards associated with aircraft hangars, fire suppression is required based upon the criteria of Table 412.4.6. The table determines the hangar classification (Group I, II, or III) to which the fire suppression must be designed in accordance with NFPA 409, *Aircraft Hangars*. The classification is based upon the hangar's type of construction and fire area size. Fire area size is based on the aggregate floor area bounded by specified fire walls that have a minimum 2-hour fire-resistance rating. For the purposes of hangar classification, ancillary uses located within the fire area are no longer required to be included in the fire area size provided they are separated from the aircraft serving area by minimum 1-hour fire barriers.

Many times there are ancillary areas associated with an aircraft hangar, such as offices, maintenance shops, and storage rooms. Unless located in a fire area different from that of the aircraft storage and servicing area, the floor area of such ancillary areas has previously been included in the hangar fire area and included in the application of Table 412.4.6. Because the fire suppression requirements of NFPA 409 are primarily for the protection of aircraft within the storage and servicing area, inclusion of the floor area of the ancillary spaces into the fire

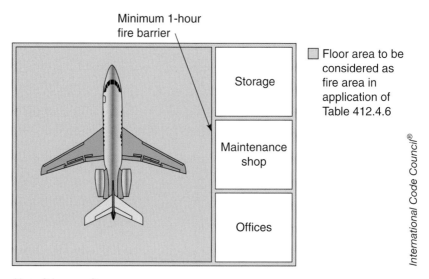

Aircraft hangar fire area

suppression criteria was thought to be inappropriate. The fire protection requirements in the ancillary areas are considered to be less extensive than those required for the aircraft servicing and storage areas. Therefore, their inclusion in the application of Table 412.4.6 for fire area size has been eliminated where a limited degree of fire separation is provided.

In order to be exempted from the fire area calculation within the aircraft hangar, it is necessary that the ancillary areas be separated from the aircraft storage and servicing areas by minimum 1-hour fire barriers. The 1-hour requirement intends to provide an acceptable fire separation without the creation of additional fire areas that would require separation by minimum 2-hour fire walls.

414.5

Inside Storage, Dispensing, and Use of Hazardous Materials

CHANGE TYPE: Clarification

CHANGE SUMMARY: The scoping provisions regarding the inside storage, dispensing, and use of hazardous materials have been revised to provide consistency with those of the *International Fire Code*.

2012 CODE: 414.5 Inside Storage, Dispensing, and Use. The inside storage, dispensing, and use of hazardous materials ~~in excess of the maximum allowable quantities per control area of Tables 307.1(1) and 307.1(2)~~ shall be in accordance with Sections 414.5.1 through ~~414.5.5~~ <u>414.5.4</u> of this code and the *International Fire Code*.

CHANGE SIGNIFICANCE: The storage, dispensing, and use of hazardous materials within buildings are strictly regulated by Section 414.5, as well as the *International Fire Code* (IFC), for explosion control; monitor control equipment; standby and emergency power; and spill control, drainage and containment. IBC Table 415.3.1, Minimum Separation Distances for Buildings Containing Explosive Materials, replicates Table 911.1 of the IFC. However, the scoping provisions for this table have previously differed between the IBC and the IFC. IFC Table 911.1 applies where an explosion hazard exists regardless of the quantity of the hazardous material involved, while IBC Table 415.3.1 has previously only applied where the maximum allowable quantities of the hazardous materials are such that a Group H occupancy is created. The provisions in the IBC have been modified for consistency with the IFC by requiring compliance with Section 414.5 and Table 415.3.1 where any amount of hazardous material is present.

Gas packaging plant

International Code Council®

CHANGE TYPE: Modification

CHANGE SUMMARY: The means of egress and plumbing facilities requirements for the nonresidential portion of a live/work unit are now regulated based upon the specific function of the nonresidential space rather than those of a Group R-2 occupancy.

2012 CODE:

202 Definitions.

LIVE/WORK UNIT.   A dwelling unit or sleeping unit in which a significant portion of the space includes a nonresidential use that is operated by the tenant.

SECTION 419
LIVE/WORK UNITS

419.1 General.   A live/work unit ~~is a dwelling unit or sleeping unit in which a significant portion of the space includes a nonresidential use that is operated by the tenant and~~ shall comply with Sections 419.1 through 419.9.

> **Exception:** Dwelling or sleeping units that include an office that is less than 10 percent of the area of the dwelling unit ~~shall not be classified as a live/work unit~~ are permitted to be classified as dwelling units with accessory occupancies in accordance with Section 508.2.

419.1.1 Limitations.   The following shall apply to all live/work areas:

1. The live/work unit is permitted to be a maximum of 3,000 square feet (279 m^2);
2. The nonresidential area is permitted to be a maximum 50 percent of the area of each live/work unit;

419, 202 continues

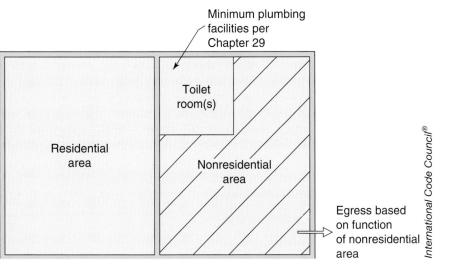

Live/work unit

419, 202 continued

3. The nonresidential area function shall be limited to the first or main floor only of the live/work unit; and

4. A maximum of five nonresidential workers or employees are allowed to occupy the nonresidential area at any one time.

419.2 Occupancies.   Live/work units shall be classified as a Group R-2 occupancy. Separation requirements found in Sections 420 and 508 shall not apply within the live/work unit when the live/work unit is in compliance with Section 419. ~~High-hazard and storage occupancies~~ <u>Nonresidential uses that would otherwise be classified as either a Group H or S occupancy</u> shall not be permitted in a live/work unit.

> **Exception:** <u>Storage shall be permitted in the live/work unit provided the aggregate area of storage in the nonresidential portion of the live/work unit shall be limited to 10 percent of the space dedicated to nonresidential activities.</u>

419.3 Means of Egress.   Except as modified by this section, ~~the provisions for Group R-2 occupancies in Chapter 10 shall apply to the entire live/work unit~~ <u>the means of egress components for a live/work unit shall be designed in accordance with Chapter 10 for the function served.</u>

419.3.1 Egress Capacity.   The egress capacity for each element of the live/work unit shall be based on the occupant load for the function served in accordance with Table 1004.1.1.

~~419.3.2 Sliding Doors.~~   ~~Where doors in a means of egress are of the horizontal –sliding type, the force to slide the door to its fully open position shall not exceed 50 pounds (220 N) with a perpendicular force against the door of 50 pounds (220 N).~~

~~419.3.3~~ <u>419.3.2</u> Spiral Stairways.   Spiral stairways that conform to the requirements of Section 1009.12 shall be permitted.

~~419.3.4 Locks.~~   ~~Egress doors shall be permitted to be locked in accordance with Exception 4 of Section 1008.1.9.3.~~

419.4 Vertical Openings.   Floor openings between floor levels of a live/work unit are permitted without enclosure.

419.5 Fire Protection.   The live/work unit shall be provided with a monitored fire alarm system where required by Section 907.2.9 and an automatic sprinkler system in accordance with Section 903.2.8.

419.6 Structural.   Floor loading for the areas within a live/work unit shall be designed to conform to Table 1607.1 based on the function within the space.

419.7 Accessibility.   Accessibility shall be designed in accordance with Chapter 11 <u>for the function served.</u>

419.8 Ventilation.   The applicable requirements of the *International Mechanical Code* shall apply to each area within the live/work unit for the function within that space.

419.9 Plumbing Facilities.   The nonresidential area of the live/work unit shall be provided with minimum plumbing facilities as specified by Chapter 29, based on the function of the nonresidential area. Where the nonresidential area of the live/work unit is required to be accessible by Section 1103.2.13, the plumbing fixtures specified by Chapter 29 shall be accessible.

CHANGE SIGNIFICANCE: Although the concept and unique regulation of live/work units is relatively new in the code, such arrangements are a throwback to 1900 era community planning where residents could walk to all of the needed services within their neighborhood. These types of units began to reemerge in the 1990s through a development style know as "traditional neighborhood design." More recently, adaptive reuse of many older urban structures in city centers incorporated the same live/work tools to provide a variety of residential unit types. Residential live/work units typically include a dwelling unit along with some public service business, such as an artist's studio, coffee shop, or chiropractor's office. There may be a small number of employees working within the unit and the public is able to enter the work area of the unit to acquire service. The special provisions of Section 419 specifically addressing live/work units recognize the uniqueness of this type of use. A number of changes to the live/work provisions were made that both clarify and modify the requirements.

Now defined in Chapter 2, a live/work unit is primarily residential in nature but has a sizable portion of the space devoted to nonresidential activities. The designation of a residential unit as a live/work unit is only applicable where it would otherwise be classified as a mixed occupancy condition. For example, a home office operating out of an extra bedroom in a residential condominium unit would typically not trigger the provisions of Section 419, assuming the local zoning and home occupation ordinances are met. As a related note, the revised exception permitting classification of a small office as an accessory occupancy would seem to have no application as the live/work provisions are intended to eliminate classification as a mixed-occupancy condition and the classification of a small office as an accessory occupancy would be counter to this approach.

There are significant changes that modify the application of the means of egress and plumbing facility provisions. The means of egress is now regulated based upon the specific function of the space rather than the egress requirements for a Group R-2 occupancy. In many instances, this will require an occupancy classification determination for the nonresidential portion for means of egress purposes without assigning such a classification to the function itself. For example, if the nonresidential portion of the live/work unit is a retail sales use, that portion of the unit would be regulated as a Group M occupancy for means of egress provisions. The same concept is also now applied in the determination of required plumbing facilities. The nonresidential area of the live/work unit is to be provided with minimum plumbing facilities as specified by Chapter 29, based on the function of the nonresidential area. Based on the previous example, plumbing fixtures requirements for the nonresidential retail sales area would be based on a Group M classification. However, in both cases, the Group M classification would not be specifically assigned to the live/work unit. The application of this same approach for accessibility purposes has also been clarified.

422

Ambulatory Care Facilities

CHANGE TYPE: Modification

CHANGE SUMMARY: In a multi-tenant or mixed-occupancy building where there are uses present other than an ambulatory care facility, a fire partition is now required between the care facility and those nonrelated spaces where the ambulatory care facility is intended to have at least four care recipients incapable of self-preservation at any one time.

2012 CODE:

SECTION 422
AMBULATORY ~~HEALTH~~ CARE FACIILITIES

422.1 General. Occupancies classified as ~~Group B~~ ambulatory ~~health~~ care facilities shall comply with the provisions of Sections 422.1 through ~~422.6~~ <u>422.7</u> and other applicable provisions of this code.

422.2 Separation. <u>Ambulatory care facilities where the potential for four or more care recipients are to be incapable of self-preservation at any time, whether rendered incapable by staff or staff accepted responsibility for a care recipient already incapable, shall be separated from adjacent spaces, corridors, or tenants with a fire partition installed in accordance with Section 708.</u>

~~422.2~~ <u>422.3</u> Smoke ~~Barriers~~ <u>Compartments</u>. ~~Smoke barriers shall be provided to subdivide every~~ <u>Where the aggregate area of one or more</u> ambulatory care facilities is greater than 10,000 square feet <u>on one story, the story shall be provided with a smoke barrier to subdivide the story into no fewer than</u> ~~into a minimum of~~ two smoke compartments ~~per story~~. <u>The area of any one such smoke compartment shall be not greater than 22,500 square feet (2092 m^2).</u> The travel distance from any point in a smoke compartment to a smoke barrier door shall be not greater than 200 feet (60 960 mm). The smoke barrier shall be installed in accordance with Section 709 <u>with the exception that smoke barriers shall be continuous from outside wall to an outside wall, a floor to a floor, or from a smoke barrier to a smoke barrier, or a combination thereof.</u>

~~422.3~~ <u>422.4</u> Refuge Area. Not less than 30 net square feet (2.8 m^2) for each nonambulatory ~~patient~~ <u>care recipient</u> shall be provided within

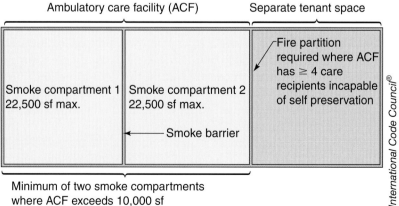

Ambulatory care facility (ACF)   Separate tenant space

Smoke compartment 1
22,500 sf max.

Smoke compartment 2
22,500 sf max.

Fire partition required where ACF has ≥ 4 care recipients incapable of self preservation

Smoke barrier

Minimum of two smoke compartments where ACF exceeds 10,000 sf

International Code Council®

Ambulatory care facility

the aggregate area of corridors, ~~patient~~ care recipient rooms, treatment rooms, lounge or dining areas, and other low-hazard areas ~~on each side of each smoke barrier~~ within each smoke compartment. Each occupant of an ambulatory care facility shall be provided with access to a refuge area without passing through or utilizing adjacent tenant spaces.

~~422.4~~ 422.5 Independent Egress. A means of egress shall be provided from each smoke compartment created by smoke barriers without having to return through the smoke compartment from which means of egress originated.

~~422.5~~ 422.6 Automatic Sprinkler Systems. Automatic sprinkler systems shall be provided for ambulatory care facilities in accordance with Section 903.2.2.

~~422.6~~ 422.7 Fire Alarm Systems. A fire alarm system shall be provided for ambulatory care facilities in accordance with Section 907.2.2.1.

CHANGE SIGNIFICANCE: Ambulatory care facilities, as regulated by Section 422, are health care facilities providing medical, surgical, psychiatric, nursing, or similar care on a less-than-24-hour basis to individuals who are rendered incapable of self-preservation. Defined as Group B occupancies, such facilities are generally regarded as moderate in hazard level due to their office-like conditions. However, additional hazards are typically present due to presence of individuals who are temporarily rendered incapable of self-preservation due to the application of nerve blocks, sedation, or anesthesia. While the occupants may walk in and walk out the same day with a quick recovery time after surgery, there is a period of time where a potentially large number of people could require physical assistance in case of an emergency that would require evacuation or relocation. In a multi-tenant or mixed-occupancy building where there are uses present other than an ambulatory care facility, a degree of fire separation between the ambulatory care facility and those nonrelated spaces is now required where the ambulatory care facility is intended to have at least four care recipients incapable of self-preservation at any one time. The minimum required separation, a fire partition complying with Section 708, is viewed as an important tool in isolating the ambulatory care portion of the building from fire hazards that may occur in other areas of the building. Where a building is considered wholly as an ambulatory care facility, or if the ambulatory care facility occupies the entire story of a multi-story building, there is no requirement for a fire partition to be constructed.

The smoke compartmentation requirements for ambulatory care facilities have also been revised. As a past requirement, ambulatory care facilities having more than 10,000 square feet of floor area were to be subdivided into at least two smoke compartments by smoke barriers in accordance with Section 709. The 10,000-square-foot threshold is now based on a story-by-story basis. As a result, a multi-story ambulatory care facility with no story exceeding 10,000 square feet in floor area would not be required to have the facility divided into smoke compartments. Another modification is the potential increase in the minimum number of smoke compartments that are required. Previously, there was no maximum size of smoke compartments, based on floor area, provided at least two compartments were provided. Consistent with similar provisions for Group I-2 occupancies, the maximum floor area of any one smoke compartment is now limited to 22,500 square feet.

424

Children's Play Structures

CHANGE TYPE: Modification

CHANGE SUMMARY: The regulations for children's play structures, previously limited in application only to covered mall buildings, are now applicable where such structures are located within any building regulated by the IBC, regardless of occupancy classification.

2012 CODE:

SECTION 424
CHILDREN'S PLAY STRUCTURES

402.12 424.1 Children's Playground Structures. ~~Structures intended as children's playgrounds~~ Children's play structures installed inside all occupancies covered by this code that exceed 10 feet (3048 mm) in height and 150 square feet (14 m^2) in area shall comply with Sections ~~402.12.1 through 402.12.4~~ 424.2 through 424.5.

402.12.1 424.2 Materials. Children's play~~ground~~ structures shall be constructed of noncombustible materials or of combustible materials that comply with the following:

1. Fire-retardant-treated wood complying with Section 2303.2.
2. Light-transmitting plastics complying with Section 2606.
3. Foam plastics (including the pipe foam used in soft-contained play equipment structures) having a maximum heat-release rate not greater than 100 kilowatts when tested in accordance with

Children's play structure

International Code Council®

UL 1975 <u>or when tested in accordance with NFPA 289, using the 20 kW ignition source.</u>

4. Aluminum composite material (ACM) meeting the requirements of Class A interior finish in accordance with Chapter 8 when tested as an assembly in the maximum thickness intended for use.

5. Textiles and films complying with the flame propagation performance criteria contained in NFPA 701.

6. Plastic materials used to construct rigid components of soft-contained play equipment structures (such as tubes, windows, panels, junction boxes, pipes, slides, and decks) exhibiting a peak rate of heat release not exceeding 400 kW/m^2 when tested in accordance with ASTM E 1354 at an incident heat flux of 50 kW/m^2 in the horizontal orientation at a thickness of 6 mm.

7. Ball pool balls, used in soft-contained play equipment structures, having a maximum heat-release rate not greater than 100 kilowatts when tested in accordance with UL 1975 <u>or when tested in accordance with NFPA 289, using the 20 kW ignition source.</u> The minimum specimen test size shall be 36 inches by 36 inches (914 mm by 914 mm) by an average of 21 inches (533 mm) deep, and the balls shall be held in a box constructed of galvanized steel poultry netting wire mesh.

8. Foam plastics shall be covered by a fabric, coating or film meeting the flame propagation performance criteria of NFPA 701.

9. The floor covering placed under the children's play~~ground~~ structure shall exhibit a Class I interior floor finish classification, as described in Section 804, when tested in accordance with NFPA 253.

~~402.12.2~~ **424.3 Fire Protection.** Children's play~~ground~~ structures ~~located within the mall~~ shall be provided with the same level of approved fire suppression and detection devices required for ~~kiosks and similar structures~~ <u>other structures in the same occupancy.</u>

~~402.12.3~~ **424.4 Separation.** <u>Children's play structures shall have a minimum horizontal separation from building walls, partitions, and from elements of the means of egress of not less than 5 feet (1524 mm).</u> Children's play~~ground~~ structures shall have a horizontal separation from other ~~structures within the mall~~ <u>children's play structures</u> of not less than 20 feet (6090 mm).

~~404.12.4~~ **424.5 Area Limits.** Children's play~~ground~~ structures shall be not greater than 300 square feet (28 m^2) in area, unless a special investigation<u>, acceptable to the building official,</u> has demonstrated adequate fire safety.

CHANGE SIGNIFICANCE: Play structures for children's activities have been regulated for some time by the IBC where such structures are located within covered mall buildings. The primary concern, consistent with that of other structures located within a covered mall building, is the combustibility of such play structures. Children's play structures must be constructed of noncombustible materials or, as an option if combustible, comply with the appropriate criteria previously established in

424 continues

424 continued   Section 424.2. Such alternative methods include the use of fire-retardant-treated wood, textiles complying with the designated flame propagation performance criteria, and plastics exhibiting an established maximum peak rate of heat release. Due to the potential fire hazards associated with children's play structures, the regulations are now applicable where such structures are located within any building regulated by the IBC, regardless of occupancy classification. As a result, the provisions have been relocated from Section 402 regulating covered mall and open mall buildings.

Included in the code change is a new allowance permitting the use of NFPA 289 as an alternative testing standard to UL 1975 for foam plastics and ball pool balls that are commonly used in soft-contained play equipment structures. NFPA 289 is widely considered to be more versatile than UL 1975 and is also likely to offer lower variability. The 20-kW gas burner ignition source in NFPA 289 was specifically designed with the intent of being a substitute for UL 1975.

501.2
Address Identification

CHANGE TYPE: Modification

CHANGE SUMMARY: The fire code official can now require that address numbers be posted in multiple locations if necessary to facilitate emergency response.

2012 CODE: 501.2 Address Identification. New and existing buildings shall be provided with approved address numbers or letters. Each character shall be not less than 4 inches (102 mm) in height and not less than 0.5 inch (12.7 mm) in width. They shall be installed on a contrasting background and be plainly visible from the street or road fronting the property. <u>When required by the fire code official, address numbers shall be provided in additional approved locations to facilitate emergency response</u>. Where access is by means of a private road and the building address cannot be viewed from the public way, a monument, pole or other approved sign or means shall be used to identify the structure. <u>Address numbers shall be maintained.</u>

CHANGE SIGNIFICANCE: Building address numbers enable responding fire, medical, or law enforcement forces to locate a building where emergency services are needed without going through a lengthy search procedure. As a fundamental requirement, the approved street numbers are to be placed in a location readily visible from the street fronting the property. In order to provide even better recognition, the fire code official can now require that the address numbers be posted in additional locations. Such locations, subject to approval of the fire code official, should be selected to help eliminate any confusion or delay in identifying the location of the emergency. Specific language has also been added to require such address markings to be maintained to allow for ongoing visibility.

International Code Council®

Address monument at high school

505.2.2

Mezzanine Means of Egress

CHANGE TYPE: Modification

CHANGE SUMMARY: The specific provisions for mezzanine means of egress have been deleted and replaced with a general reference to Chapter 10.

2012 CODE: ~~505.3~~ **505.2.2.** The means of egress for mezzanines shall comply with the applicable provisions of Chapter 10. ~~Each occupant of a mezzanine shall have access to at least two independent means of egress where the common path of egress travel exceeds the limitations of Section 1014.3. Where an unenclosed stairway provides a means of exit access from a mezzanine, the maximum travel distance includes the distance traveled on the stairway measured in the plane of the tread nosing. Accessible means of egress shall be provided in accordance with Section 1007.~~

> **Exception:** ~~A single means of egress shall be permitted in accordance with Section 1015.1.~~

CHANGE SIGNIFICANCE: Defined as "an intermediate level or levels between the floor and ceiling of any story," a mezzanine must also comply with the special conditions established in Section 505. By virtue of the conditions placed on such elevated floor levels, a mezzanine is not considered to create additional building area or an additional story for the purpose of limiting building size. In addition, a relaxation of the means of egress requirements has previously been granted in regard to the minimum required number of independent egress paths from the mezzanine. This special allowance for mezzanine means of egress has been deleted and the provisions have been replaced with a general reference to Chapter 10. As a result, the means of egress requirements for a mezzanine are consistent

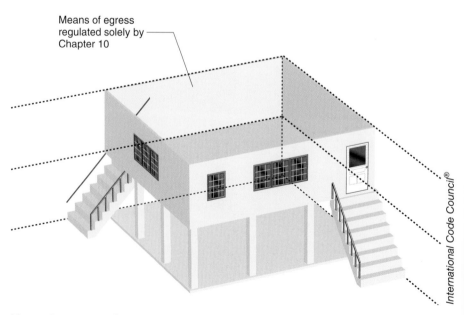

Means of egress regulated solely by Chapter 10

Mezzanine means of egress

International Code Council®

with those for other portions of the building regulated as the exit access. The consideration of an elevated floor level as a mezzanine no longer provides any special allowances for means of egress purposes.

The inclusion of specific mezzanine means of egress provisions caused inconsistent application in two ways. One, the incomplete statement of certain means of egress design requirements erroneously alluded to the notion that any provisions that were not stated did not apply. The new reference to Chapter 10 clarifies that mezzanines must comply with the general means of egress provisions as applicable. Secondly, the provision addressing required access to at least two means of egress did not indicate required compliance with the occupant load limitations of Table 1015.1. It was often assumed that a single means of egress was permitted as long as the common path of egress travel was compliant, regardless of the mezzanine's occupant load. This confusion has also been addressed through the blanket reference to Chapter 10.

506.2

Allowable Area Frontage Increase

CHANGE TYPE: Clarification

CHANGE SUMMARY: The method of calculating the appropriate allowable area increase for buildings fronting on public ways and/or open space has been clarified.

2012 CODE: 506.2 Frontage Increase.   Every building shall adjoin or have access to a public way to receive a building area increase for frontage. Where a building has more than 25 percent of its perimeter on a public way or open space having a minimum width of 20 feet (6096 mm), the frontage increase shall be determined in accordance with ~~the following~~ Equation 5-2:

$$I_f = [F/P - 0.25]W/30 \qquad \textbf{(Equation 5-2)}$$

where
  I_f = Area increase due to frontage.
  F = Building perimeter that fronts on a public way or open space having 20 feet (6096 mm) open minimum width (feet).
  P = Perimeter of entire building (feet).
  W = Width of public way or open space (feet) in accordance with Section 505.2.1.

506.2.1 Width Limits.   To apply this section, the value of W shall be at least 20 feet (6096 mm). Where the value of W varies along the perimeter of the building, the calculation performed in accordance with Equation 5-2 shall be based on the weighted average calculated in accordance with

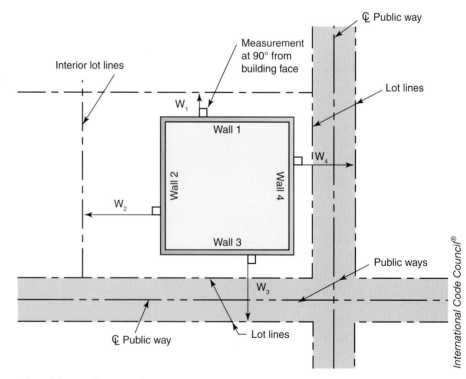

Allowable area frontage increase

Equation 5-3 for portions of the exterior perimeter walls ~~of each portion of exterior wall and open space~~ where the value of W is greater than or equal to 20 feet (6096 mm). Where the value of W exceeds 30 feet (9144 mm), a value of 30 feet (9144 mm) shall be used in calculating the weighted average, regardless of the actual width of the open space. W shall be measured perpendicular from the face of the building to the closest interior lot line. Where the building fronts on a public way, the entire width of the public way shall be used. Where two or more buildings are on the same lot, W shall be measured from the exterior face of ~~a~~ each building to the opposing exterior face of ~~an opposing~~ each adjacent building, as applicable.

$$\text{Weighted average } W = (L_1 \times w_1 + L_2 \times w_2 + L_3 \times w_3 \ldots)/F$$

(Equation 5-3)

where
  L_n = Length of a portion of the exterior perimeter wall.
  w_n = Width of open space associated with that portion of the exterior perimeter wall.
  F = Building perimeter that fronts on a public way or open space having a width of 20 feet (6096 mm) or more.

Exception: ~~The value of W divided by 30 shall be permitted to be a maximum of 2 when~~ Where the building meets ~~all~~ the requirements of Section 507, as applicable, except for compliance with the 60-foot (18 288-mm) public way or yard requirement, and the value of W is greater than 30 feet (9144 mm), the value of W divided by 30 shall be limited to a maximum of 2, ~~as applicable~~.

CHANGE SIGNIFICANCE: In the calculation of allowable building area, an increase is provided to the tabular areas established in Table 503 where there is a significant amount of open space adjacent to the building's exterior walls. If a building is provided with frontage consisting of public ways and/or open space for an increased portion of its perimeter, some benefit is accrued based on better access for the fire department. In addition, if the open space and public ways are wide enough, there is a benefit that is due to the decreased exposure from adjoining properties. The method of calculating the appropriate allowable area increase has been clarified in hopes of achieving more consistent interpretation, application, and enforcement of the provisions. The clarification focuses on three basic issues: (1) the amount of public way that can be utilized for a frontage increase, (2) the appropriate method of measuring the available width, and (3) the definition of the term "weighted average."

When calculating the frontage increase for a building that fronts on a public way, the entire width of the public way is to be used. Always the intent, the confusion evolves from Section 702.1, which states that fire separation distance is measured from the building face to the centerline of a street, alley, or public way. Although there was no reference to applying the fire separation distance for the frontage increase, the lack of any specific methodology created questions as to the appropriate method. The code now clearly states the intended application. As for the method of measuring the actual distance to be utilized in the frontage increase calculation, the measurement is taken at a right angle from the building face toward an interior lot line or public way as applicable. Again, the

506.2 continues

506.2 continued   new language provides a consistent approach to calculating the frontage increase that previously caused some confusion.

Where the open space or public way adjacent to the building's perimeter is between 20 feet (6096 mm) and 30 feet (9144 mm) in width, the code permits the use of the weighted average method in the determination of the allowable area frontage increase. This approach allows for the width W in Equation 5-2 to be somewhat representative of the actual open area that is available. A definition of "weighted average" intends to allow for a more consistent application of the provisions. In all cases, the additional and revised text is intended to simply clarify the intent of the provisions and does not modify its application.

CHANGE TYPE: Clarification

CHANGE SUMMARY: The allowance for occupancy groups not specifically scoped under the unlimited area building provisions of Section 507 to be located in such buildings under the accessory occupancies provisions of Section 508.2 is now contained within the code text.

507.1

Unlimited Area Buildings—Accessory Occupancies

Example:

Given: A 120,000 square-foot retail sales building housing a 1,080 square foot cafe with an occupant load of 72 persons. The building is fully sprinklered, is of Type IIB construction, and qualifies for unlimited area under the provisions of Section 507.

Determine: How the accessory occupancy provisions of Section 508.2 are applicable to the unlimited building provisions of Section 507.

1. Is the accessory occupancy subsidiary to the building's major occupancy?

   Yes, the cafe is intended as an extension of the sales function if they are part of the same tenant.

2. Is the accessory occupancy no more than 10% of the floor area of the story?

   Yes, 10% of 120,000 sq. ft. = 12,000 sq. ft. maximum; the cafe is 1,080 sq. ft.

Cafe
Group A-2

No physical or fire-resistance-rated separation required

Retail Sales
Group M

International Code Council®

Accessory occupancy in unlimited area building

3. Is the accessory occupancy no larger than the tabular values in Table 503?

   Yes, the tabular value for a Group A-2 of IIB construction is 9500 sq. ft.; the cafe is 1,080 sq. ft.

4. What is the occupancy classification of the cafe?

   Group A-2, based on the individual classification of the use.

5. How are the other requirements of the IBC applied?

   The provisions for each occupancy are applied to only that specific occupancy.

6. What is the allowable height and area of the building?

   The building's allowable height and area are based on the major occupancy involved, in this case it is Group M. Based upon the criteria of Section 507, the building is permitted to be unlimited in area and limited to two stories in height.

7. What is the allowable height for the lunchroom?

   Two stories, based on Table 503 for Group A-2 in a Type IIB building.

8. What is the minimum required separation between the cafe and the manufacturing area?

   There is no fire-resistive or physical separation required due to compliance with the provisions of Section 508.2 for accessory occupancies.

507.1 continues

507.1 continued

2012 CODE: 507.1 General.   The area of buildings of the occupancies and configurations specified ~~herein~~ <u>Sections 507.1 through 507.12</u> shall not be limited.

> **Exception:**   <u>Other occupancies shall be permitted in unlimited area buildings in accordance with the provisions of Section 508.2.</u>

CHANGE SIGNIFICANCE: In other than Type I construction, buildings are typically limited in allowable floor area based on occupancy classification and type of construction. However, Section 507 permits a variety of buildings to be unlimited in floor area where specified safeguards are present. Historically, structures constructed under the provisions for unlimited area buildings have performed quite well in regard to fire and life safety. In general, only those occupancies and configurations that are specifically identified in Section 507 are subject to the unlimited area allowance. A commonly applied method of allowing occupancy groups not specifically mentioned in Section 507 to be located in such buildings under the accessory occupancies provisions of Section 508.2 is now contained within the code text.

As an example, Group I occupancies are not mentioned in Section 507 as an occupancy group permitted to use the unlimited area building allowance. If the Group I complies with the provisions of Section 508.2 as an accessory occupancy, it is permitted to be located in an unlimited area building complying with Section 507. The basis for this allowance is found in Section 508.2.3 which indicates that the allowable area for an accessory occupancy is to be based upon the allowable area of the main occupancy. If the main occupancy is permitted by Section 507 to be in an unlimited area building, the accessory occupancy also enjoys the same benefit.

CHANGE TYPE: Clarification

CHANGE SUMMARY: In the determination of the required open space width for unlimited area building compliance, information has been added to clarify the measurement method to allow for consistent application.

2012 CODE: 507.1 General. The area of buildings of the occupancies and configurations specified in Sections 507.1 through 507.12 shall not be limited.

> **Exception:** Other occupancies shall be permitted in unlimited area buildings in accordance with the provisions of Section 508.2.

Where Sections 507.2 through 507.12 require buildings to be surrounded and adjoined by public ways and yards, those open spaces shall be determined as follows:

1. Yards shall be measured from the building perimeter in all directions to the closest interior lot lines or to the exterior face of an opposing building located on the same lot, as applicable.
2. Where the building fronts on a public way, the entire width of the public way shall be used.

507.1 continues

507.1
Unlimited Area Buildings—Open Space

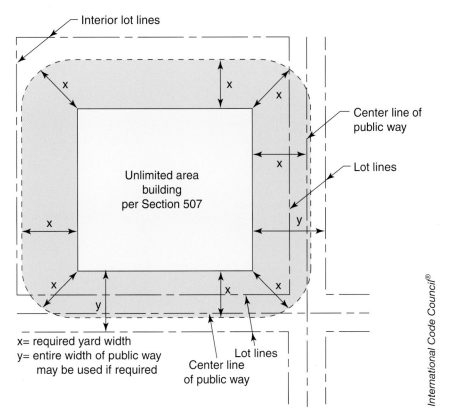

x= required yard width
y= entire width of public way may be used if required

Open space measurement for unlimited area building.

507.1 continued

CHANGE SIGNIFICANCE: All buildings regulated under the unlimited area building provisions of Section 507 are required to be surrounded by public ways and yards of substantial width. This continuous open space provides a means for the fire service to access the building as necessary from the exterior, while at the same time maintain a sizable separation from any other structures on the site. In the determination of the required open space width, information has been added to clarify the measurement method to allow for consistent application.

The required open space is to be measured from all points along the building's exterior wall in all directions, ensuring that the full perimeter of the building is provided with continuous open space. This differs somewhat from the right-angle method established for an allowable area frontage increase as set forth in Section 506.2 where such continuity of open space is not required at the building corners. Consistent with the measurement method of Section 506.2 for gaining a frontage increase, the open space width adjacent to those exterior walls fronting on a public way is permitted to include the entire width of the public way.

CHANGE TYPE: Clarification

CHANGE SUMMARY: The limitations placed on Group H occupancies permitted in unlimited area buildings have been clarified and reformatted to aid in their consistent application.

507.8 continues

507.8

Unlimited Area Buildings—Group H Occupancies

Unlimited Area Group F or S Building with Group H Occupancies

Example:

Given: A 130,000 square-foot Group F-1 of Type IIB construction having unlimited area under the provisions of Section 507.3. One Group H-3 storage room is located on the building's perimeter. Multiple Group H-3 storage rooms are located such that they are not located along an exterior wall.

Determine: The maximum allowable floor areas for the Group H-3 storage rooms.

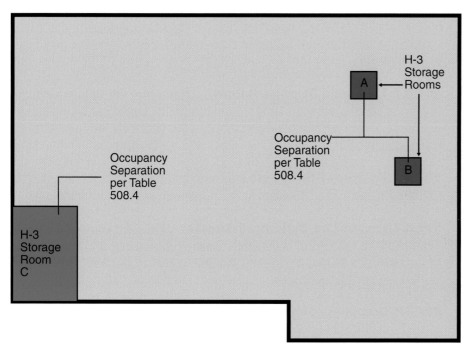

Group H occupancies in unlimited area building

Aggregate of Group H: (rooms A, B and C)	10% of 130,000 13,000 sf max, *nor*
	Table 503 with frontage 14,000 + 3,500 = 17,500 sf
	∴ Aggregate limit of 13,000 sf
Located within building: (rooms A & B)	25% of 14,000 ∴ Aggregate limit of 3,500 sf
Located on perimeter: (room C)	13,000 − 3,500 ∴ Limit of 9,500 sf

International Code Council®

507.8 continued

2012 CODE: 507.8 Group H Occupancies. Group H-2, H-3, and H-4 occupancies shall be permitted in unlimited areas buildings containing Group F and S occupancies, in accordance with Sections 507.3 and 507.4 and the <u>provisions</u> ~~limitations~~ of ~~this section~~ <u>Sections 507.8.1 through 507.8.4.</u>

507.8.1 Allowable Area. The aggregate floor area of ~~the~~ Group H occupancies located ~~at the perimeter of the~~ <u>in an</u> unlimited area building shall not exceed 10 percent of the area of the building nor the area limitations for the Group H occupancies as specified in Table 503 as modified by Section 506.2 based upon ~~the percentage of~~ the perimeter of each Group H floor area that fronts on a ~~street or other unoccupied~~ <u>public way or open</u> space.

507.8.1.1 Located within the Building. The aggregate floor area of Group H occupancies not located at the perimeter of the building shall not exceed 25 percent of the area limitations for the Group H occupancies as specified in Table 503.

507.8.1.1.1 Liquid Use, Dispensing, and Mixing Rooms. <u>Liquid use, dispensing, and mixing rooms having a floor area of not more than 500 square feet (46.5 m^2) need not be located on the outer perimeter of the building where they are in accordance with the *International Fire Code* and NFPA 30.</u>

507.8.1.1.2 Liquid Storage Rooms. <u>Liquid storage rooms having a floor area of not more than 1000 square feet (93 m^2) need not be located on the outer perimeter where they are in accordance with the *International Fire Code* and NFPA 30.</u>

507.8.1.1.3 Spray Paint Booths. <u>Spray paint booths that comply with the *International Fire Code* need not be located on the outer perimeter.</u>

507.8.2 Located on Building Perimeter. <u>Except as provided for in Section 508.8.1.1, Group H occupancies shall be located on the perimeter of the building. In Group H-2 and H-3 occupancies, not less than 25 percent of the perimeter of such occupancies shall be an exterior wall.</u>

507.8.3 Occupancy Separations. <u>Group H occupancies shall be separated from the remainder</u> ~~rest~~ of the unlimited area building and from each other in accordance with Table 508.4.

507.8.4 Height Limitations. For two-story unlimited area buildings, ~~the~~ Group H occupancies shall not be located more than one story above grade plane unless permitted <u>based on</u> ~~by~~ the allowable height in stories and feet as set forth in Table 503 <u>for</u> ~~based on~~ the type of construction of the unlimited area building.

CHANGE SIGNIFICANCE: Because many large industrial operations, both manufacturing and warehousing, have a need to utilize a limited quantity of high-hazard materials in some manner, it has for some time been deemed acceptable to allow limited Group H occupancies in Group F and S unlimited area buildings. Because of the allowances given to buildings of unlimited area under the provisions of Section 507, it is critical that any high-hazard Group H occupancies be strictly limited in floor area and adequately separated by fire-resistance-rated construction from the remainder of the building. The limitations placed on Group H occupancies in such buildings have been clarified and reformatted to aid in their consistent application, and related requirements from Section 415.3 addressing various types of hazardous material rooms have been replicated. There was no intent to change the application of the requirements related to those Group H occupancies located within unlimited area warehouses and factories regulated by Section 507.8.

A simple procedure allows the code user to establish the maximum amount of Group H occupancies permitted within the unlimited area Group F or S building, based upon the location of the Group H occupancies. The first step of the process now requires that the aggregate floor area limitation for all of the Group H occupancies throughout building be determined. Secondly, the maximum permitted floor area for those Group H occupancies not located at the building's perimeter is calculated. The third and final step is simply identifying the remaining floor area available for Group H uses located at the perimeter of the building.

509

Incidental Uses—General Provisions

CHANGE TYPE: Clarification

CHANGE SUMMARY: The concept of incidental uses has been clarified by eliminating the previous relationship with the mixed-occupancy provisions.

2012 CODE: ~~508.2.5~~ 509.1 ~~Separation of Incidental Uses~~. **General.** ~~The incidental accessory occupancies listed in Table 508.2.5 shall be separated from the remainder of the building or equipped with an automatic fire-extinguishing system, or both, in accordance with Table 508.2.5.~~ Incidental uses located within single occupancy or mixed occupancy buildings shall comply with the provisions of this section. Incidental uses are ancillary functions associated with a given occupancy that generally pose a greater level of risk to that occupancy and are limited to those uses listed in Table 509.

> **Exception:** Incidental ~~accessory occupancies~~ uses within and serving a dwelling unit are not required to comply with this section.

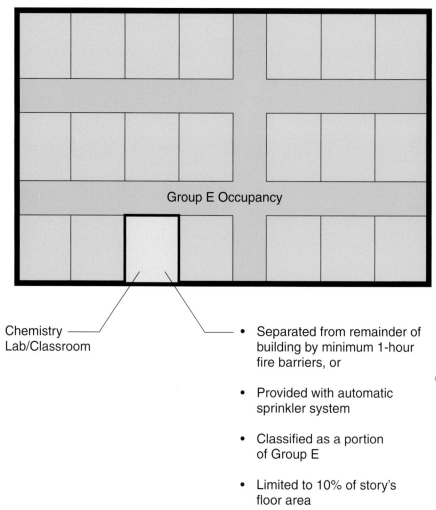

Group E Occupancy

Chemistry Lab/Classroom

- Separated from remainder of building by minimum 1-hour fire barriers, or
- Provided with automatic sprinkler system
- Classified as a portion of Group E
- Limited to 10% of story's floor area

School lab/classroom regulated as incidental use

International Code Council®

509.2 Occupancy Classification.   Incidental uses shall not be individually classified in accordance with Section 302.1. Incidental uses shall be included in the building occupancies within which they are located.

509.3 Area Limitations.   Incidental uses shall not occupy more than 10 percent of the building area of the story in which they are located.

CHANGE SIGNIFICANCE: There are occasionally one or more rooms or areas in a building that pose risks not typically addressed by the provisions for the occupancy group(s) under which the building is classified. However, such rooms or areas may functionally be an extension of the primary use. These types of spaces were previously considered in the 2009 IBC to be "incidental accessory occupancies" and evaluated according to their hazard level. As incidental accessory occupancies, such rooms or areas were inappropriately regulated under the mixed-occupancy provisions relating to accessory occupancies. The guiding concept has been restored with new terminology, "incidental uses," and a separate and distinct code section, Section 509. The term "incidental uses" more accurately reflects the long-held intent on this subject. Table 509, which lists those types of rooms and areas considered as incidental uses, addresses potential hazards based upon the specific use of the space, not the occupancy classification. The introduction of a new code section, outside of the scope of Section 508 addressing mixed occupancies, further emphasizes that incidental uses have no relationship to mixed-occupancy conditions.

New Section 509 begins with general applicability requirements that establish the scope of the incidental use provisions. It is now clearly stated that incidental uses can be located within both single-occupancy and mixed-occupancy buildings, eliminating the past confusion caused by the location of the requirements within the mixed-occupancy portion of the code. It is further stated that incidental uses are ancillary functions included within those occupancy groups that have been established. The scoping provisions also now reinforce the concept that it is the incidental uses that pose a risk to the remainder of the occupancy in which they are located. It is the intent to protect surrounding areas from the hazards that exist due to the incidental uses. Table 509 lists all rooms or areas that are to be regulated as incidental uses.

An important new provision now expressly states that incidental uses are not considered as separate and distinct occupancy classifications, but rather are classified the same as the building occupancies in which they are located. In the past, it was permissible to regulate the listed uses as incidental or, as an alternative, classify them as unique occupancy groups exempt from the incidental use requirements. This option is no longer available, as all rooms and areas identified in Table 509 must be regulated as incidental uses and comply with the requirements of Section 509.

The floor area limitation for incidental uses has been retained to place emphasis on the ancillary function of an incidental use. Each incidental use is limited to a maximum floor area of 10 percent of the floor area of the story in which it is located. Where there are two or more tenants located on the same story, the 10 percent limitation would presumably be based upon the floor area of each individual tenant space rather than that of the entire story. The application of the limit on a tenant-by-tenant basis is consistent with the concept of incidental uses typically being ancillary only to a portion of the building, the specific tenant occupancy.

509

Incidental Uses—Separation and Protection

CHANGE TYPE: Modification

CHANGE SUMMARY: An automatic sprinkler system is now the only fire-extinguishing system specifically permitted as a means of providing any fire protection required for incidental use rooms and areas.

2012 CODE: **509.4 Separation and Protection.**   The incidental uses listed in Table 509 shall be separated from the remainder of the building or equipped with an automatic sprinkler system, or both, in accordance with the provisions of that table.

~~508.2.5.1~~ **509.4.1 Fire-resistance-rated Separation.**   Where Table ~~508.2.5~~ 509 specifies a fire-resistance-rated separation, the incidental ~~accessory occupancies~~ uses shall be separated from the remainder of the building by a fire barrier constructed in accordance with Section 707 or a horizontal assembly constructed in accordance with Section 711, or both. Construction supporting 1-hour fire-resistance-rated fire barriers or horizontal assemblies used for incidental ~~accessory occupancy~~ use separations in buildings of Type IIB, IIIB, and VB construction is not required to be fire-resistance rated unless required by other sections of this code.

~~508.2.5.2~~ **509.4.2 Nonfire-resistance-rated Separation and Protection.**   Where Table ~~508.2.5~~ 509 permits an automatic ~~fire-extinguishing~~ sprinkler system without a fire barrier, the incidental ~~accessory occupancies~~ uses shall be separated from the remainder of the building by construction capable of resisting the passage of smoke. The walls shall extend from the top of the foundation or floor assembly below to the underside of the ceiling that is a component of a fire-resistance-rated floor assembly or roof assembly above or to the underside of the floor or roof sheathing, deck or slab above. Doors shall be self- or automatic closing

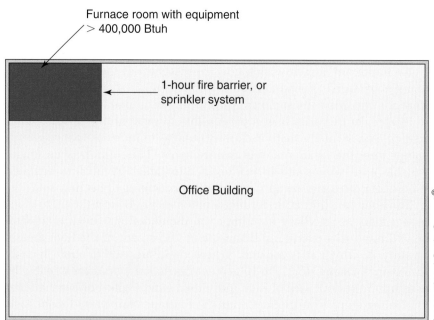

Furnace room with equipment > 400,000 Btuh

1-hour fire barrier, or sprinkler system

Office Building

International Code Council®

Separation/protection of incidental use

upon detection of smoke in accordance with Section 716.5.9.3. Doors shall not have air transfer openings and shall not be undercut in excess of the clearance permitted in accordance with NFPA 80. Walls surrounding the incidental use shall not have air transfer openings unless provided with smoke dampers in accordance with Section 710.7.

~~508.2.5.3~~ 509.4.2.1 Protection Limitation. Except as specified in Table ~~508.2.5~~ 509 for certain incidental ~~accessory occupancies~~ uses, where an ~~automatic fire-extinguishing system or an~~ automatic sprinkler system is provided in accordance with Table ~~508.2.5~~ 509, only the space occupied by the incidental ~~accessory occupancy~~ use need be equipped with such a system.

CHANGE SIGNIFICANCE: Where a building contains one or more incidental uses as established in Table 509, such uses must be specifically addressed due to the increased risk their presence poses to the remainder of the building. The IBC continues to recognize two methods of protection: construction of fire-resistance-rated separation elements and/or installation of a fire protection system. The acceptable protection methods are set forth in Table 509 for each of the incidental uses listed. Where a fire protection system is utilized, the code now mandates that an automatic sprinkler system be provided. Previously, any appropriate fire-extinguishing system was acceptable for protecting most of the incidental use rooms or areas. The change in terminology was made to eliminate any confusion regarding the anticipated results from the fire protection system. It was determined that the term "fire-extinguishing system" could lead the code user to think that a sprinkler system that is designed to control a fire is not adequate. The result of this modification also limits the type of fire-extinguishing permitted for incidental use purposes to an automatic sprinkler system. In order to satisfy the requirements, other types of fire-extinguishing systems are no longer acceptable unless specifically approved by the fire code official as set forth in Section 904.2.

Table 509

Incidental Uses—Rooms or Areas

CHANGE TYPE: Modification

CHANGE SUMMARY: The list of incidental uses now includes waste and linen collection rooms in Group B ambulatory care facilities and such rooms must be separated from the remainder of the building by minimum 1-hour fire-resistance-rated fire barriers and/or horizontal assemblies.

2012 CODE:

TABLE ~~508.2.5~~ 509 Incidental ~~Accessory Occupancies~~ Uses

Room or Area	Separation and/or Protection
Furnace room where any piece of equipment is over 400,000 Btu per hour input	1 hour or provide automatic ~~fire-extinguishing~~ sprinkler system
Rooms with boilers where the largest piece of equipment is over 15 psi and 10 horsepower	1 hour or provide automatic ~~fire-extinguishing~~ sprinkler system
Refrigerant machinery room	1 hour or provide automatic sprinkler system
Hydrogen cutoff rooms, not classified as Group H	1 hour in Group B, F, M, S, and U occupancies; 2 hours in Group A, E, I, and R occupancies.
Incinerator rooms	2 hours and automatic sprinkler system
Paint shops, not classified as Group H, located in occupancies other than Group F	2 hours or 1 hour and provide automatic ~~fire-extinguishing~~ sprinkler system
Laboratories and vocational shops, not classified as Group H, located in a Group E or I-2 occupancy	1 hour or provide automatic ~~fire-extinguishing~~ sprinkler system
Laundry rooms over 100 square feet	1 hour or provide automatic ~~fire-extinguishing~~ sprinkler system
Group I-3 cells equipped with padded surfaces	1 hour
~~Group I-2~~ Waste and linen collection rooms located in either Group I-2 occupancies or ambulatory care facilities	1 hour
Waste and linen collection rooms over 100 square feet	1 hour or provide automatic ~~fire-extinguishing~~ sprinkler system
Stationary storage battery systems having a liquid electrolyte capacity of more than 50 gallons for flooded lead-acid, nickel cadmium or VRLA, or ~~a lithium-ion capacity of more than~~ 1000 pounds for lithium-ion and lithium metal polymer used for facility standby power, emergency power, or ~~uninterrupted~~ uninterruptable power supplies	1 hour in Group B, F, M, S, and U occupancies; 2 hours in Group A, E, I, and R occupancies.
~~Rooms containing fire pumps in nonhigh-rise buildings~~	~~2 hours; or 1 hour and provide automatic sprinkler system throughout the building~~
~~Rooms containing fire pumps in high-rise buildings~~	~~2 hours~~

CHANGE SIGNIFICANCE: The purpose of Table 509 is twofold: (1) it identifies the rooms or areas that are specifically regulated as "Incidental Uses," and (2) it establishes the type and degree of fire protection that is to be afforded the remainder of the building. The listed rooms and areas have been selected for inclusion because of the increased hazard they present to the other building uses. The intent of the fire separation and fire sprinkler requirements is to provide safeguards because of the increased hazard level presented by the incidental use. Minor changes have been made to the listing of incidental uses, as well as to the type of fire protection systems that are to be installed.

The list of incidental uses now includes all waste and linen collection rooms in Group B ambulatory care facilities. Such rooms have previously only been regulated as incidental uses if they were located in a Group I-2 occupancy or over 100 square feet in floor area. A minimum 1-hour fire-resistance-rated separation, previously required for such rooms in Group I-2 occupancies, is now also applicable to all waste and linen collection rooms that are provided in ambulatory care facilities as regulated by Section 422. Several changes throughout the code, this one included, were made to regulate ambulatory care facilities consistent with the higher level of protection required when some occupants rely on staff for assisted evacuation, similar to nursing homes and hospitals. In another change to the table, rooms containing fire pumps were removed from the listing of incidental uses because they do not meet the criteria for incidental uses. The separation/protection requirements for fire pump rooms comprehensively addressed in Section 913 are intended to protect the fire pump room from risks found in the remainder of the building, the opposite of the intent of the incidental use provisions.

Table 509 also reflects the new requirement that automatic sprinkler systems be utilized for fire protection purposes, where appropriate, rather than automatic fire-extinguishing systems.

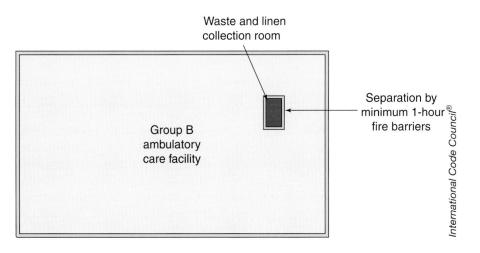

Incidental use in ambulatory care facility

Table 602, Note h

Fire Ratings of Exterior Walls

CHANGE TYPE: Modification

CHANGE SUMMARY: Nonbearing exterior walls that are permitted to have unlimited unprotected opening based on Table 705.8 are no longer required to have a fire-resistance rating due to fire separation distance.

2012 CODE:

TABLE 602 Fire-Resistance Rating Requirements for Exterior Walls Based on Fire Separation Distance[a,e,h]

Fire Separation Distance = X (feet)	Type of Construction	Occupancy Group H[f]	Occupancy Group F-1, M, S-1[g]	Occupancy Group A, B, E, F-2, I, R, S-2[g], U[b]
X < 5[c]	All	3	2	1
5 ≤ X < 10	IA	3	2	1
	Others	2	1	1
10 ≤ X < 30	IA, IB	2	1	1[d]
	IIB, VB	1	0	0
	Others	1	1	1[d]
X ≥ 30	All	0	0	0

For SI: 1 foot = 304.8 mm

a. - g. (no changes to text)

h. Where Table 705.8 permits nonbearing exterior walls with unlimited area of unprotected openings, the required fire resistance rating for the exterior walls is 0 hours.

CHANGE SIGNIFICANCE: As far as exterior wall protection is concerned, the IBC is based on the philosophy that an owner can have no control over what occurs on an adjacent lot and, therefore, the location of buildings on the owner's lot must be regulated relative to the lot line. The lot-line concept provides a convenient means of protecting one building from another insofar as potential radiant heat transmission and direct flame impingement is concerned. The regulations for protection of the wall itself based on proximity to the lot line are contained in Table 602, while the requirements for openings such as doors and windows in the wall are regulated by Table 705.8. The addition of footnote h to Table 602 addresses a previous conflict with Table 705.8.

Table 705.8 allows for an unlimited amount of unprotected openings in exterior walls of a sprinklered building that has a fire separation distance of 20 feet or greater. However, in certain buildings, Table 602 has required those same exterior walls to be fire-resistance-rated for a minimum of 1 hour. It was concluded that any nonbearing exterior wall that is permitted to be entirely open due to the unlimited unprotected opening allowances of Table 705.8 need not be required to have a fire-resistance rating due to fire separation distance.

Example:

Given: A Type IIA building housing a Group M occupancy

Determine: The minimum required fire-resistance rating for nonbearing exterior wall "C" based on fire separation distance.

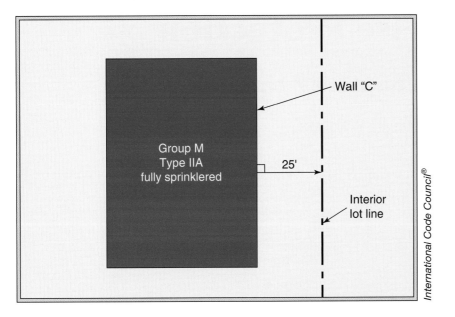

Solution: Table 602 initially requires a minimum 1-hour rating; however, in applying footnote "h", Table 705.8 has no limit on unprotected openings where the building is sprinklered and the fire separation distance is 20 feet or more. Therefore, there is no fire-resistance rating required for wall "C".

There are only a small percentage of buildings where this new footnote is applicable. It has no effect on:

- Exterior bearing walls
- Group H-1, H-2, and H-3 occupancies
- Nonsprinklered buildings
- Buildings of Type IIB and VB construction, other than Groups H-4 and H-5
- Exterior walls with a fire separation distance of less than 20 feet
- Exterior walls with a fire separation distance of 30 feet or more

PART 3

Fire Protection

Chapters 7 through 9

The fire-protection provisions of the International Building Code are found primarily in Chapters 7 through 9. There are two general categories of fire protection: active and passive. The fire and smoke resistance of building elements and systems in compliance with Chapter 7 provides for passive protection. Chapter 9 contains requirements for various active systems often utilized in the creation of a safe building environment, including automatic sprinkler systems, standpipe systems, and fire alarm systems. To further address the rapid spread of fire, the provisions of Chapter 8 are intended to regulate interior-finish materials, such as wall and floor coverings. ■

701.2

Multiple-Use Fire Assemblies

703.4

Establishing Fire Resistance Ratings

703.7

Identification of Fire and Smoke Separation Walls

704.11

Fire Protection of Bottom Flanges

705.2

Extent of Projections beyond Exterior Walls

705.2.3

Protection of Combustible Projections

705.3

Projections from Buildings on the Same Lot

706.2

Double Fire Walls

706.6, 706.6.2

Fire Wall Height at Sloped Roofs

707.8, 707.9

Intersections of Fire Barriers at Roof Assemblies

701.2

Multiple-Use Fire Assemblies

CHANGE TYPE: Clarification

CHANGE SUMMARY: Where a single fire assembly serves multiple purposes, such as a wall being utilized as both a fire barrier and a fire partition, it has been clarified that all of the applicable requirements for both types of fire separation walls must be met.

2012 CODE: **701.2 Multiple-Use Fire Assemblies.** Fire assemblies that serve multiple purposes in a building shall comply with all of the requirements that are applicable for each of the individual fire assemblies.

CHANGE SIGNIFICANCE: Fire assemblies are utilized throughout the code for a variety of purposes. Fire walls, fire barriers, fire partitions, smoke barriers, and smoke partitions are selectively mandated in order to provide the appropriate fire separation necessary for the situation under consideration. There are often times where the placement of one type of fire assembly coincides with that of another type. For example, a corridor wall required to be constructed as a fire partition may also serve as the boundary of a fire area, necessitating the use of a fire barrier. It is important to recognize that where such multiple uses exist, the requirements for each of the fire assemblies must be met. This approach is consistent with the accepted concept of applying the most restrictive provisions where multiple requirements are applicable and clarifies the method to deal with such situations.

The most common characteristic that must be evaluated is the minimum required fire-resistance rating of the fire assembly. A single wall assembly serving as both a 3-hour fire wall and a 2-hour fire barrier must be designed and constructed to the higher fire-resistance level, in this case 3 hours. However, the primary intent of separation in regard to fire and/or smoke protection must also be evaluated. In the example of a corridor fire partition also being utilized as fire barrier, the primary purpose of the fire partition can be viewed as smoke containment. On the other hand, the fire barrier is considered more of a fire separation feature. Where both types of wall assemblies occur at the same location, the necessary elements to provide both smoke resistance and fire resistance must be in place.

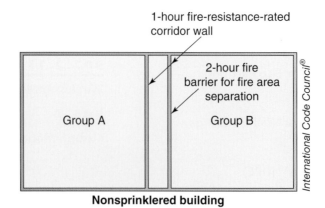

Nonsprinklered building

Wall assembly used for multiple separations

The variation in minimum requirements is typically most evident in the opening protectives that maintain the integrity of the fire assembly. Although the fire door assembly located in a 2-hour fire barrier must have a minimum fire-protection rating of 1½ hours, only a minimum 20-minute fire-protection rating is required for a corridor door installed in a fire partition. However, the corridor door must also meet the requirements for a smoke and draft control assembly tested in accordance with UL 1784 with a maximum air leakage rate. In the single fire assembly under discussion, a door opening would need to be protected with a minimum 1½-hour fire door assembly that also qualifies as a complying smoke and draft control door assembly.

703.4

Establishing Fire Resistance Ratings

ASTM E 119 vertical furnace

International Code Council®

CHANGE TYPE: Clarification

CHANGE SUMMARY: Specific language has been added to clarify that a fire suppression system is not permitted to be included as part of a tested building element, component, or assembly in order to establish the fire-resistance rating.

2012 CODE: 703.4 Automatic Sprinklers. Under the prescriptive fire resistance requirements of the *International Building Code*, the fire resistance rating of a building element, component, or assembly shall be established without the use of automatic sprinklers or any other fire suppression system being incorporated as part of the assembly tested in accordance with the fire exposure, procedures, and acceptance criteria specified in ASTM E 119 or UL 263. However, this section shall not prohibit or limit the duties and powers of the building official allowed by Sections 104.10 and 104.11.

CHANGE SIGNIFICANCE: As a general rule, the fire-resistance ratings of building elements, components, and assemblies established throughout the IBC are to be determined in accordance with the test procedures as established in ASTM E 119 or UL 263. In addition, alternative methods for determining fire-resistance set forth in Section 703.3 are acceptable where such methods are based on the fire exposure and acceptance criteria specified in ASTM E 119 or UL 263. Specific language has now been added to clarify that a fire suppression system is not permitted to be included as part of the tested element, component, or assembly in order to establish the fire-resistance rating. It has been generally accepted that the various fire resistance ratings mandated throughout the code have been established based on an assumption that the fire assembly would pass the standardized tests without the assistance of water cooling during fire exposure. The new provision clarifies this assumption.

It is important to note that these provisions are not intended to limit the use of Section 104 by building officials for the approval of alternative methods on a case-by-case basis. While the prescriptive provisions of the code are based upon fire-resistance ratings established without the benefit of any automatic fire suppression system, the building official continues to maintain the authority to evaluate and approve alternative materials, designs, and methods of construction that meet the intent and purpose of the code.

703.7

Identification of Fire and Smoke Separation Walls

CHANGE TYPE: Modification

CHANGE SUMMARY: The size and location of identifying markings required on vertical fire assemblies in accessible above-ceiling spaces have been modified to increase the potential for such markings to be seen.

2012 CODE: 703.7 Marking and Identification. Fire walls, fire barriers, fire partitions, smoke barriers, and smoke partitions or any other wall required to have protected openings or penetrations shall be effectively and permanently identified with signs or stenciling. Such identification shall:

1. Be located in accessible concealed floor, floor/ceiling, or attic spaces

2. Be <u>located within 15 feet (4572 mm) of the end of each wall and</u> ~~repeated~~ at intervals not exceeding 30 feet (9144 mm) measured horizontally along the wall or partition

3. Include lettering not less than ~~0.5 inch (12.7 mm)~~ <u>3 inches (76 mm)</u> in height <u>with a minimum ⅜-inch (9.5-mm) stroke in a contrasting color</u> incorporating the suggested wording: "FIRE AND/OR SMOKE BARRIER—PROTECT ALL OPENINGS" or other wording.

Exception: Walls in Group R-2 occupancies that do not have a removable decorative ceiling allowing access to the concealed space.

703.7 continues

Minimum
3 in. high with ⅜"
stroke lettering

Sign or stenciling
at maximum
30-ft. intervals and
within 15 ft. of
end of wall

Concealed space

Floor or roof deck

Fire barrier
Protect all openings

Ceiling

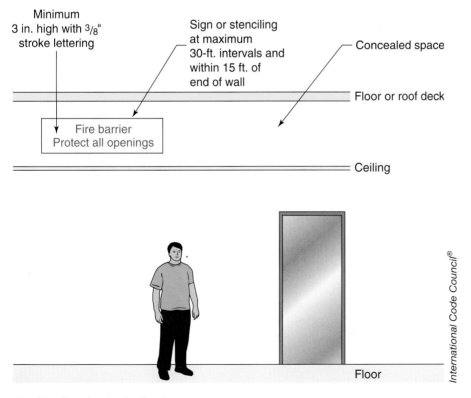

International Code Council®

Identification sign for fire barrier

703.7 continued

CHANGE SIGNIFICANCE: The integrity of fire and/or smoke separation walls is subject to compromise during the life of a building. During maintenance and remodel activities, it is not uncommon for new openings and penetrations to be installed in a fire separation wall without the recognition that the integrity of the construction must be maintained or that some type of fire or smoke protective is required. Provisions mandating the appropriate identification of such walls under certain conditions have been modified to better ensure that tradespeople, maintenance workers, and inspectors will recognize the required level of protection that must be maintained.

It is intended that the identification marks be located in areas not visible to the general public. Specific locations set forth in the provisions indicate that the identification is to be provided above any lay-in panel ceiling or similar concealed space that is deemed to be accessible. In addition to previous requirements for locating the identifying markings at maximum 30-foot intervals, it is now also necessary that such markings be provided no more than 15 feet from the end of each wall requiring such identification. This additional requirement increases the possibility that the identifying markings will be visible during any work on the wall assemblies. The minimum required letter height has also been increased from ½ inch to 3 inches to make the markings much more visible. In addition, a minimum stroke width has been established at ⅜ inch and the lettering must be of a color that contrasts with its background. All of the code modifications are intended to increase the possibility that the identification of the information will be achieved.

The requirements apply to all wall assemblies where openings or penetrations are required to be protected. This would include exterior fire-resistance-rated walls as well as fire walls, fire barriers, fire partitions, smoke barriers, and smoke partitions.

CHANGE TYPE: Modification

CHANGE SUMMARY: In buildings required to be fire-resistance-rated, the permitted span of a lintel, shelf angle, or plate whose bottom flange has no fire protection has been increased slightly to accommodate an opening containing a pair of 3-foot doors.

2012 CODE: 704.11 Bottom Flange Protection.   Fire protection is not required at the bottom flange of lintels, shelf angles, and plates, spanning not more than ~~6 feet (1829 mm)~~ 6 feet 4 inches (1931 mm) whether part of the primary structural frame or not, and from the bottom flange of lintels, shelf angles, and plates not part of the structural frame, regardless of span.

CHANGE SIGNIFICANCE: Where a building is required to be fire-resistance-rated, the fire protection need not necessarily extend to the bottom flange of lintels, shelf angles, and plates that occur over wall openings. Where such elements are not regarded as part of the primary structural frame, there is no maximum length on the span of the member with the unprotected bottom flange. Where the lintel, shelf angle, or plate is considered part of the building's primary structural frame, the member span with the unprotected flange is limited. Typically applicable to exposed steel lintels or angles over openings in masonry walls, the span limitation of 6 feet has been increase to 6 feet 4 inches to accommodate a pair of 3-foot 0-inch doors in a single door frame.

Although a portion of the structural frame is unprotected, it is assumed that the arching action of the masonry or concrete above the short-span lintel will prevent anything other than a localized failure. Furthermore, only the bottom flange is permitted to be unprotected, and as a result, the wall supported by the lintel will act as a heat sink to draw heat away from the lintel and thereby increase the length of time until failure occurs that is due to heat.

704.11

Fire Protection of Bottom Flanges

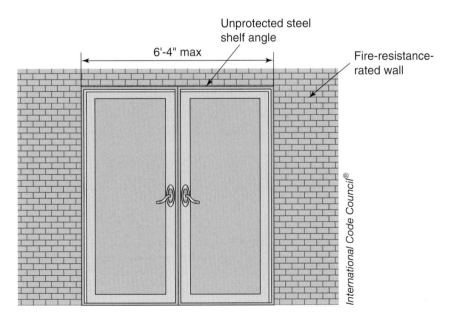

Unprotected shelf angle above door assembly

705.2

Extent of Projections beyond Exterior Walls

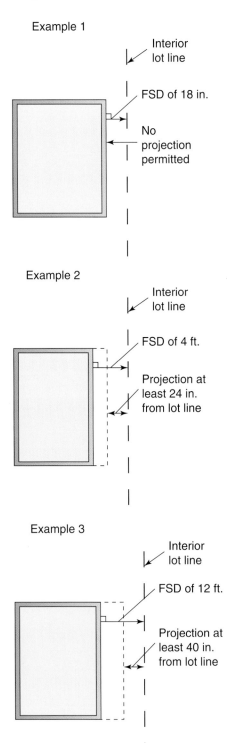

Example 1

Interior lot line

FSD of 18 in.

No projection permitted

Example 2

Interior lot line

FSD of 4 ft.

Projection at least 24 in. from lot line

Example 3

Interior lot line

FSD of 12 ft.

Projection at least 40 in. from lot line

Maximum permitted projections

International Code Council®

CHANGE TYPE: Modification

CHANGE SUMMARY: The permitted extent of projections beyond exterior walls is now regulated in a straightforward manner that establishes a minimum clear distance that is required between the leading edge of the projection and the line used to establish the fire separation distance.

2012 CODE: 705.2 Projections. Cornices, eave overhangs, exterior balconies, and similar projections extending beyond the exterior wall shall conform to the requirements of this section and Section 1406. Exterior egress balconies and exterior exit stairways shall also comply with Sections 1019 and 1026, respectively. Projections shall not extend ~~beyond the distance determined by the following three methods, whichever results in the lesser projection:~~ <u>any closer to the line used to determine the fire separation distance than shown in Table 705.2.</u>

1. ~~A point one-third the distance from the exterior face of the wall to the lot line where protected openings or a combination of protected and unprotected openings are required in the exterior walls.~~

2. ~~A point one-half the distance from the exterior face of the wall to the lot line where all openings in the exterior wall are permitted to be unprotected or the building is equipped throughout with an automatic sprinkler system installed under the provisions of Section 705.8.2.~~

3. ~~More than 12 inches (305 mm) into areas where openings are prohibited.~~

Exception: Buildings on the same lot and considered as portions of one building in accordance with Section 705.3 are not required to comply with this section.

TABLE 705.2   Minimum Distance of Projection

Fire Separation Distance (FSD)	Minimum Distance From Line Used To Determine FSD
0 feet to less than 2 feet	Projections not permitted
2 feet to less than 5 feet	24 inches
5 feet or greater	40 inches

CHANGE SIGNIFICANCE: Where projections like cornices, eave overhangs, and balconies are located such that there is a limited fire separation distance, they create problems due to the trapping of convected heat from a fire in an adjacent building. Therefore, the extent of such projections is addressed. By providing some degree of physical horizontal separation from adjacent buildings and lots, the concerns associated with such trapped heat can be substantially reduced. The permitted extent of projections has been modified and simplified from previous requirements that were viewed as complex, confusing, and inconsistent. The permitted extent of projections is now regulated in a manner that establishes a minimum required clear distance between the leading edge of the projection and the line used to establish the fire separation distance. This methodology is in contrast to the previous approach, which limited the distance a projection could extend beyond the building's exterior wall.

The permitted extent of projections is established by new Table 705.2 and based on the distance between the building's exterior wall and an interior lot line, centerline of a public way, or assumed imaginary line between two buildings on the same lot. Under these new limitations, no projections are permitted beyond the exterior wall where the wall has a fire separation distance of less than 2 feet. Where the exterior wall is located such that the fire separation distance is at least 2 feet, a projection is permitted but its extent is regulated. For example, if a building's exterior wall is located 54 inches from an adjacent to an interior lot line, at least 24 inches must be maintained between the edge of the projection and the lot line. This results in a maximum permitted projection beyond the exterior wall of 30 inches. If an exterior wall is located 96 inches from an interior lot line, the projection may not extend within 40 inches of the lot line. In this case, a projection of up to 56 inches is permitted.

705.2.3

Protection of Combustible Projections

CHANGE TYPE: Modification

CHANGE SUMMARY: The threshold at which combustible projections must be protected for fire exposure has been modified to include projections with greater fire separation distances than previously regulated.

2012 CODE: 705.2.3 Combustible Projections. Combustible projections <u>extending to within 5 feet of the line used to determine the fire separation distance, or</u> located where openings are not permitted, or where protection of <u>some</u> openings is required shall be of at least 1-hour fire-resistance-rated construction, Type IV construction, fire-retardant-treated wood, or as required by Section 1406.3.

> **Exception:** Type V<u>B</u> construction shall be allowed <u>for combustible projections</u> in Group R-3 <u>and U</u> occupancies <u>with a fire separation distance greater than or equal to 5 feet</u>.

CHANGE SIGNIFICANCE: Projections from buildings are regulated in order to prevent a fire hazard from the inappropriate use of combustible materials extending beyond exterior walls. Thus, the IBC requires that projections from walls of Type I and II buildings be of noncombustible materials. For buildings permitted to be of combustible construction (Type III, IV, and V construction), both combustible and noncombustible materials are permitted. Where combustible projections are utilized, they are regulated based upon the potential for a severe fire exposure hazard. This exposure potential is directly related to the distance between the leading edge of the projection and the line used to determine the fire separation distance. The threshold at which combustible projections must

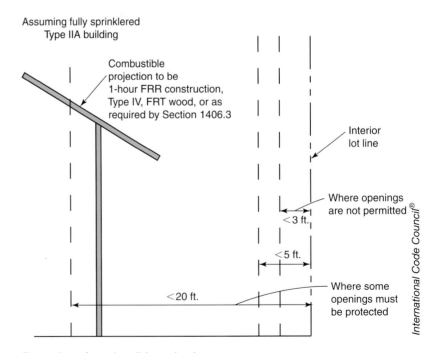

Protection of combustible projections

be protected for fire exposure has been modified to include projections with greater fire separation distances than previously regulated.

The new criteria require that combustible projections be of 1-hour construction, Type IV construction, or fire-retardant treated wood, or alternatively comply with Section 1406.3, where any one of the three following listed conditions exists:

1. The projection extends within a distance of 5 feet to the line where fire separation distance is measured (interior lot line, centerline of a public way, or assumed imaginary line between two buildings on the same lot).

2. The projection extends into the zone where exterior wall openings are prohibited (as regulated by Table 705.8).

3. The projection extends into the zone where unlimited unprotected exterior wall openings are prohibited (as regulated by Table 705.8).

The zone described in item 3 is always inclusive of those described in items 1 and 2, therefore it becomes the regulating provision. In order to determine the point at which a combustible projection is no longer regulated by Section 705.2.3, it is necessary to apply the information established in Table 705.8. Only where the table provides for no limit on the allowable area of exterior wall openings where the degree of opening protection is considered to be unprotected is a combustible projection outside the scope of Section 705.3.2.

A change to the exception provides the same level of protection to Group R-3 and U occupancies as established in the IRC. In addition, the reference to Type VB construction removes any ambiguity as to whether rated or nonrated projections are required.

705.3

Projections from Buildings on the Same Lot

CHANGE TYPE: Modification

CHANGE SUMMARY: Projections extending beyond opposing exterior walls of two buildings on the same lot must now comply with the projection provisions of Section 705.2 based upon the location of the projections in relation to the assumed imaginary line placed between the buildings.

2012 CODE: 705.3 Buildings on the Same Lot. For the purposes of determining the required wall and opening protection, <u>projections,</u> and roof-covering requirements, buildings on the same lot shall be assumed to have an imaginary line between them.

Where a new building is to be erected on the same lot as an existing building, the location of the assumed imaginary line with relation to the existing building shall be such that the exterior wall and opening protection of the existing building meet the criteria as set forth in Sections 705.5 and 705.8.

> **Exception:** Two or more buildings on the same lot shall either be regulated as separate buildings or shall be considered as portions of one building if the aggregate area of such buildings is within the limits specified in Chapter 5 for a single building. Where the buildings contain different occupancy groups or are of different types of construction, the area shall be that allowed for the most restrictive occupancy or construction.

CHANGE SIGNIFICANCE: As a fundamental concept, multiple buildings on the same lot are to be regulated as separate and distinct buildings for applying code requirements. In the determination of required exterior wall protection, exterior opening protection, and roof-covering requirements, it is necessary that an imaginary line be assumed between the buildings to properly evaluate the necessary degree of fire protection. As a portion of this evaluation, the projections extending beyond opposing exterior walls of these buildings must now also be considered. The provisions of Section 705.2 regulating the extent, construction, and protection of

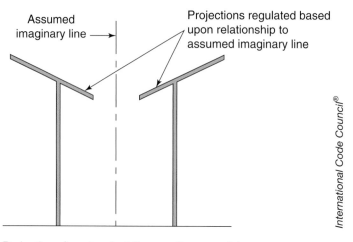

Projections from two buildings on the same lot

International Code Council®

projections are now applicable where Section 705.3 is applicable. If the exception to Section 705.3 is utilized in order to consider the multiple buildings as a single building, the imaginary line is not to be assumed, and those projections beyond opposing exterior walls are not regulated as indicated in the exception to Section 705.2.

706.2

Double Fire Walls

CHANGE TYPE: Addition

CHANGE SUMMARY: In order to satisfy the intended objective of structural stability, the use of a double fire wall complying with NFPA 221 is now permitted as an alternative to a single fire wall.

2012 CODE: 706.2 Structural Stability. Fire walls shall have sufficient structural stability under fire conditions to allow collapse of construction on either side without collapse of the wall for the duration of time indicated by the required fire-resistance rating <u>or shall be constructed as double fire walls in accordance with NFPA 221</u>.

CHANGE SIGNIFICANCE: Fire walls are fire-resistance-rated building elements constructed within a structure that are utilized to create two or more smaller-area buildings. Each portion of the structure so separated may be considered a separate and unique building for all purposes of the code. One of the key criteria to the design and construction of a fire wall is that it performs structurally under fire conditions in a manner that will maintain the integrity of the fire separation. A new allowance permits the use of a double fire wall in lieu of a single fire wall that satisfies the intended objective of structural stability.

Double fire walls are simply two back-to-back walls, each having an established fire-resistive rating. While acceptable for use in a new structure, double fire walls are most advantageous where an addition is being constructed adjacent to an existing building and the intent is to regulate the addition as a separate building under the fire wall provisions. The exterior wall of the existing building, if compliant, can be utilized as one wall of the double wall system, with the new wall of the addition providing the second wall.

Double fire wall assemblies are to comply with the applicable provisions of NFPA 221, *Standard for High Challenge Fire Walls, Fire Walls,*

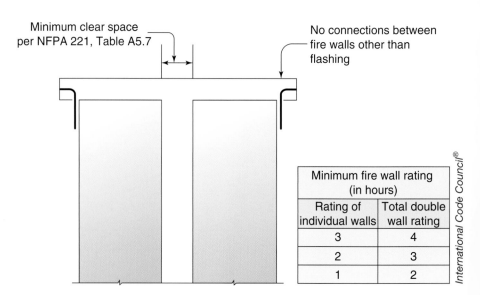

Minimum clear space per NFPA 221, Table A5.7

No connections between fire walls other than flashing

Minimum fire wall rating (in hours)	
Rating of individual walls	Total double wall rating
3	4
2	3
1	2

International Code Council®

Double fire wall

and Fire Barrier Walls. This standard addresses a number of criteria for double fire walls, including fire-resistance rating, connections, and structural support. In order to meet the minimum fire-resistance rating for a fire wall as set forth in IBC Table 706.4, each individual wall of a double fire wall assembly is permitted to be reduced to 1 hour less than the minimum required rating for a single fire wall. For example, where IBC Table 706.4 requires the use of a minimum 3-hour fire wall, two 2-hour fire-resistance-rated (double) fire walls can be utilized. Similarly, two 3-hour fire walls in a double wall system can be considered as a single 4-hour fire wall, and two 1-hour fire walls used as a double wall qualify as a single 2-hour fire wall.

Because the intended goal of fire wall construction is to allow collapse of a building on either side of the fire wall while maintaining an acceptable level of fire separation, the only connection permitted by NFPA 221 between the two walls that make up the double fire wall is the flashing, if provided. Illustrated in the explanatory material to the standard, the choice of flashing methods must provide for separate flashing sections in order to maintain a complete physical separation between the walls. Each individual wall of the double wall assembly must be supported laterally without any assistance from the adjoining building. In addition, a minimum clear space between the two walls is recommended by NFPA 221 in order to allow for thermal expansion between unprotected structural framework, where applicable, and the wall assemblies that make up the double fire wall.

706.6, 706.6.2

Fire Wall Height at Sloped Roofs

CHANGE TYPE: Addition

CHANGE SUMMARY: In the application of the minimum parapet height requirement for fire walls, provisions have been added to address conditions where a sloped roof occurs on one or both sides of the fire wall parapet.

2012 CODE: **706.6 Vertical Continuity.** Fire walls shall extend from the foundation to a termination point at least 30 inches (762 mm) above both adjacent roofs.

Exceptions:

1.-5. (no changes to text)
6. Buildings with sloped roofs in accordance with Section 706.6.2.

706.6.2 Buildings with Sloped Roofs. Where a fire wall serves as an interior wall for a building, and the roof on one side or both sides of the fire wall slopes toward the fire wall at a slope greater than 2 units vertical in 12 units horizontal (2:12), the fire wall shall extend to a height equal to the height of the roof located 4 feet (1219 mm) from the fire wall plus 30 inches (762 mm). In no case shall the extension of the fire wall be less than 30 inches (762 mm).

CHANGE SIGNIFICANCE: The fire-resistive separation provided by a fire wall must typically extend above the roof surface in order to minimize any potential for fire spread at or above the roof. A minimum parapet height of 30 inches has historically been required to ensure that any fire that reaches the roof of a structure divided by one or more fire walls will not travel beyond the fire wall separation to the adjacent building. In the application

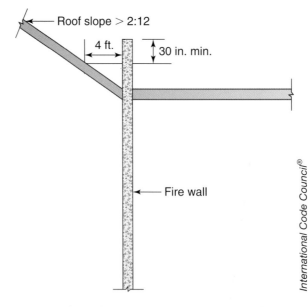

Minimum fire wall height adjacent to sloped roof

of the minimum parapet height requirement, it has always been based on the assumption that the roof surfaces on each side of the parapet are relatively flat. Provisions have been added to address conditions where a sloped roof occurs on one or both sides of the fire wall parapet.

Where the roof surface slopes significantly upward adjacent to the parapet, there is an increased potential for fire spread over the top of the fire wall. Therefore, it may be necessary to increase the parapet height in order to address this concern. A similar provision has been in the code for some time regarding sloping roofs at parapets of exterior walls. This new requirement regarding the vertical continuity of fire walls is based on those provisions previously established in Section 705.11.1 for exterior wall continuity.

The increase in required parapet height is only applicable where the adjacent roof slope exceeds 2:12. As the roof slope increases, the parapet height must also be increased. For example, where the fire wall parapet adjoins a roof having a slope of 3:12 upward from its intersection with the fire wall, the minimum required parapet height is 42 inches. Where the roof slope is 12:12, the parapet must be at least 78 inches in height. The minimum parapet height is based upon the roof height at a point 48 inches, measured horizontally, from the parapet wall. The following table indicates the minimum fire wall parapet height based on varying adjacent roof slopes.

Roof slope adjacent to parapet	3:12	4:12	6:12	9:12	12:12
Minimum parapet height (inches)	42	46	54	66	78

As the table indicates, for each 1:12 increase in roof slope, the parapet must be increased at least 4 inches in height. And in all cases, a minimum 30-inch parapet height is required.

707.8, 707.9

Intersections of Fire Barriers at Roof Assemblies

CHANGE TYPE: Modification

CHANGE SUMMARY: The void at the intersection between a fire barrier and a nonfire-resistance rated roof assembly now need only be protected with an approved material rather than a fire-resistant joint system.

2012 CODE: 707.8 Joints. Joints made in or between fire barriers and joints made at the intersection of fire barriers with the underside of ~~the~~ a fire-resistance rated floor or roof sheathing, slab, or deck above, and the exterior vertical wall intersection shall comply with Section 715.

707.9 Voids at Intersections. The voids created at the intersection of a fire barrier and a non-fire-resistance-rated roof assembly shall be filled. An approved material or system shall be used to fill the void, shall be securely installed in or on the intersection for its entire length so as not to dislodge, loosen, or otherwise impair its ability to accommodate expected building movements and to retard the passage of fire and hot gases.

CHANGE SIGNIFICANCE: A fire barrier is one of several specific elements established in the IBC to provide a fire-resistance-rated separation of adjacent spaces to safeguard against the spread of fire and smoke. Limited to fire-resistance-rated wall assemblies, fire barriers must extend from the floor to the bottom of the floor or roof sheathing, deck, or slab directly above. This high degree of required continuity minimizes the potential for fire spread from one area to another over the top of the wall. Historically, where a head-of-wall or similar joint was created at the intersection of the fire barrier and the floor or roof sheathing, deck, or slab above, a fire-resistant joint system complying with ASTM E 1966 or UL 2079 has been required. New language addressing the void at the intersection between a

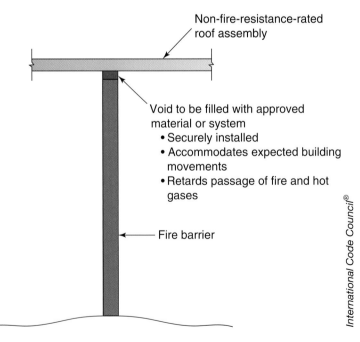

Joint protection at fire barrier/roof assembly intersection

fire barrier and a non-fire-resistance-rated roof assembly now allows for a reduced degree of protection.

The two conditions of top-of-wall joints at fire barriers are now addressed differently based upon the type of floor or roof construction involved. Where a fire barrier intersects with a floor or a fire-resistance-rated roof assembly above, the joint must continue to comply with the provisions of Section 715 addressing fire-resistant joint systems. However, a reduced degree of joint protection is now afforded where a fire barrier intersects with a non-fire-resistance-rated roof assembly. The void at the joint need only be an approved material that is securely installed and capable of retarding the passage of fire and hot gases. It is important to note that the allowance for use of an approved material rather that a complying fire-resistant joint system is not applicable where the joint occurs at a non-fire-resistance-rated floor assembly.

709.4
Continuity of Smoke Barriers

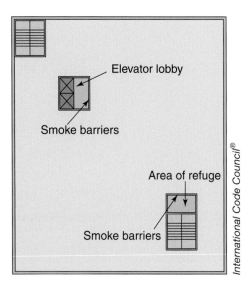

Continuity of smoke barriers

CHANGE TYPE: Clarification

CHANGE SUMMARY: Smoke barrier walls used for elevator lobbies and areas of refuge are no longer required to extend from outside wall to outside wall.

2012 CODE: 709.4 Continuity. Smoke barriers shall form an effective membrane continuous from outside wall to outside wall and from the top of the foundation or floor/ceiling assembly below to the underside of the floor or roof sheathing, deck, or slab above, including continuity through concealed spaces, such as those found above suspended ceilings, and interstitial structural and mechanical spaces. The supporting construction shall be protected to afford the required fire-resistance rating of the wall or floor supported in buildings of other than Type IIB, IIIB, or VB construction.

Exceptions:

1. Smoke-barrier walls are not required in interstitial spaces where such spaces are designed and constructed with ceilings that provide resistance to the passage of fire and smoke equivalent to that provided by the smoke-barrier walls.

2. Smoke barriers used for elevator lobbies in accordance with Section 405.4.3, 3007.4.2, or 3008.11.2 are not required to extend from outside wall to outside wall.

3. Smoke barriers used for areas of refuge in accordance with Section 1007.6.2 are not required to extend from outside wall to outside wall.

CHANGE SIGNIFICANCE: Smoke barriers are occasionally mandated by the code to resist the passage of smoke from one area to another. In addition, smoke barriers must be of 1-hour fire-resistance-rated construction. By using smoke barriers to create compartments within a building, it is anticipated that smoke and fire will not travel to areas outside of the compartment of fire origin. The use of smoke barriers was initially limited to construction elements utilized to create smoke compartments in Group I-2 and I-3 occupancies. Due to the difficulty in evacuating such institutional buildings under fire conditions, the concept focused on the creation of protected compartments where occupants could be relocated rather than evacuated. As such, the smoke barriers were required to extend from outside wall to outside wall. Two exceptions have been added that now permit smoke barrier walls to be used in a manner where they create fully interior spaces without the need to extend to the exterior walls of the building.

Special requirements mandate that lobbies for underground building elevators, fire service access elevators, and occupant evacuation elevators be constructed with smoke barriers. Areas of refuge required under the provisions for accessible means of egress must also be enclosed by smoke barrier construction. In all of these situations, it is often impractical that the smoke barrier walls extend from exterior wall to exterior wall. In addition, no additional degree of fire and smoke protection is afforded by doing so. Therefore, the requirement that smoke barriers extend from outside wall to outside wall no longer applies to the specified elevator lobbies and areas of refuge.

712
Vertical Openings

CHANGE TYPE: Clarification

CHANGE SUMMARY: A significant reformatting in Chapter 7 now places the emphasis on the presence of vertical openings rather than on shaft enclosures, recognizing that the use of shaft enclosures is just one of many acceptable protective measures that can be utilized to address the hazards related to vertical openings.

2012 CODE:

SECTION ~~708~~ 712
~~SHAFT ENCLOSURES~~ VERTICAL OPENINGS

~~708.1~~ 712.1 General.   The provisions of this section shall apply to the <u>vertical opening applications listed in Sections 712.1.1 through 712.1.18.</u> ~~shafts required to protect openings and penetrations through floor/ceiling and roof/ceiling assemblies. Shaft enclosures shall be constructed as fire barriers in accordance with Section 707 or horizontal assemblies in accordance with Section 712, or both.~~

~~708.2 Shaft Enclosure Required.~~   ~~Openings through a floor-ceiling assembly shall be protected by a shaft enclosure complying with this section.~~

> **~~Exceptions:~~**   (Exceptions 1 through 16 have been reformatted as Sections 712.1.2 through 712.1.18 with limited editorial changes.)

712.1.1 Shaft Enclosures.   <u>Vertical openings contained entirely within a shaft enclosure complying with Section 713 shall be permitted.</u>

712 continues

Piping extending through opening in floor

International Code Council®

712 continued

SECTION 713
SHAFT ENCLOSURES

713.1 General. The provisions of this section shall apply to shafts required to protect openings and penetrations through floor/ceiling and roof/ceiling assemblies. Exit access stairways and exit access ramps shall be protected in accordance with the applicable provisions of Section 1009. Interior exit stairways and interior exit ramps shall be protected in accordance with the requirements of Section 1022.

713.2 Construction. Shaft enclosures shall be constructed as fire barriers in accordance with Section 707 or horizontal assemblies in accordance with Section 711, or both.

(remainder of section remains relatively unchanged from 2009 IBC Section 708)

CHANGE SIGNIFICANCE: In multi-story buildings, the upward transmission of fire, smoke, and toxic gases through openings in the floor/ceiling assemblies continues to be a hazard of the highest degree. Historically, the provisions of the code intended to address such concerns have primarily been located under the requirements for shaft enclosures. The fundamental premise has been that a shaft enclosure is mandated to protect openings within a floor/ceiling assembly. Other methods of protection were simply identified as exceptions to the shaft enclosure approach. The code has been reformatted in a manner that now places the emphasis on the presence of vertical openings, while identifying the use of shaft enclosures as one of many protective measures that can be utilized to address the concern.

The criteria for shaft enclosures have been maintained as Section 713 for those situations where a shaft enclosure is used as the desired method of opening protection. Limited technical changes were made to the shaft enclosure provisions.

713.13
Refuse and Laundry Chutes in Group I-2 Occupancies

CHANGE TYPE: Modification

CHANGE SUMMARY: The specific IBC requirements addressing refuse and laundry chutes are no longer applicable in Group I-2 occupancies, because chutes in such institutional occupancies are now regulated by Chapter 5 of NFPA 82, *Standard on Incinerators and Waste and Linen Handling Systems and Equipment.*

2012 CODE: ~~708.13~~ 713.13 **Refuse and Laundry Chutes.**   In other than Group I-2, refuse and laundry chutes, access and termination rooms, and incinerator rooms shall meet the requirements of Section 713.13.1 through 713.13.6.

Exceptions:

1. Chutes serving and contained within a single dwelling unit.
2. Refuse and laundry chutes in Group I-2 shall comply with the provisions of NFPA 82, Chapter 5.

CHANGE SIGNIFICANCE: The installation of refuse- and laundry-handling systems in multi-story buildings can potentially create conditions that provide for rapid vertical fire spread. The emphasis on the provisions regulating refuse and laundry chutes is because such systems interconnect multiple stories and can contain combustible material. Additionally, these systems all too often receive an ignition source. Primary areas of regulation include the construction, fire suppression, and termination requirements. These specific IBC requirements addressing refuse and laundry chutes are no longer applicable in Group I-2 occupancies, because chutes in such institutional occupancies are now regulated by Chapter 5 of NFPA 82, *Standard on Incinerators and Waste and Linen Handling Systems and Equipment.*

713.13 continues

Refuse and laundry chutes in Group I-2 occupancies to comply with NFPA 82, Chapter 5

International Code Council®

Opening to refuse chute

713.13 continued   The construction of hospitals accredited by the Joint Commission must comply not only the IBC but also with NFPA 101, *Life Safety Code.* NFPA 101 further references NFPA 82 for provisions applicable to linen and rubbish chutes. In order to eliminate any inconsistencies between the IBC and NFPA 82 regarding such chutes, the standard is now directly referenced in lieu of the IBC for the applicable construction requirements in regard to hospitals, nursing homes, and other Group I-2 occupancies. Of particular note is a requirement in NFPA 82 regarding the venting of linen and rubbish chutes to the exterior of the building, which was not previously mandated by the IBC. By requiring that refuse and linen chutes in Group I-2 occupancies meet the requirements of NFPA 82, the provisions are consistent with the current Joint Commission rules.

713.13.4
Fire Protection of Termination Rooms

CHANGE TYPE: Modification

CHANGE SUMMARY: The level of fire protection required for a refuse or laundry chute termination room has been modified to provide consistency with those requirements mandated for the shaft that encloses the chutes.

2012 CODE: ~~708.13.4~~ **713.13.4 Termination Room.** Refuse, recycling, and laundry chutes shall discharge into an enclosed room separated from the remainder of the building by ~~not less than 1-hour~~ fire barriers constructed in accordance with Section 707 or horizontal assemblies constructed in accordance with Section 711, or both. Openings into the termination room shall be protected by opening protectives having a fire protection rating ~~of not less than ¾ hour~~ equal to the protection required for the shaft enclosure. Doors shall be self- or automatic-closing upon the detection of smoke in accordance with Section 716.5.9.3. Refuse chutes shall not terminate in an incinerator room. Refuse, recycling, and laundry rooms that are not provided with chutes need only comply with Table 509.

CHANGE SIGNIFICANCE: The provisions of Section 713.13 are intended to further strengthen the shaft enclosure requirements where chutes and termination rooms for refuse, recycled materials, or laundry are constructed. Refuse, recycling, and laundry areas are often poorly maintained, with a greater potential for a fire incident than most other areas of the building. Therefore, it is important that refuse, recycling, and laundry chutes discharge into a room adequately separated from the remainder of the building. The level of fire-resistance required for the termination room separation elements has been modified to provide consistency with the fire-resistance-rated separation mandated for the shaft enclosure itself.

Previously, the termination room was required to be enclosed by minimum 1-hour fire barriers and/or horizontal assemblies. No increased level of separation was required if the room under consideration was at the termination of a 2-hour shaft enclosure. The revised text sends the code user to Section 707 for fire barriers where the path to determining the minimum fire-resistance requirements of shaft enclosures is provided. Under the modified provisions, a refuse, recyling, or laundry room must be separated from the remainder of the building by minimum 2-hour fire

713.13.4 continues

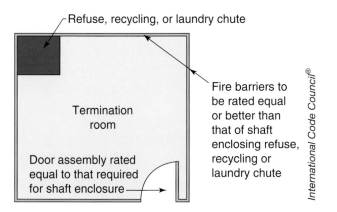

Refuse chute termination room

713.13.4 continued   barriers and/or horizontal assemblies where the refuse, recycling, or laundry shaft is required to be a 2-hour shaft enclosure. The required level of opening protection has also been revised accordingly, such that minimum 1½-hour opening protectives are required for openings into a 2-hour termination room.

713.14.1

High-Rise Buildings—Elevator Lobbies

CHANGE TYPE: Modification

CHANGE SUMMARY: Elevator hoistways in a high-rise building that serve more than three stories but do not serve any stories located more than 75 feet above the lowest level of fire department access no longer require elevator lobby protection.

2012 CODE: 708.14.1 713.14.1 Elevator Lobby.   An enclosed elevator lobby shall be provided at each floor where an elevator shaft enclosure connects more than three stories. The lobby enclosure shall separate the elevator shaft enclosure doors from each floor by fire partitions. In addition to the requirements in Section 708 for fire partitions, doors protecting openings in the elevator lobby enclosure walls shall also comply with Section 716.5.3 as required for corridor walls and penetrations of the elevator lobby enclosure by ducts, and air transfer openings shall be protected as required for corridors in accordance with Section 717.5.4.1. Elevator lobbies shall have at least one means of egress complying with Chapter 10 and other provisions within this code.

Exceptions:

  1. Enclosed elevator lobbies are not required at the ~~street floor,~~ level(s) of exit discharge, provided the ~~entire street floor~~ level(s) of exit discharge is equipped with an automatic sprinkler system in accordance with Section 903.3.1.1.

  2. Elevators not required to be located in a shaft in accordance with Section 712.1 are not required to have enclosed elevator lobbies.

  3. Enclosed elevator lobbies are not required where additional doors are provided at the hoistway opening in accordance

713.14.1 continues

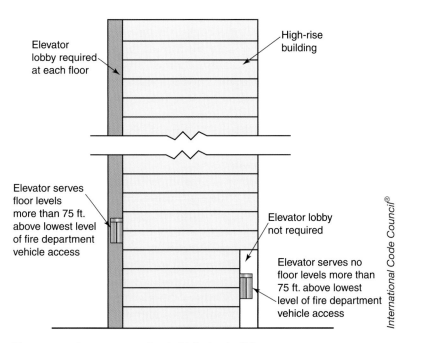

Elevator lobby required at each floor

High-rise building

Elevator serves floor levels more than 75 ft. above lowest level of fire department vehicle access

Elevator lobby not required

Elevator serves no floor levels more than 75 ft. above lowest level of fire department vehicle access

International Code Council®

Elevator enclosure protection in high-rise building

713.14.1 continued

with Section 3002.6. Such doors shall ~~be~~ comply with the smoke and draft control door assembly requirements in Section 716.5.3.1 when tested in accordance with UL 1784 without an artificial bottom seal.

4. Enclosed elevator lobbies are not required where the building is protected by an automatic sprinkler system installed in accordance with Section 903.3.1.1 or 903.3.1.2. This exception shall not apply to the following:
   4.1 Group I-2 occupancies,
   4.2 Group I-3 occupancies, and
   4.3 ~~High-rise buildings~~ Elevators serving floor levels over 75 feet above the lowest level of fire department vehicle access in high-rise buildings.

5. Smoke partitions shall be permitted in lieu of fire partitions to separate the elevator lobby at each floor where the building is equipped throughout with an automatic sprinkler system installed in accordance with Section 903.3.1.1 or 903.3.1.2. In addition to the requirements in Section 710 for smoke partitions, doors protecting openings in the smoke partitions shall also comply with Sections 710.5.2.2, 710.5.2.3 and 716.5.9, and duct penetrations of the smoke partitions shall be protected as required for corridors in accordance with Section 717.5.4.1.

6. Enclosed elevator lobbies are not required where the elevator hoistway is pressurized in accordance with Section 909.21.

7. Enclosed elevator lobbies are not required where the elevator serves only open parking garages in accordance with Section 406.5.

CHANGE SIGNIFICANCE: To reduce the potential for smoke to travel from the story of fire origin to any other story of the building by way of an elevator hoistway shaft enclosure, elevator lobbies are identified as a means to provide the necessary smoke separation. Typical elevator doors, although fire rated, cannot provide the necessary barrier required to keep smoke from passing from story to story by way of the elevator shaft. To restrict such smoke movement, elevator lobbies are the points where each story can be adequately separated from the elevator shaft. Although there are multiple exceptions eliminating or reducing the general elevator lobby requirement, high-rise buildings have been specifically identified as requiring the protection afforded by elevator lobbies. The provisions addressing elevator lobbies in high-rise buildings have been modified to indicate which elevators are regulated.

As a general rule, elevator lobbies are mandated for elevators that connect four or more stories unless the building is provided with an automatic sprinkler system. Fully sprinklered buildings are exempt from the lobby requirement unless the elevators serve Group I-2 or I-3 occupancies. The restricted use of the exception for sprinklered buildings also previously applied to all high-rise buildings. Although all high-rise buildings require sprinkler protection, elevator lobbies were previously mandated for all elevators located in high-rise buildings where the elevator shaft enclosure connected four or more stories. Now only those elevators that serve floor levels that occur above the 75-foot threshold require lobby protection.

In a high-rise building, any elevators that only serve lower floor levels based on the 75-foot threshold no longer require elevator lobbies. The stack effect concern at elevator hoistways does not become an issue until the building height is significant.

The scope of the revised provisions is limited because it only affects those elevator hoistways that serve more than three stories but do not serve any stories located more than 75 feet above the lowest level of fire department access. Using a 50-story fully sprinklered hotel as an example, the elevators that serve the upper floors of the building must continue to be protected by elevator lobbies. However, an elevator that travels only between the first and fourth stories needs no elevator lobby protection provided the floor level of the fourth story is no more than 75 feet above the lowest level of fire department access. An elevator that connects only the first and second stories continues to be exempt from the lobby requirements. It should be noted that where an elevator lobby is required, alternative methods established in the exceptions to Section 713.14.1 may continue to be utilized.

714.4.1.1.2

Floor Penetrations of Horizontal Assemblies

CHANGE TYPE: Modification

CHANGE SUMMARY: An approved through-penetration firestop system used to protect floor penetrations of horizontal assemblies due to the presence of floor, tub, and shower drains is no longer required to have a T rating.

2012 CODE: ~~713.4.1.1.2~~ 714.4.1.1.2 **Through-Penetration Firestop System.** Through penetrations shall be protected by an approved through-penetration firestop system installed and tested in accordance with ASTM E 814 or UL 1479, with a minimum positive pressure differential of 0.01 inch of water (2.49 Pa). The system shall have an F rating/T rating of not less than 1 hour but not less than the required rating of the floor penetrated.

Exceptions:

1. Floor penetrations contained and located within the cavity of a wall above the floor or below the floor do not require a T rating.

2. Floor penetrations by floor drains, tub drains, or shower drains contained and located within the concealed space of a horizontal assembly do not require a T rating.

CHANGE SIGNIFICANCE: Through penetrations are permitted in a horizontal assembly where such penetrations are protected by an approved through-penetration firestop system. The firestop system must

Shower drain penetration of horizontal assembly

be tested and installed in accordance with either ASTM E 814, or UL 1479. The firestop system must always have an appropriate F rating to indicate that it is capable of stopping fire, flame, and hot gases from passing through the horizontal assembly at the point of penetration. In addition, as a general rule the system must also have an appropriate T rating demonstrating that the firestop system adequately limits the temperature transfer to the unexposed side of the assembly. A second exception to the requirement for T ratings now allows those floor penetrations of horizontal assemblies due to the presence of floor, tub, and shower drains to be provided with only an F rating.

It is common for most floor penetrations of horizontal assemblies to occur within the cavity of a wall above or below the floor. In such cases, a T rating is not required for the penetration due to the allowance granted by Exception 1. However, drain piping from floor drains, bathtubs, and showers typically penetrates at a point where no such wall exists. Thus, those drains cannot be addressed under the previous T rating exception. It was determined that the concealed space in a horizontal assembly is comparable in construction and protection to that of a wall cavity, thus providing a degree of protection equivalent to that of Exception 1.

714.4.1.2

Interruption of Horizontal Assemblies

CHANGE TYPE: Modification

CHANGE SUMMARY: The ceiling membrane of a 1-hour or 2-hour fire-resistance-rated floor/ceiling or roof/ceiling assembly is now permitted to be interrupted by a double wood top plate of a fire-resistance-rated wall.

2012 CODE: ~~713.4.1.2~~ __714.4.1.2__ **Membrane Penetrations.** Penetrations of membranes that are part of a horizontal assembly shall comply with Section 714.4.1.1.1 or 714.4.1.1.2. Where floor/ceiling assemblies are required to have a fire-resistance rating, recessed fixtures shall be installed such that the required fire resistance will not be reduced.

Exceptions:

1.-5. (no changes to text)

__6.__ Noncombustible items that are cast into concrete building elements and that do not penetrate both top and bottom surfaces of the element.

__7.__ The ceiling membrane of 1-hour and 2-hour fire-resistance-rated horizontal assemblies is permitted to be interrupted with the double wood top plate of a fire-resistance wall assembly, provided that all penetrating items through the double top plates are protected in accordance with Section 714.4.1.1.1 or 714.4.1.1.2. The fire-resistance rating of the wall shall not be less than the rating of the horizontal assembly.

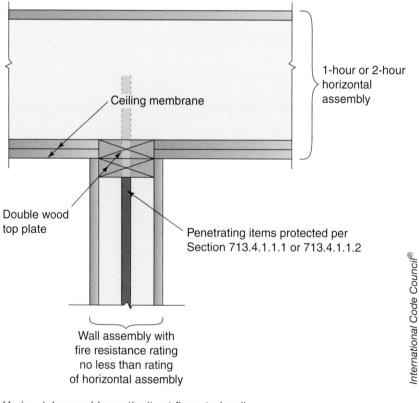

Ceiling membrane

1-hour or 2-hour horizontal assembly

Double wood top plate

Penetrating items protected per Section 713.4.1.1.1 or 713.4.1.1.2

Wall assembly with fire resistance rating no less than rating of horizontal assembly

International Code Council®

Horizontal assembly continuity at fire-rated wall

CHANGE SIGNIFICANCE: A horizontal assembly, defined as "a fire-resistance-rated floor or roof assembly of materials designed to restrict the spread of fire in which continuity is maintained," is required to be continuous without openings, penetrations, or joints unless specifically permitted. Based on the recognition that some degree of penetrations is necessary, acceptable membrane penetration methods are established in Section 714.4.1.2. In addition to five existing specific allowances that permit various membrane penetrations, a new exception allows the ceiling membrane of a 1-hour or 2-hour fire-resistance-rated floor/ceiling or roof/ceiling assembly to be interrupted by a double wood top plate of a fire-resistance-rated wall. There are two conditions of use: (1) the wall must have fire-resistance rating no less than that of the horizontal assembly, and (2) any penetrations of the double wood top plate must be adequately addressed.

The new exception allows for the practical application of the code where wood-framed walls extend up and attach directly to the underside of wood floor joists/trusses or roof joists/trusses for structural requirements. However, there are limits to its use. Non-fire-rated wall top plates are not allowed to interrupt the gypsum board membrane of the floor/ceiling or roof/ceiling membrane. The allowance is only permitted where the horizontal assembly has a required fire-resistance-rating of 2 hours or less, and the intersecting wall must have a fire-resistance rating equal or greater than that of the horizontal assembly. Piping, conduit, and similar items within the fire-resistance-rated wall must be adequately protected where they penetrate the double wood top plate.

Compliance with the established criteria was deemed to provide for an equivalent degree of fire resistance at the discontinuous portion of the ceiling membrane at the intersection of a horizontal assembly and a fire-resistance-rated wall with a double wood top plate.

Another new exception addresses the partial penetration of noncombustible items into fire-resistance-rated concrete building elements.

714.5, 715.6, 202

L Ratings

CHANGE TYPE: Clarification

CHANGE SUMMARY: An "L" rating identifying the air leakage rate—newly defined in Chapter 2—is now mandated for penetration firestop systems and fire-resistant joint systems that are utilized in smoke barrier construction.

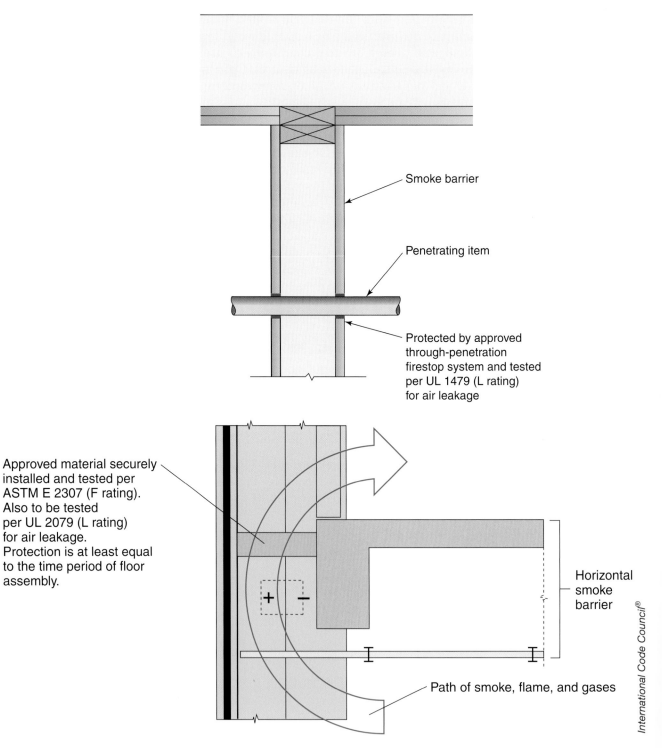

Smoke barrier

Penetrating item

Protected by approved through-penetration firestop system and tested per UL 1479 (L rating) for air leakage

Approved material securely installed and tested per ASTM E 2307 (F rating). Also to be tested per UL 2079 (L rating) for air leakage. Protection is at least equal to the time period of floor assembly.

Horizontal smoke barrier

Path of smoke, flame, and gases

International Code Council®

L rating locations

2012 CODE:

202 Definitions.

L RATING. The air leakage rating of a through-penetration firestop system or a fire-resistant joint system when tested in accordance with UL 1479 or UL 2079, respectively.

~~713.5~~ 714.5 Penetrations in Smoke Barriers. Penetrations in smoke barriers shall be protected by approved through-penetration firestop systems installed and tested in accordance with the requirements of UL 1479 for air leakage. The ~~air leakage rate~~ L rating of the ~~penetration assemblies~~ system measured at 0.30 inch (7.47 Pa) of water in both the ambient temperature and elevated temperature tests, shall not exceed:

1. 5.0 cfm per square foot (0.025 m^3/s m^2) of penetration opening for each through-penetration firestop system; or
2. A total cumulative leakage of 50 cfm (0.024 m^3/s) for any 100 square feet (9.3 m^2) of wall area or floor area.

~~714.6~~ 715.6 Fire-Resistant Joint Systems In Smoke Barriers. Fire-resistant joint systems in smoke barriers, and joints at the intersection of a horizontal smoke barrier and an exterior curtain wall, shall be tested in accordance with the requirements of UL 2079 for air leakage. The ~~air leakage rate~~ L rating of the joint system shall not exceed 5 cfm per lineal foot (0.00775 m^3/s m) of joint at 0.30 inch (7.47 Pa) of water for both the ambient temperature and elevated temperature tests.

CHANGE SIGNIFICANCE: Smoke barriers are utilized to create compartments with a building so that smoke and fire will not travel to areas outside of the compartment of fire origin. The most common use of smoke barriers is to create smoke compartments in Group I-2 and I-3 occupancies. In addition, lobbies for underground building elevators, fire service access elevators, and occupant evacuation elevators must be constructed with smoke barriers. Areas of refuge required under the provisions for accessible means of egress must also be enclosed by smoke barrier construction. Where penetrations and/or joints occur in smoke barriers, it is necessary to maintain the integrity of the separation under both fire and smoke conditions. Applicable provisions have now been modified to require an L rating for penetration firestop systems and fire-resistant joint systems that are utilized in smoke barrier construction. In addition, a definition for L rating has been introduced in Chapter 2.

715.4

Exterior Curtain Wall/ Floor Intersection

CHANGE TYPE: Modification

CHANGE SUMMARY: The use of ASTM E 119 test criteria is now recognized as an acceptable evaluation method for addressing voids at the intersection of fire-resistance-rated floor assemblies and exterior curtain wall assemblies, but only for those curtain wall assemblies where the vision glass extends down to the finished floor level.

2012 CODE: ~~714.4~~ 715.4 Exterior Curtain Wall/Floor Intersection. Where fire resistance-rated floor or floor/ceiling assemblies are required, voids created at the intersection of the exterior curtain wall assemblies and such floor assemblies shall be sealed with an approved system to prevent the interior spread of fire. Such systems shall be securely installed and tested in accordance with ASTM E 2307 to ~~prevent the passage of flame~~ provide an F rating for ~~the~~ a time period at least equal to the fire-resistance rating of the floor assembly ~~and prevent the passage of heat and hot gases sufficient to ignite cotton waste~~. Height and fire-resistance requirements for curtain wall spandrels shall comply with Section 705.8.5.

> **Exception:** Voids created at the intersection of the exterior curtain wall assemblies and such floor assemblies where the vision glass extends to the finished floor level shall be permitted to be sealed with an approved material to prevent the interior spread of fire. Such material shall be securely installed and capable of preventing the passage of flame and hot gases sufficient to ignite cotton waste where subjected to ASTM E 119 time–temperature fire conditions under a minimum positive pressure differential of 0.01 inch (0.254 mm) of water column (2.5 Pa) for the time period at least equal to the fire-resistance rating of the floor assembly.

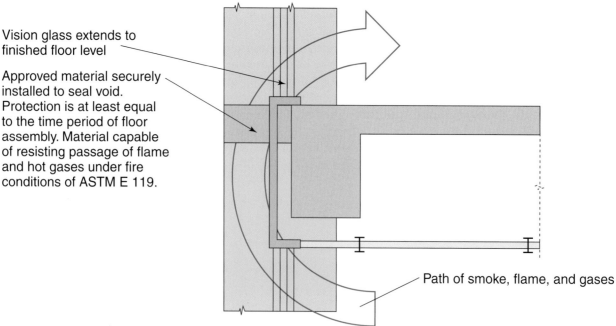

Vision glass extends to finished floor level

Approved material securely installed to seal void. Protection is at least equal to the time period of floor assembly. Material capable of resisting passage of flame and hot gases under fire conditions of ASTM E 119.

Path of smoke, flame, and gases

International Code Council®

Joint protection at exterior curtain wall/floor intersection

CHANGE SIGNIFICANCE: Vertical passages—including those that occur at the intersection of an exterior curtain wall and floor system—allow fire, smoke, and hot gases to quickly travel from story to story if they are not appropriately addressed. Where such openings occur in relationship to a fire-resistance-rated floor or floor assembly, the provisions mandate an approved barrier at the intersection at least equal to the required fire resistance of the floor or floor assembly. The necessary degree of protection has been clarified by specifying that the perimeter fire barrier system provide an F rating equal to that of the fire-resistance rating of the floor assembly. Accordingly, the performance language has been removed because the "F" rating as determined in accordance with ASTM E 2307 evaluates the effectiveness of the material or assembly for passage of flame, heat, and hot gases. The revision also clarifies that the provisions only call for an F rating and a T rating is not required. Previous performance language addressed the prevention of heat passage which potentially brought in the issue of regulating temperature rise.

The use of ASTM E 2307, *Standard Test Method for Determining Fire Resistance of Perimeter Fire Barrier Systems Using Intermediate-scale, Multistory Test Apparatus*, as the test method for perimeter fire barrier systems first occurred in the 2006 edition of the IBC. Prior to that time, the materials and systems were required to comply with the ASTM E 119, *Test Methods for Fire Tests of Building Construction and Materials*, time–temperature fire conditions. Although the 2006 IBC allowed perimeter fire barrier systems to comply with either ASTM E 119 or ASTM E 2307, the 2009 edition of the IBC only recognized those systems complying with ASTM E 2307. The use of ASTM E 119 test criteria has now again been viewed as an acceptable evaluation method, but only for those curtain wall assemblies where the vision glass extends down to the finished floor level. Where the curtain wall consists of full height vision panels, acceptance under the criteria of ASTM E 2307 cannot be attained, although compliance in accordance with the traditional ASTM E 119 criteria is possible. Therefore, the previous IBC allowance utilizing ASTM E 119 has been reinstated for this specific condition. The language in the exception is consistent with the text in previous code editions recognizing the E 119 test method.

716.3, 202

Marking of Fire-Rated Glazing Assemblies

CHANGE TYPE: Clarification

CHANGE SUMMARY: Table 716.3 has been added to define and relate the various test standards for fire-rated glazing, now defined in Chapter 2, to the designations used to mark such glazing.

2012 CODE:

202 Definitions.

Fire-Rated Glazing. Glazing with either a fire protection rating or a fire resistance rating.

716.3 Marking Fire-Rated Glazing Assemblies. Fire-rated glazing assemblies shall be marked in accordance with Tables 716.3, 716.5, and 716.6.

716.3.1 Fire-Rated Glazing That Exceeds the Code Requirements. Fire-rated glazing assemblies marked as complying with hose stream requirements (H) shall be permitted in applications that do not require compliance with hose stream requirements. Fire-rated glazing assemblies marked as complying with temperature rise requirements (T) shall be permitted in applications that do not require compliance with temperature rise requirements. Fire-rated glazing assemblies marked with ratings (XXX) that exceed the ratings required by this code shall be permitted.

TABLE 716.3 **Marking Fire-Rated Glazing Assemblies**

Fire Test Standard	Marking	Definition Of Marking
ASTM E 119 or UL 263	W	Meets wall assembly criteria.
NFPA 257 or UL 9	OH	Meets fire window assembly criteria including the hose stream test.
NFPA 252 or UL 10B or UL 10C	D	Meets fire door assembly criteria.
	H	Meets fire door assembly "Hose Stream" test.
	T	Meets 450° F temperature rise criteria for 30 minutes
	XXX	The time in minutes of the fire resistance or fire protection rating of the glazing assembly

CHANGE SIGNIFICANCE: Fire separation elements such as fire barriers and fire walls will often include glazing in some form, such as glazed wall assemblies, fire windows, and/or glazed fire doors. A definition of "fire-rated glazing" has been added to Chapter 2 that encompasses both types of such glazing addressed by the code: fire-resistance-rated glazing and fire-protection-rated glazing. Fire-resistance-rated glazing, introduced in Section 703.6, must be tested in accordance with ASTM E 119 or UL 263 as a wall assembly. Fire-protection-rated glazing, established for use by Section 715.6, is to be tested in accordance with NFPA 257 or UL 9 as an opening protective. Both types of glazing are now collectively referred to as fire-rated glazing.

Glazing to be labeled with 4-part identifier:

- •"D": applicable for fire-door assemblies and meets applicable fire-resistance requirements

- • "H": meets hose stream requirements (if applicable)

- • "T": meets temperature requirements (if applicable)

- • "XXX": fire-protection rating in minutes

International Code Council®

Marking of fire-rated glazing in fire door

Both fire-resistance-rated glazing and fire-protection-rated glazing must be appropriately identified for verification of its appropriate application. These markings will establish compliance with hose-stream and temperature rise requirements, while also identifying the minimum assembly rating in minutes. It is not unusual for such glazing to be marked to indicate a higher degree of protection than mandated by the code. A new provision clarifies that the use of glazing marked to indicate a higher level of compliance is permitted for use where such compliance is not required.

Table 716.3 has been added to define and relate the various test standards for fire-rated glazing to the designations used to mark such glazing. The marking of fire-rated glazing has been simplified by deleting the "NH" (not hose stream tested) and NT (not temperature rise tested) designations, because these designations correspond with test standards, not end uses. The table reflects the continued use of the designations "W," "OH," "D," "DT," "DH," and "XXX" as markings for fire-rated glazing. Tables 716.5 and 716.6 set forth the markings required for acceptance in specified applications.

Table 716.5
Opening Protection Ratings and Markings

CHANGE TYPE: Clarification

CHANGE SUMMARY: The information previously available in Table 715.4 addressing the minimum required fire-protection ratings of fire door and fire shutter assemblies has been extensively expanded to also include the maximum size and marking requirements for door vision panels and the minimum assembly rating and glazing marking requirements for sidelights and transoms.

2012 CODE:

TABLE ~~715.4~~ 716.5 ~~Fire Door and Fire Shutter Fire Protection Ratings~~
Opening Fire-Protection Assemblies, Ratings, and Markings

Type of Assembly	Required Wall Assembly Rating (Hours)	Minimum Fire Door and Fire Shutter Assembly Rating (Hours)	Door Vision Panel Size	Fire-Rated Glazing Marking Door Vision Panel[−e]	Minimum Sidelight/ Transom Assembly Rating (Hours)		Fire-Rated Glazing Marking Sidelite/ Transom Panel	
					Fire protection	Fire resistance	Fire protection	Fire resistance
Fire walls and fire barriers having a required fire-resistance rating greater than 1 hour	4	3	Not Permitted	Not Permitted	Not Permitted	4	Not Permitted	W-240
	3	3[a]	Not Permitted	Not Permitted	Not Permitted	3	Not Permitted	W-180
	2	1½	100 sq. in.[c]	≤100.in.^{2} = D-H--90 >100 in.^{2} = D-H-W-90 ≤100 in.^{2} = D-H-90	Not Permitted	2	Not Permitted	W-120
	1½	1½	100 sq. in.[c]	>100.in.^{2} = D-H-W-90 <100 in.^{2} = D-H-90	Not Permitted	1½	Not Permitted	W-90
Shaft, exit enclosures, and exit passageway walls	2	1½	100 in.^{2} [c, d]	≤100 in.^{2} = D-H -T-or D-H-T-W-90	Not Permitted	2	Not Permitted	W-120

continued

Table 716.5 continued

Type of Assembly	Required Wall Assembly Rating (Hours)	Minimum Fire Door and Fire Shutter Assembly Rating (Hours)	Door Vision Panel Size	Fire-Rated Glazing Marking Door Vision Panel[e]	Minimum Sidelight/Transom Assembly Rating (Hours)		Fire-Rated Glazing Marking Sidelite/Transom Panel	
					Fire protection	Fire resistance	Fire protection	Fire resistance
Fire barriers having a required fire-resistance rating of 1 hour: Enclosures for shafts, eixt access stairways, exit access ramps, interior exit stairways, interior exit ramps, and exit passageway walls	1	1	100 in.$^{2\ c,\ d}$	≤100 in.^{2} = D-H-60 >100 in.^{2} = D-H-T-60 or D-H-T-W-60	Not Permitted	1	Not Permitted	W-60
					Fire protection			
Other fire barriers	1	¾	Maximum size tested	D-H-NT-45	¾		D-H-NT-45	
	1	⅓^b	Maximum size tested	D-20	¾^b		D-H-OH-45	
Fire partitions Corridor walls	0.5	⅓^b	Maximum size tested	D-20	⅓		D-H-OH-20	
Other fire partitions	1	¾	Maximum size tested	D-H-45	¾		D-H-45	
	0.5	⅓	Maximum size tested	D-H-20	⅓		D-H-20	
					Fire protection	Fire resistance	Fire protection	Fire resistance
Exterior walls	3	1½	100 in.$^{2\ c}$	≤100 in.^{2} = D-H-90 >100 in.^{2} = D-H-W-90	Not Permitted	3	Not Permitted	W-180
	2	1½	100 in.$^{2\ c}$	≤100 in.^{2} = D-H-90 >100 in.^{2} = D-H-W-90	Not Permitted	2	Not Permitted	W-120
					Fire protection			
	1	¾	Maximum size tested	D-H-45	¾		D-H-45	
					Fire protection			
Smoke barriers	1	⅓^b	Maximum size tested	D-20	¾		D-H-OH-45	

a. Two doors, each with a fire protection rating of 1-½ hours, installed on opposite sides of the same opening in a fire wall, shall be deemed equivalent in fire protection rating to one 3-hour fire door.

b. For testing requirements, see Section 716.6.3.

c. Fire-resistance-rated glazing tested to ASTM E 119 per Section 716.2 shall be permitted, in the maximum size tested.

d. Except where the building is equipped throughout with an automatic sprinkler and the fire-rated glazing meets the criteria established in Section 716.5.5.

e. Under the column heading "Fire-Rated Glazing Marking Door Vision Panel," W refers to the fire-resistance rating of the glazing, not the frame.

Table 716.5 continued

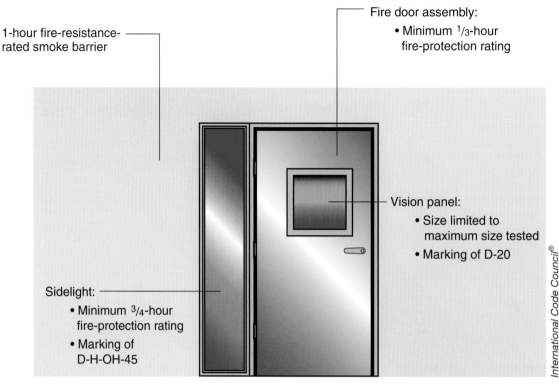

1-hour fire-resistance-rated smoke barrier

Fire door assembly:
• Minimum ⅓-hour fire-protection rating

Vision panel:
• Size limited to maximum size tested
• Marking of D-20

Sidelight:
• Minimum ¾-hour fire-protection rating
• Marking of D-H-OH-45

International Code Council®

Markings for fire door assembly, vision panel, and sidelight

CHANGE SIGNIFICANCE: Intended to maintain the integrity of the fire separation elements in which they are located, fire door and fire shutter assemblies used as opening protectives in fire-resistance-rated walls are uniquely regulated. Under certain conditions, door assemblies must provide a high degree of smoke resistance as well. Vision panels, sidelights, and transoms have their own specific requirements. It is critical that all aspects related to the installation of a fire door or fire shutter assembly be evaluated in order to provide a complete fire separation. The information previously available in Table 715.4 addressing the minimum required fire-protection ratings of fire door and fire shutter assemblies has been extensively expanded to also include the maximum size and marking requirements for door vision panels and the minimum assembly rating and glazing marking requirements for sidelights and transoms. The inclusion of this information in the table retains the technical requirements while making them more convenient for the code user. As a result, all text provisions used to define and relate test standards to marking designations have been deleted in favor of the tabular format.

CHANGE TYPE: Modification

CHANGE SUMMARY: The allowance for glazing in fire door assemblies in interior stairways and ramps and exit passageways has been revised in regard to the maximum permitted size of the glazing and the limitations where the building is fully sprinklered.

2012 CODE: ~~715.4.4~~ 716.5.5 **Doors in** ~~Exit Enclosures~~ **Interior Exit Stairways and Ramps and Exit Passageways.**   Fire door assemblies in ~~exit enclosures~~ interior exit stairways and ramps and exit passageways shall have a maximum transmitted temperature rise of not more than 450°F (250°C) above ambient at the end of 30 minutes of standard fire test exposure.

> **Exception:**   The maximum transmitted temperature rise is not required in buildings equipped throughout with an automatic sprinkler system installed in accordance with Section 903.3.1.1 or 903.3.1.2.

~~715.4.4.1~~ 716.5.5.1 **Glazing in Doors.**   Fire-protection-rated glazing in excess of 100 square inches (0.065 m²) is not permitted. Fire-resistance rated glazing in excess of 100 square inches (0.065 m²) shall be permitted in fire door assemblies when tested as components of the door assemblies, and not as glass lights, and shall have a maximum transmitted temperature rise of 450°F (250°C) in accordance with Section 716.5.5.

716.5.5.1. continues

716.5.5.1

Glazing in Exit Enclosure and Exit Passageway Doors

• Fire-protection-rated glazing limited to 100 square inches

• Fire-resistance-rated glazing permitted in excess of 100 square inches when:

> • Tested as component of door assembly

> • Limited in maximum transmitted temperature rise to 450°F

International Code Council®

Glazing in interior exit stairway or ramp or exit passageway door

716.5.5.1. continued

Exception: ~~The maximum transmitted temperature rise is not required in buildings equipped throughout with an automatic sprinkler system installed in accordance with Section 903.3.1.1 or 903.3.1.2.~~

CHANGE SIGNIFICANCE: Interior exit stairways and ramps and exit passageways are intended to provide a high degree of occupant protection within the means of egress system. As such, fire door assemblies in such exit elements of nonsprinklered buildings must provide for a maximum temperature rise of 450°F above ambient after 30 minutes of standard fire test exposures. The end-point limitation on temperature transmission through the fire door assembly is to protect the person inside the enclosure from excessive heat radiation at the fire door as he or she passes through the fire floor. The allowance for glazing in such fire door assemblies has been revised in regard to two issues: (1) the maximum permitted size of the glazing and (2) the limitations where the building is fully sprinklered.

Fire-protection-rated glazing in a fire door assembly in an interior exit stairway or ramp or exit passageway is no longer permitted to exceed 100 square inches in area. This limitation is now consistent with those for fire-protection-rated glazing installed in 1-hour and 1½-hour fire door assemblies in other types of fire separation elements, such as horizontal exits, control area separations, and occupancy separations. Previously, there was no maximum permitted amount of fire-protection-rated glazing provided the glazing was tested as a component of the door assembly. Where the glazing is fire-resistance-rated, the limit of 100 square inches does not apply if tested as a fire door component.

CHANGE TYPE: Clarification

CHANGE SUMMARY: In addition to fire window assembly fire-protection ratings, Table 716.6 now identifies the markings required on the fire-rated glazing for acceptance in specified applications.

2012 CODE: ~~715.5~~ **716.6 Fire-Protection-Rated Glazing.** Glazing in fire window assemblies shall be fire-protection rated in accordance with this section and Table 716.6. Glazing in fire door assemblies shall comply with Section 716.5.8. Fire-protection-rated glazing <u>in fire window assemblies</u> shall be tested in accordance with and shall meet the acceptance criteria of NFPA 257 or UL 9. Fire-protection-rated glazing shall also comply with NFPA 80. Openings in nonfire-resistance-rated exterior wall assemblies that require protection in accordance with Section 705.3, 705.8, 705.8.5, or 705.8.6 shall have a fire-protection rating of not less than ¾ hour. <u>Fire protection-rated glazing in 0.5-hour fire-resistance-rated partitions is permitted to have a 0.33-hour fire-protection rating.</u>

~~Exceptions:~~

~~2.~~ ~~Fire protection-rated glazing in 0.5-hour fire-resistance-rated partitions is permitted to have an 0.33-hour fire-protection rating.~~

716.6 continues

Table 716.6
Fire-Protection-Rated Glazing

TABLE ~~715.5~~ <u>716.6</u> Fire Window Assembly Fire-Protection Ratings

Type of <u>Wall</u> Assembly	Required <u>Wall</u> Assembly Rating (Hours)	Minimum Fire Window Assembly Rating (Hours)	<u>Fire-Rated Glazing Marking</u>
Interior walls			
Fire walls	All	NP[a]	<u>W-xxx</u>[b]
Fire barriers	>1	NP[a]	<u>W-xxx</u>[b]
	1	NP[a]	<u>W-xxx</u>[b]
<u>Incidental-use areas (707.3.6)</u>	1	¾	<u>OH-45 or W-60</u>
<u>Mixed-occupancy separations (707.3.8)</u>			
Fire partitions	1	¾	<u>OH-45 or W-60</u>
	0.5	⅓	<u>OH-20 or W-30</u>
Smoke barriers	1	¾	<u>OH-45 or W-60</u>
Exterior walls	>1	1½	<u>OH-90 or W-XXX</u>[b]
	1	¾	<u>OH-45 or W-60</u>
	0.5	⅓	<u>OH-20 or W-30</u>
Party wall	All	NP	<u>Not applicable</u>

NP – Not Permitted

a. Not permitted except <u>fire-resistance-rated glazing assemblies tested to ASTM E 119 or UL 263,</u> as specified in Section 716.2.

b. <u>XXX = The fire rating duration period in minutes, which shall be equal to the fire resistance rating required for the wall assembly.</u>

716.6 continued

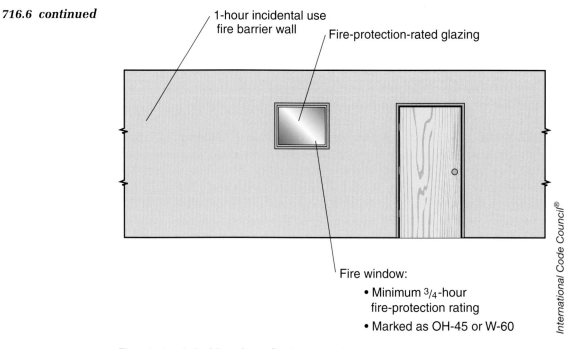

1-hour incidental use
fire barrier wall

Fire-protection-rated glazing

Fire window:

• Minimum ³/₄-hour
  fire-protection rating

• Marked as OH-45 or W-60

International Code Council®

Fire window in incidental-use fire barrier wall

CHANGE SIGNIFICANCE: In many situations, it is necessary to provide glazed openings in fire-resistance-rated walls. Fire window assemblies satisfy this need as opening protectives in fire partitions, smoke barriers, exterior walls, and specified fire barriers. Table 716.6 has historically identified the minimum fire-protection rating required for fire windows based upon the type of wall assembly and the required wall assembly rating. The table now also identifies the marking required on the fire-rated glazing for acceptance in specified applications. By inserting the marking information into Table 716.6, it is intended to provide building and fire code officials with easy access to all of the information needed when inspecting fire window installations, including required marking designations.

As part of the table's expansion, the allowance for 3/4-hour fire windows in fire barriers utilized as incidental use separations and occupancy separations has been relocated from the text of the IBC. In addition, fire window requirements for ½-hour fire-resistance-rated exterior walls have been included, however the IBC currently has no requirement for the use of such walls.

716.6.4
Wired Glass in Fire Window Assemblies

CHANGE TYPE: Deletion

CHANGE SUMMARY: The allowance for the use of wired glass without compliance with the appropriate test standards has been deleted.

2012 CODE: ~~715.5~~ __716.6__ **Fire-Protection-Rated Glazing.** Glazing in fire window assemblies shall be fire-protection rated in accordance with this section and Table 716.6. Glazing in fire door assemblies shall comply with Section 716.5.8. Fire-protection-rated glazing in fire window assemblies shall be tested in accordance with and shall meet the acceptance criteria of NFPA 257 or UL 9. Fire-protection-rated glazing shall also comply with NFPA 80. Openings in non-fire-resistance-rated exterior wall assemblies that require protection in accordance with Section 705.3, 705.8, 705.8.5, or 705.8.6 shall have a fire-protection rating of not less than ¾ hour. Fire protection-rated glazing in 0.5-hour fire-resistance-rated partitions is permitted to have a 0.33-hour fire-protection rating.

Exceptions:

> 1. ~~Wired glass in accordance with Section 715.5.4.~~

~~**715.5.4 Wired glass.** Steel window frame assemblies of 0.125-inch (3.2 mm) minimum solid section or of not less than nominal 0.048-inch-thick (1.2 mm) formed sheet steel members fabricated by pressing, mitering, riveting, interlocking or welding and having provision for glazing with ¼-inch (6.4 mm) wired glass where securely installed in the building construction and glazed with ¼-inch (6.4 mm) labeled wired glass shall be deemed to meet the requirements for a ¾-hour fire window assembly. Wired glass panels shall conform to the size limitations set forth in Table 715.5.4.~~

716.6.4 continues

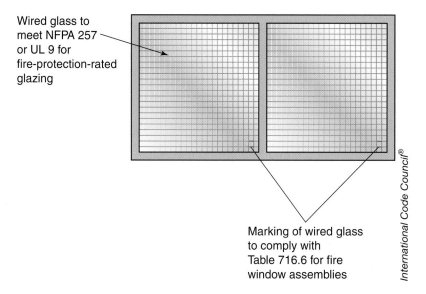

Wired glass to meet NFPA 257 or UL 9 for fire-protection-rated glazing

Marking of wired glass to comply with Table 716.6 for fire window assemblies

International Code Council®

Wired glass used in a fire window assembly

716.6.4 continued

~~TABLE 715.5.4~~   ~~Limiting Sizes Of Wired Glass Panels~~

~~Opening Fire Protection Rating~~	~~Maximum Area (Square Inches)~~	~~Maximum Height(Inches)~~	~~Maximum Width (Inches)~~
~~3 hours~~	~~0~~	~~0~~	~~0~~
~~1½-hour doors in exterior walls~~	~~0~~	~~0~~	~~0~~
~~1 and 1½ hours~~	~~100~~	~~33~~	~~10~~
~~¾ hours~~	~~1,296~~	~~54~~	~~54~~
~~20 minutes~~	~~Not Limited~~	~~Not Limited~~	~~Not Limited~~
~~Fire window assemblies~~	~~1,296~~	~~54~~	~~54~~

~~715.5.5 Nonwired glass.~~ 716.6.4 Glass and Glazing.   Glazing ~~other than wired glass~~ in fire window assemblies shall be fire-protection-rated glazing installed in accordance with and complying with the size limitations set forth in NFPA 80.

CHANGE SIGNIFICANCE: Where glazing occurs in walls that require openings to have a fire-protection rating, such glazing (fire windows) must be tested in accordance with either NFPA 257, *Standard for Fire Test for Window and Glass Block Assemblies*, or UL 9, *Fire Tests of Window Assemblies*. Other than fire-resistance-rated glazing, the only glazing permitted without such a fire-protection rating has historically been wired glass installed within a steel frame in accordance with specific prescriptive provisions established by the code. The allowance for the use of wired glass without compliance with the appropriate test standards has been removed, along with the companion Table 715.5.4, which addressed the maximum size of wired glass panels. Specific reference to the use of wired glass in fire window assemblies has also been deleted from NFPA 80, *Fire Doors and Other Opening Protectives,* which regulates the installation and size limitations of such assemblies. With the removal of Exception 1 to Section 715.5, all glazing in fire-window assemblies must now be fire protection rated, including wired glass.

The use of traditional wired glass has been prohibited for some time in fire doors because it does not meet the CPSC safety glazing requirements of IBC Section 2406.1. Table 715.5.4 has been confusing to many code users because it appears to prescribe permitted size limits for wired glass in doors which are no longer allowed of any significant size. The only accepted application for wired glass is in fire assemblies in nonhazardous locations, and it was determined that a table was not needed to prescribe those size limitations.

717.5.4
Fire Damper Exemption for Fire Partitions

CHANGE TYPE: Modification

CHANGE SUMMARY: The omission of fire dampers in fire partitions is now permitted under the same criteria that have been previously established for fire barriers.

2012 CODE: ~~716.5.4~~ **717.5.4 Fire Partitions.**   Ducts and air transfer openings that penetrate fire partitions shall be protected with listed fire dampers installed in accordance with their listing.

> **Exceptions:**   In occupancies other than Group H, fire dampers are not required where any of the following apply:
>
> **1.-3.** (no changes to text)
>
>   4. Such walls are penetrated by ducted HVAC systems, have a required fire-resistance rating of 1 hour or less, and are in buildings equipped throughout with an automatic sprinkler system in accordance with Section 903.3.1.1 or 903.3.1.2. For the purposes of this exception, a ducted HVAC system shall be a duct system for conveying supply, return, or exhaust air as part of the structure's HVAC system. Such a duct system shall be constructed of sheet steel not less than 26 gage thickness and shall be continuous from the air-handling appliance or equipment to the air outlet and inlet terminals.

CHANGE SIGNIFICANCE: Where a vertical fire assembly, such as a fire barrier or fire partition, is penetrated by a duct or air transfer opening, the integrity of the assembly must typically be maintained through the installation of a listed fire damper. Exception 3 to Section 717.5.2 has

717.5.4 continues

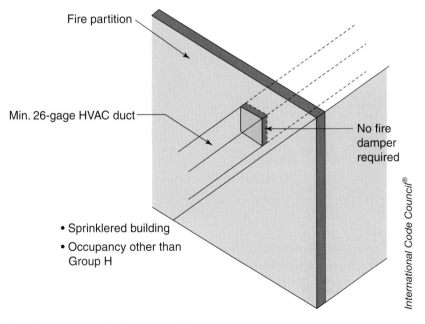

Min. 26-gage HVAC duct

Fire partition

No fire damper required

• Sprinklered building
• Occupancy other than Group H

International Code Council®

Fire damper omission in fire partition

717.5.4 continued historically allowed the omission of fire dampers at penetrations of fire barriers where the specific conditions of the exception are met; however, such an allowance has not previously been available for fire partitions. The new exception to Section 717.5.4 now permits the omission of fire dampers in fire partitions under the same criteria that have been established for fire barriers. Fire dampers are no longer required in duct and air transfer openings that penetrate fire partitions provided:

• the penetration consists of a duct that is a portion of a ducted HVAC system.
• the fire-resistance rating of the fire partition is 1 hour or less.
• the area is not a Group H occupancy.
• the building is fully protected by an automatic fire-sprinkler system.

The limitations established continue to provide for an acceptable alternative to fire dampers in fully sprinklered buildings.

There are two important considerations regarding the application of this new exception. First, the exception is applicable only to the omission of fire dampers and does not eliminate any smoke damper requirements that may be imposed by the code. Second, the exception has no application to fire-resistance-rated corridors because there is no requirement for fire dampers in such corridors in fully sprinklered buildings.

CHANGE TYPE: Modification

CHANGE SUMMARY: In combustible construction, the installation of fireblocking within concealed spaces of exterior wall coverings is no longer required where the wall covering is tested and installed in conformance with NFPA 285.

2012 CODE: ~~717.2.6 Architectural trim.~~ **718.2.6 Exterior Wall Coverings.** Fireblocking shall be installed within concealed spaces of exterior wall ~~finish~~ <u>coverings</u> and other exterior architectural elements where permitted to be of combustible construction as specified in Section 1406 or where erected with combustible frames. <u>Fireblocking shall be installed</u> at maximum intervals of 20 feet (6096 mm) <u>in either dimension</u> so that there will be no ~~open~~ <u>concealed</u> space exceeding 100 square feet (9.3 m^3) <u>between fireblocking</u>. Where wood furring strips are used, they shall be of approved wood of natural decay resistance or preservative-treated wood. If noncontinuous, such elements shall have closed ends, with at least 4 inches (102 mm) of separation between sections.

Exceptions:

1. Fireblocking of cornices is not required in single-family dwellings. Fireblocking of cornices of a two-family dwelling is required only at the line of dwelling unit separation.

718.2.6 continues

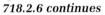

NFPA 285 test furnace

International Code Council®

718.2.6
Fireblocking within Exterior Wall Coverings

2. Fireblocking shall not be required where <u>the exterior wall covering is</u> installed on noncombustible framing and the face of the exterior wall ~~finish~~ <u>covering</u> exposed to the concealed space is covered by one of the following materials:
   2.1. Aluminum having a minimum thickness of 0.019 inch (0.5 mm).
   2.2. Corrosion-resistant steel having a base of metal thickness not less than 0.016 inch (0.4 mm) at any point.
   2.3. Other approved noncombustible materials.

3. <u>Fireblocking shall not be required where the exterior wall covering has been tested in accordance with, and complies with the acceptance criteria of, NFPA 285. The exterior wall covering shall be installed as tested in accordance with NFPA 285.</u>

CHANGE SIGNIFICANCE: Concealed spaces of combustible exterior wall construction are required to be fireblocked in order to limit the spread of fire and smoke. The maximum size of such concealed spaces cannot exceed 100 square feet with the maximum distance between fireblocking of 20 feet either vertically or horizontally. In addition to long-standing exceptions for one- and two-family dwellings and where noncombustible wall framing is used, the installation of fireblocking is also no longer required where the exterior wall covering is tested and installed in conformance with NFPA 285.

NFPA 285, *Standard Method of Test for the Evaluation of Flammability Characteristics of Exterior Non-Load-Bearing Wall Assemblies Containing Combustible Components,* addresses the performance for resisting exterior fire and flame spread along the face of and within the interior cavities of a combustible exterior wall system. Compliance with the criteria of NFPA 285 for a combustible non-load-bearing wall is deemed to demonstrate the ability of the wall to resist the spread of fire within the concealed space. The anticipation of limited fire spread makes the need for periodic fireblocking of the exterior wall unnecessary.

CHANGE TYPE: Modification

CHANGE SUMMARY: Polypropylene plastics used as interior finishes must now be tested using the room corner burn test versus the typical Steiner tunnel test, because the room corner test gives a more accurate evaluation of the flame spread hazards for this type of plastic.

2012 CODE: 803.12 High-Density Polyethylene (HDPE) and Polypropylene (PP). Where high density polyethylene <u>or polypropylene</u> is used as an interior finish it shall comply with Section 803.1.2.

CHANGE SIGNIFICANCE: High-density polyethylene (HDPE) as well as polypropylene (PP) is used extensively in toilet room privacy partitions as well as in other locations. Polypropylene plastics behave in a very similar manner to HDPE when they are exposed to a fire. Therefore, it is appropriate that they be treated the same way and comply with the same flame spread test requirements.

The primary concern is that both of these plastics give off considerable heat energy and will melt and drip flaming droplets that produce a pooling flammable liquids fire on the floor beneath the material. This pooling liquids fire can lead to additional flame spread along the floor and also lead to a greater exposure to the partition above it, creating somewhat of a self-feeding fire.

When these materials are tested using the NFPA 286 room corner burn test, the significant hazards can be seen; yet, when the same material is tested in accordance with the ASTM E 84, Steiner tunnel test, they are often given a flame-spread-index rating of 25 or less. Therefore, because the safety of the material cannot be guaranteed when testing is based only on the ASTM E 84 test, this material needed to be included with this section so the NFPA 286 test will be used.

Code users do need to realize that not all PP will perform in a hazardous manner. Depending on the formulations, it is possible to include fire-retarders into the material, which will greatly improve the fire performance to the point that significant flaming can be eliminated or reduced.

803.12

High-Density Polyethylene (HDPE) and Polypropylene (PP)

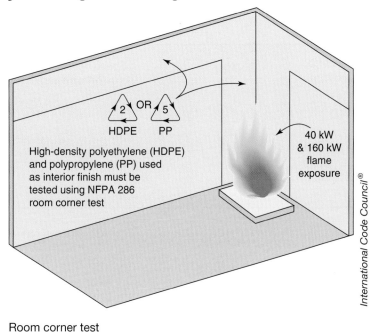

Room corner test

804.4

Interior Floor Finish Requirements

CHANGE TYPE: Clarification

CHANGE SUMMARY: Where fibrous floor finishes are used, it has been clarified that rooms or spaces that are not separated from the corridor by full-height walls must meet the same requirements as the corridor regarding floor finish material.

2012 CODE: **804.4 Interior Floor Finish Requirements.** Interior floor covering materials shall comply with Sections 804.4.1 and 804.4.2, and interior floor finish materials shall comply with Section 804.4.2. ~~In all occupancies, interior floor finish and floor covering materials in exit enclosures, exit passageways, corridors and rooms or spaces not separated from corridors by full-height partitions extending from the floor to the underside of the ceiling shall withstand a minimum critical radiant flux as specified in Section 804.4.1.~~

804.4.1 Minimum Critical Radiant Flux. ~~Interior floor finish and floor covering materials in exit enclosures, exit passageways and corridors shall not be less than Class I in Groups I-1, I-2 and I-3 and not less than Class II in Groups A, B, E, H, I-4, M, R-1, R-2 and S. In all areas, floor covering materials shall comply with the DOC FF-1 "pill test" (CPSC 16 CFR, Part 1630).~~

804.4.1 Test Requirement.   In all occupancies, interior floor covering materials shall comply with the requirements of the DOC FF-1 "pill test" (CPSC 16 CFR, Part 1630) or with ASTM D 2859.

804.4.2 Minimum Critical Radiant Flux.   In all occupancies, interior floor finish and floor covering materials in enclosures for stairways and ramps, exit passageways, corridors, and rooms or spaces not separated from corridors by partitions extending from the floor to the underside of the ceiling shall withstand a minimum critical radiant flux. The minimum critical radiant flux shall not be less than Class I in Groups I-1, I-2, and I-3 and not less than Class II in Groups A, B, E, H, I-4, M, R-1, R-2, and S.

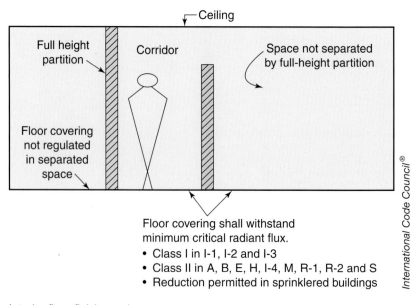

Floor covering shall withstand minimum critical radiant flux.
- Class I in I-1, I-2 and I-3
- Class II in A, B, E, H, I-4, M, R-1, R-2 and S
- Reduction permitted in sprinklered buildings

International Code Council®

Interior floor finish requirements

Exception: Where a building is equipped throughout with an automatic sprinkler system in accordance with Section 903.3.1.1 or 903.3.1.2, Class II materials are permitted in any area where Class I materials are required and materials complying with DOC FF-1 "pill test" (CPSC 16 CFR, Part 1630) or with ASTM D 2859 are permitted in any area where Class II materials are required.

CHANGE SIGNIFICANCE: Primarily, this revision will clarify how the "critical radiant flux" requirements are to be applied to floor finishes in rooms or spaces that are not separated from corridors by full-height partitions. Looking at the wording that has been deleted in Section 804.4 will show that for "rooms or spaces not separated from corridors by full-height partitions extending from the floor to the underside of the ceiling," they were expected to comply with the requirements of the previously existing Section 804.4.1. However, once a user looked at Section 804.4.1, that text did not distinguish how the spaces that were open to the corridor were regulated because the provisions only addressed exit enclosures, exit passageways, and corridors. The new language clarifies that the rooms or spaces that are not separated from the corridor need to meet the same requirements as those for the corridor. From a flame spread requirement standpoint, it is logical that these open rooms or spaces need to meet the same requirements as the corridors from which they are not separated.

These revisions do not have as broad an application as what it may seem when first reading the requirements. Based on the scoping of Section 804.1 and the exception to that section, it is clear that "traditional-type" floor finishes are not regulated and that the requirements only apply to materials that are comprised of fibers. Traditional finish floors and floor coverings, such as wood flooring and resilient floor coverings, have not proved to present an unusual hazard and are known to pass the "pill test"; they are thus exempted by the exception in Section 804.1.

Within the United States, the revisions to this section will not change the way the provisions are enforced or impose any additional requirements. All carpets and carpet-like floor materials have been regulated by the federal government and have been required to comply with the pill test since the 1970s. Therefore, all U.S. carpeting materials are tested and regulated through this process. The revisions to this section have added a referenced standard, ASTM D 2859, that is an equivalent test standard and could be used internationally where the pill test may not be used.

901.8

Pump and Riser Room Size

Sprinkler riser room

International Code Council®

CHANGE TYPE: Addition

CHANGE SUMMARY: Where provided, rooms housing fire protection systems must be adequately sized to facilitate maintenance.

2012 CODE: <u>**901.8 Pump and Riser Room Size.** Fire pump and automatic sprinkler system riser rooms shall be designed with adequate space for all equipment necessary for the installation, as defined by the manufacturer, with sufficient working room around the stationary equipment. Clearances around equipment to elements of permanent construction, including other installed equipment and appliances, shall be sufficient to allow inspection, service, repair, or replacement without removing such elements of permanent construction or disabling the function of a required fire-resistance-rated assembly. Fire pump and automatic sprinkler system riser rooms shall be provided with a door(s) and unobstructed passageway large enough to allow removal of the largest piece of equipment.</u>

CHANGE SIGNIFICANCE: Given the costs associated with the design and installation of fire protection systems, it is important that these systems be maintained. IFC Section 901.6 has extensive requirements for the maintenance of such systems. However, maintenance can be difficult to perform if adequate space is not provided to allow personnel access for the removal of large, cumbersome, or heavy components such as pump casings, control valves, or alarm check valves.

Section 901.8 establishes new requirements to ensure rooms housing fire protection system risers or fire pumps and their components have adequate space to facilitate their maintenance. This section does not require the construction of a room to house fire protection systems—however, if a room is provided, it is now mandated that it be adequately sized to allow for maintenance.

Instead of prescribing arbitrary dimensions, this provision bases the minimum room area on clearances specified by the equipment manufacturers to ensure adequate space is available for its installation or removal. The design must provide enough floor area so that walls, finish materials, and doors are not required to be removed during maintenance activities. The door serving a riser or pump room must also be of a size to accommodate the removal of the largest piece of equipment.

Given that the design of fire protection systems generally commences during the period that building construction drawings and specifications are being reviewed by the jurisdiction, it will be especially important for the building's designer to establish dialogue with the fire-protection system contractor early in the design process to ensure the room and at least one door that can accommodate the largest equipment and provide the space needed for maintenance.

CHANGE TYPE: Modification

CHANGE SUMMARY: Automatic sprinkler requirements for Group B ambulatory care facilities are now regulated on a floor-by-floor basis.

2012 CODE: 903.2.2 ~~Group B~~ Ambulatory ~~Health~~ Care Facilities. An automatic sprinkler system shall be installed throughout ~~all fire areas~~ <u>the entire floor</u> containing <u>an</u> ~~Group B~~ ambulatory ~~health~~ care facility, where either of the following conditions exist at any time:

1. Four or more care recipients are incapable of self-preservation, <u>whether rendered incapable by staff or staff has accepted responsibility for care recipients already incapable.</u>

2. One or more care recipients that are incapable of self-preservation are located at other than the level of exit discharge serving such ~~an~~ <u>facility</u> ~~occupancy~~.

<u>In buildings where care is provided on levels other than the level of exit discharge, an automatic sprinkler system shall be installed throughout the entire floor where such care is provided as well as all floors below, and all floors between the level of ambulatory care and the nearest level of exit discharge, including the level of exit discharge.</u>

CHANGE SIGNIFICANCE: Requirements for ambulatory care facilities were introduced in the 2009 code and previously defined as Ambulatory Health Care Facilities. Ambulatory care facilities, also known as ambulatory surgery centers, are designed so health care practitioners can deliver

903.2.2 continues

903.2.2
Sprinklers in Ambulatory Care Facilities

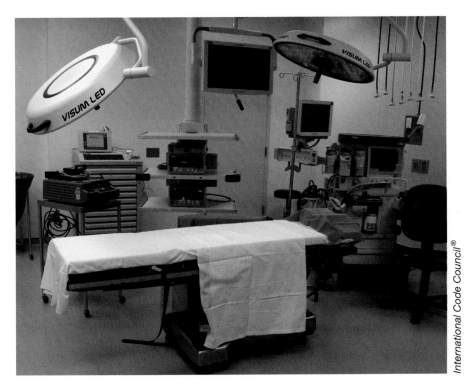

International Code Council®

Operating suite in an ambulatory care facility

903.2.2 continued

surgical procedures that do not require the patient to have such a treatment within a hospital. Within ambulatory care facilities (ACF), the patient is capable of entering and leaving the building on the same day of the procedure. Conversely, hospitalization of an individual normally requires a 24-hour stay, which is the basis for the institutional (Group I) occupancy classification and code requirements. Many of the procedures performed in an ACF require that the patient be incapacitated by anesthesia or sedation of the body's central nervous system, which means the individual is no longer capable of self-rescue and preservation in the event a fire or other emergency occurs within the building. Patients undergoing treatment in an ACF are capable of being placed under home health care within a few hours of the treatment and do not require an overnight stay in a hospital.

ACFs remain classified as Group B occupancies in the 2012 IBC. The 2009 IBC required automatic sprinkler protection for an ACF based on its location in relation to the level of exit discharge or the number of patients who were incapable of self-preservation. However, the requirement for the sprinkler system was limited only to the fire area containing the ACF. Under the 2012 IBC, automatic sprinkler protection is now required to be extended throughout the entire story where the ACF is located, not just within its fire area. In addition, where the ACF is located on a story other than the level of exit discharge, the automatic sprinkler system is required on the level of exit discharge and all of the stories between it and the ACF.

Because occupants are incapable of self-preservation, it is important that the sprinkler protection be extended over the entire floor because occupant evacuation times will be greater when compared to buildings where occupants are capable of self-preservation and able to initiate self-rescue. That same rationale about the increased evacuation times is also one of the reasons protecting the stories between the ACF and the level of exit discharge is important. This will provide a safe route for evacuating the occupants to the exterior of the building and will ensure they do not leave a sprinklered area to egress through a nonsprinklered area.

CHANGE TYPE: Modification

CHANGE SUMMARY: Automatic sprinkler systems are now required in occupancies where upholstered furniture or mattresses are manufactured, stored, or displayed.

2012 CODE: 903.2.4 Group F-1.   An automatic sprinkler system shall be provided throughout all buildings containing a Group F-1 occupancy where one of the following conditions exists:

1. A Group F-1 fire area exceeds 12,000 square feet (1115 m^2).

2. A Group F-1 fire area is located more than three stories above grade plane.

3. The combined area of all Group F-1 fire areas on all floors, including any mezzanines, exceeds 24,000 square feet (2230 m^2).

4. <u>A Group F-1 occupancy used for the manufacture of upholstered furniture or mattresses exceeds 2,500 square feet (232 m^2).</u>

903.2.7 Group M.   An automatic sprinkler system shall be provided throughout buildings containing a Group M occupancy where one of the following conditions exists:

1. A Group M fire area exceeds 12,000 square feet (1115 m^2).

2. A Group M fire area is located more than three stories above grade plane.

3. The combined area of all Group M fire areas on all floors, including any mezzanines, exceeds 24,000 square feet (2230 m^2).

4. A Group M occupancy used for the display and sale of upholstered furniture<u> or mattresses exceeds 5,000 square feet (464 m^2)</u>.

903.2.9 Group S-1.   An automatic sprinkler system shall be provided throughout all buildings containing a Group S-1 occupancy where one of the following conditions exists:

1. A Group S-1 fire area exceeds 12,000 square feet (1115 m^2).

2. A Group S-1 fire area is located more than three stories above grade plane.

3. The combined area of all Group S-1 fire areas on all floors, including any mezzanines, exceeds 24,000 square feet (2230 m^2).

4. A Group S-1 fire area used for the storage of commercial trucks or buses where the fire area exceeds 5,000 square feet (464 m^2).

5. <u>A Group S-1 occupancy used for the storage of upholstered furniture or mattresses exceeds 2,500 square feet (232 m^2).</u>

CHANGE SIGNIFICANCE: The 2009 IBC introduced a new requirement that prescribed the installation of an automatic sprinkler system in any Group M occupancy that displayed and sold upholstered furniture,

903.2.4, 903.2.7, 903.2.9 continues

903.2.4, 903.2.7, 903.2.9

Furniture Storage and Display in Group F-1, M, and S-1 Occupancies

Storage area containing upholstered furniture

903.2.4, 903.2.7, 903.2.9 continued

regardless of fire area size. The provision was not tied to the amount or height of furniture storage and it was unclear whether the requirement could be applied to bedding such as mattresses or box springs. Mattresses and box springs are not considered to be "upholstered furniture" under current Consumer Products Safety Commission regulations found in 16 CFR Part 1633, which is a performance standard that measures the ignition resistance of mattresses. Therefore, further refinement was deemed necessary for the requirement to be effective.

New limits have now been established for the presence of upholstered furniture and mattresses in Group F-1, M, and S-1 occupancies. Sections 903.2.4 and 903.2.9 addressing Group F-1 and Group S-1 occupancies, respectively, now establish a threshold of 2500 square feet for the storage or manufacturing of upholstered furniture and mattresses. In Group M occupancies, Section 903.2.7 establishes a threshold of 5000 square feet. These floor area values are arbitrary but are intended to reduce the burden on the regulated businesses while providing reasonable thresholds as to when automatic sprinkler protection is required.

The requirements in Section 903.2.4, 903.2.7, and 903.2.9 are tied to the floor area devoted to the manufacture, display, or storage of upholstered furniture rather than building fire area. Jurisdictions may want to develop some type of policy on these provisions because the exceptions all are tied to the area "used for" manufacturing, display, sale, or storage of the upholstered furniture or mattresses. The code does not clearly state how the storage or display area's size and quantity of the materials are to be measured. For example, can the occupancy have multiple areas within it, provided each area is below the size threshold, or would a single sofa in a large retail store trigger the requirements? Using a Group M occupancy in a nonsprinklered 11,000-square-foot space as an example, is it permissible to divide the display and storage of upholstered furniture or mattresses into areas of 4,900 square feet, each separated by exit access aisles and consider that each area is beneath the 5,000-square-foot threshold? Or, on the other hand, could a single piece of upholstered furniture in the store trigger the requirement because the store itself is over the area limitation? Jurisdictions should consider these scenarios and develop a policy to address how the floor area and quantity of the materials will be measured for the purpose of applying these requirements to determine when automatic sprinkler protection is required.

Another consideration when applying these provisions is the height of storage. Upholstered furniture or mattresses are commonly classified as high-hazard commodities in accordance with IFC Chapter 32 because they commonly are composed of large amounts of expanded Group A plastics. If the height of storage exceeds 6 feet and the area of storage exceeds 500 square feet in buildings accessible to the public or 2500 square feet in buildings that are not accessible to the public, IFC Table 3206.2 requires automatic sprinkler protection designed and installed in accordance with Section 903.3.1.1.

CHANGE TYPE: Modification

CHANGE SUMMARY: Basements provided with walls, partitions, or fixtures that can obstruct water from hose streams now require automatic sprinkler protection.

2012 CODE: 903.2.11.1.3 Basements.   Where any portion of a basement is located more than 75 feet (22 860 mm) from openings required by Section 903.2.11.1, <u>or where walls, partitions, or other obstructions are installed that restrict the application of water from hose streams</u>, the basement shall be equipped throughout with an approved automatic sprinkler system.

CHANGE SIGNIFICANCE: Interior structural firefighting is a high-risk operation for firefighters. Numerous complications can arise when commencing an interior fire attack, including (but not limited to), problems with the water supply, protective clothing, breathing apparatus, or the structure. One area of buildings that can complicate interior firefighting is basements. IBC Section 202 defines a basement as a *story that is not a story above grade plane.* Basements can be partially or completely underground. Basements present some of the more challenging complications for firefighters because entering the area is analogous to entering a building through the chimney of a fireplace. All of the heat will collect at the highest point, which can be the entry doorway into the basement, so firefighters must push their way through these fire gases

903.2.11.1.3 continues

903.2.11.1.3

Sprinkler Protection for Basements

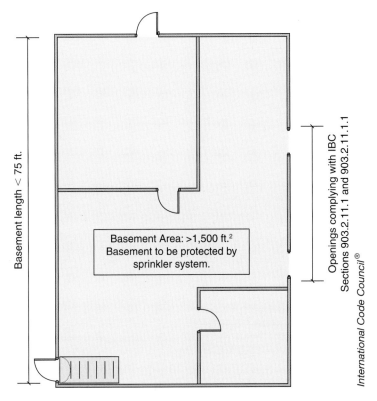

Basement length < 75 ft.

Basement Area: >1,500 ft.²
Basement to be protected by
sprinkler system.

Openings complying with IBC
Sections 903.2.11.1 and 903.2.11.1.1

International Code Council®

Sprinkler requirements for basements

903.2.11.1.3 continued

before commencing the application of water. Basements almost always contain building load-bearing elements so a fire involving this area can adversely affect structural stability when the area is involved in fire.

One concern during interior firefighting operations is obstruction of fire streams. Obstructions such as walls or partitions may prevent the application of water onto the area of fire involvement. The installation of an automatic sprinkler system in basements over 1500 square feet in floor area is now required when obstructions such as walls, partitions or similar elements are introduced which could obstruct the application of hose streams. It should be noted that whether the wall contains door openings or not has no effect on the application of the provision. While some code requirements such as exit access travel distance (Section 1016) and the location of Class II standpipes (Section 905.5) allow measuring along an available route through the building and through doors, the presence of doorways has no bearing on the code's application. Because a wall of any size has the potential to "restrict the application of water," the building official should be consulted if the design indicates anything other than a wide-open, unfurnished space and sprinklers are not intended to be installed.

CHANGE TYPE: Modification

CHANGE SUMMARY: Automatic sprinkler protection requirements for rubbish and linen chutes have been clarified for consistency of application.

2012 CODE: 903.2.11.2 Rubbish and Linen Chutes.   An automatic sprinkler system shall be installed at the top of rubbish and linen chutes and in their terminal rooms. Chutes ~~extending through three or more floors~~ shall have additional sprinkler heads installed ~~within such chutes~~ at alternate floors <u>and at the lowest intake. Where a rubbish chute extends through a building more than one floor below the lowest intake, the extension shall have sprinklers installed that are recessed from the drop area of the chute and protected from freezing in accordance with Section 903.3.1.1. Such sprinklers shall be installed at alternate floors, beginning with the second level below the last intake and ending with the floor above the discharge.</u> Chute sprinklers shall be accessible for servicing.

903.2.11.2 continues

903.2.11.2

Sprinkler Protection of Rubbish and Linen Chutes

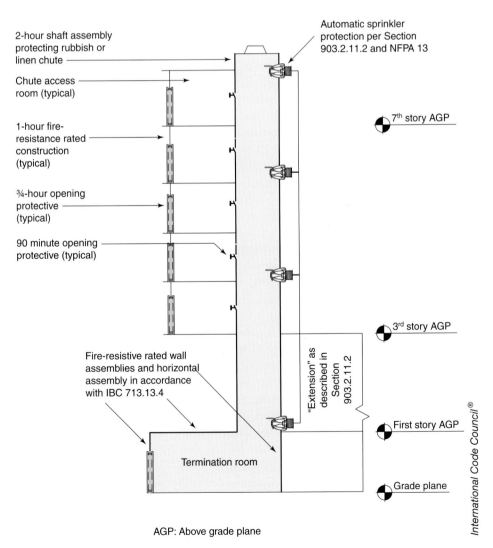

AGP: Above grade plane

Rubbish and linen chute requirements

903.2.11.2 continued

CHANGE SIGNIFICANCE: Gravity rubbish and linen chutes can present a significant hazard to building occupants if they are not properly installed and protected. Generally, these systems are installed in buildings where the occupants will be sleeping or are incapable of self-rescue—Group R-1, R-2, and I-2 occupancies. Secondly, for occupant convenience, openings to the chutes are commonly provided in areas accessible to the public, and in older buildings, the chute opening may be located in a corridor. In comparison to other building shafts, gravity rubbish and linen chutes always contain fuel. As bags of waste debris or linen fall through the chute, they can deposit fluids such as waste cooking oil, which adheres to the shaft surface. This waste material and other debris provide fuel that can support and accelerate vertical fire spread. The greatest accumulation of fuel will be in the termination room; however, a significant amount of fuel that covers the interior surface area of the chute will be found in the sections of chutes closest to the collection or "termination" room. Therefore, it is important that the automatic sprinklers be properly placed and protected so they are available in the event of a fire in the termination room and, if equipped, its waste compaction equipment.

Installation of gravity chutes for rubbish or linen requires compliance with the IBC and IFC. Under the IBC, permanent rubbish and linen chutes are constructed inside of a fire-resistance-rated shaft assembly with a minimum 1-hour fire-resistance rating in buildings less than four stories in height; in buildings four or more stories in height, the fire-resistance rating is increased to 2 hours by IBC Section 713.4. The design of the shaft system and its openings must also comply with the requirements in Sections 713.11 and 713.13, which require the termination room receiving the discharged material to be separated from the building by a fire-resistance rating equivalent to that of the shaft that it serves.

In Group I-2 occupancies, Exception 2 of Section 713.13 now requires gravity rubbish or linen chutes comply with the requirements in Chapter 5 of NFPA 82, *Standard on Incinerators and Waste and Linen Handling Systems and Equipment.* The requirements in NFPA 82 are essentially equivalent to the shaft and fire-resistive construction provisions in IBC Chapter 7. When dealing with Group I-2 occupancies, code users should be aware of the requirement to comply with NFPA 82, and a review of the discussion of Section 713.13 is suggested.

IBC Section 713.13.6 requires the installation of an automatic sprinkler system in rubbish and linen chutes to comply with the requirements of Section 903.2.11.2. Section 903.2.11.2 has been revised so it is more closely correlated to the requirements in Chapter 21 of NFPA 13, *Standard for the Installation of Sprinkler Systems.* Chapter 21 of the NFPA standard contains the special occupancy requirements for all buildings, including gravity waste and linen chutes. The changes shown introduce new requirements for sprinkler spacing and address chutes installed in buildings with pedestal construction in which the chute is routed through areas of a building intended for low-occupant uses such as parking garages. Its intent is to more closely align the IBC and IFC requirements with those in NFPA 82 and NFPA 13.

A critical term in this code change is "extension." This specific term was selected to address chutes installed in buildings of pedestal construction or other designs in which the fire-resistance-rated shaft and chute pass through a less-hazardous occupancy, such as a Group S-2 parking garage or other floors that do not have access to the shaft. In these areas, chute openings are not generally provided. As a result, this section now contains a specific provision that may impose a requirement for sprinklers in the portion of the chute that serves as an "extension" beyond the last intake and the termination room or discharge area.

Because objects will be falling through the chute, the code requires the chute sprinklers be recessed and protected from impact. Sprinklers are not required at every story housing a chute. The code requires automatic sprinklers at the top of the chute and at their termination. In addition, sprinkler heads are required at alternate floors within the chute with a head being installed at the floor level with the lowest intake point into the chute. Previously, these additional sprinkler heads were only required where the shaft extended through three or more floors. These revisions, plus the previously discussed requirements for "extensions," may result in additional sprinkler heads within some shafts as compared to the previous requirements.

Sprinklers in chutes that are in locations subject to freezing require freeze protection in accordance with the requirements of Section 903.3.1.1 and, therefore, the NFPA 13 standard. This can be accomplished using a dry-pendant sprinkler or constructing a dry-pipe sprinkler system.

903.3.5.2

Secondary Water Supply

Tank providing secondary water supply

CHANGE TYPE: Modification

CHANGE SUMMARY: Secondary water supplies must now be designed to operate automatically.

2012 CODE: 903.3.5.2 Secondary Water Supply. ~~A~~ An automatic secondary on-site water supply ~~equal to~~ having a capacity not less than the hydraulically calculated sprinkler demand, including the hose stream requirement, shall be provided for high-rise buildings assigned to Seismic Design Category C, D, E, or F as determined by the *International Building Code*. An additional fire pump shall not be required for the secondary water supply unless needed to provide the minimum design intake pressure at the suction side of the fire pump supplying the automatic sprinkler system. The secondary water supply shall have a duration of not less than 30 minutes as determined by the occupancy hazard classification in accordance with NFPA 13.

Exception: Existing buildings.

CHANGE SIGNIFICANCE: Any high-rise building constructed in accordance with the IFC and IBC requires a secondary water supply when it is located on property classified as a Seismic Design Category (SDC) C, D, E, or F. SDC is a classification assigned to a building based on its structural occupancy category and the severity of the design earthquake ground motion at the site. Buildings located with SDCs categorized as C, D, E, or F are susceptible to damage as a result of soil liquefaction or the level of ground motion it may be subjected to during an earthquake.

Because an earthquake can break underground water pipes, Section 903.3.5.2 requires high-rise buildings within the indicated SDCs to have a secondary water supply. The secondary water supply must be sized to provide the hydraulic demand of the building's automatic sprinkler system, including hose streams, for a minimum flow duration of 30 minutes. In most high-rise buildings, the hydraulic demand is based on an ordinary hazard group I or II occupancy classification in mechanical rooms or similar spaces and the hose stream.

Section 903.3.5.2 was revised by prescribing automatic operation of the secondary water supply; in other words, switchover to the secondary water source cannot be manually activated. This change is consistent with definitions of "automatic sprinkler system" and "classes of standpipe systems" in that both systems are required to be connected to a reliable water supply. This code change ensures that if an earthquake disables the primary water supply, the secondary source is available for service.

The second revision to this provision clarifies the requirements for a second fire pump. Section 903.3.5.2 does not require a second fire pump in high-rise buildings located in the indicated SDCs unless the water supply cannot provide the minimum suction pressure necessary to supply the hydraulic demand. In such a case, the installation of a second fire pump is now mandated to ensure that a sufficient volume of water at the required pressure is available at the primary fire pump.

CHANGE TYPE: Modification

CHANGE SUMMARY: When two or more alternative automatic fire-extinguishing systems are required to protect a hazard, all of the systems must now be designed to simultaneously operate.

2012 CODE: 904.3.2 Actuation.   Automatic fire-extinguishing systems shall be automatically actuated and provided with a manual means of actuation in accordance with Section 904.11.1. Where more than one hazard could be simultaneously involved in a fire due to their proximity, all hazards shall be protected by a single system designed to protect all hazards that could become involved.

> **Exception:**   Multiple systems shall be permitted to be installed if they are designed to operate simultaneously.

CHANGE SIGNIFICANCE: Section 904.3.2 requires alternative fire-extinguishing systems to be designed for automatic activation. Activation commonly occurs when a heat, fire, or smoke detection system operates. In Type I commercial kitchen hoods, Section 904.11 requires a manual and automatic means of activating the fire-extinguishing system. Designing a fire-extinguishing system to only operate upon manual actuation is prohibited by the IBC and many of the NFPA fire-protection system standards.

The requirements for fire-extinguishing system actuation in Section 904.3.2 have been revised to correlate the requirements in the code with existing provisions in NFPA 17, *Standard for Dry Chemical Extinguishing Systems,* and NFPA 17A, *Standard for Wet Chemical*

904.3.2 continues

904.3.2
Actuation of Multiple Fire-Extinguishing Systems

International Code Council®

Activated hood fire-extinguishing system

904.3.2 continued

Extinguishing Systems. The new requirement prescribes that when a hazard is protected by two or more fire-extinguishing systems, all of the systems must be designed to operate simultaneously. The reason for the revision is that a typical alternative automatic fire-extinguishing system has a limited amount of fire-extinguishing agent. The amount of agent that is available is based on the area or volume of the hazard and the fire behavior of the fuel. Because the amount of agent is limited, the simultaneous operation of all the fire-extinguishing systems ensures that enough agent is applied to extinguish the fire and prevent its spread from the area of origin.

It is fairly common for a single hazard to be protected by two or more alternative automatic fire-extinguishing systems. For example, protection of a spray booth used for the application of flammable finishes using dry chemical commonly requires two or three alternative automatic fire-extinguishing systems since many dry chemical and all wet chemical systems are preengineered systems. Utilizing listed nozzles, preengineered systems are designed and constructed based on the manufacturer's installation requirements. Because these systems are assembled using listed nozzles and extinguishing agents, one system may not be able to protect the spraying space and exhaust plenum. As a result, two or more systems may be required as a provision of an extinguishing system's listing to protect certain hazards.

Another example is commercial kitchen cooking operations. Consider a flat grill broiler and a deep fat fryer located beneath the same Type I hood. It is quite common for each of these commercial cooking appliances to be protected by separate automatic fire-extinguishing systems. Based on the revision to Section 904.3.2, both extinguishing systems must simultaneously operate in the event a fire involves either of the example appliances.

CHANGE TYPE: Modification

CHANGE SUMMARY: Requirements for roof hose connections on Class I standpipes have been clarified.

2012 CODE: 905.4 Location of Class I Standpipe Hose Connections. Class I standpipe hose connections shall be provided in all of the following locations:

Items 1 through 3 are unchanged.

4. In covered mall buildings, adjacent to each exterior public entrance to the mall and adjacent to each entrance from an exit passageway or exit corridor to the mall. <u>In open mall buildings, adjacent to each public entrance to the mall at the perimeter line and adjacent to each entrance from an exit passageway or exit corridor to the mall.</u>

5. Where the roof has a slope less than four units vertical in 12 units horizontal (33.3 percent slope), ~~each standpipe shall be provided with~~ a hose connection <u>shall be</u> located <u>to serve</u> ~~either on~~ the roof or at the highest landing of a stairway with stair access to the roof provided in accordance with Section 1009.16. ~~An additional hose connection shall be provided at the top of the most hydraulically remote standpipe for testing purposes.~~

Item 6 is unchanged.

CHANGE SIGNIFICANCE: The IBC requires Class I or III standpipe systems in all buildings where the highest story has a floor level that is more than 30 feet above the lowest level of fire department access, as

905.4 continues

905.4
Location of Class I Standpipe Hose Connections

Hose connection for Class I standpipe

International Code Council®

905.4 continued

well as for large assembly occupancies, covered mall buildings, stages, underground buildings, helistops, heliports, marinas, and boatyards. Standpipe systems required by the code are to be designed, erected, and tested in accordance with NFPA 14, *Standard for the Installation of Standpipe and Hose Systems.* Standpipe systems provide a means of flowing water inside of a building using a network of pipes and valves to help expedite manual firefighting operations.

The revision to item 4 helps coordinate with the open mall requirements that were added in the 2009 code. This item and the provisions of Section 905.3.3 state the requirement for standpipes in a mall building and requires placement near the exits. Because the open mall does not have "exits" per se between the mall and the outside, the building perimeter line is used in lieu of the exits to specify the standpipe locations.

Item 5 was revised to reflect changes made in the NFPA 14 standard regarding the installation of Class I standpipe hose connections on the roofs of buildings. The most significant change is that in a building with two or more vertical standpipes, the IBC no longer prescribes that a hose connection be provided for "each standpipe" either at the roof or at the top landing. This will help coordinate with IBC Section 1009.16, which only requires one stairway to extend to the roof. As previously written, the code could have required a hose connection at the top of a standpipe even though there was no means to access the roof from the stairway or to reach the connection that was located on the roof. Roof hose connections can be a source of maintenance problems because of the potential for water freezing in the pipe. Additionally, two or more roof hose connections are not required to facilitate acceptance or maintenance testing, especially when testing combination automatic sprinkler and standpipe systems.

The second change to item 5 is referencing the provisions in IBC Section 1009.16 for building roof access. IBC Section 1009.16 provisions are based on:

- The roof slope.
- Whether the roof is occupied.
- Height of the building above grade plane.

In general, the IBC requires the construction of a single stairway to a roof when the building is four or more stories above grade plane, the roof slope is 4 units vertical in 12 units horizontal or less (33.3 percent slope), or if the roof or penthouse houses an elevator machinery room. Where a stairway is provided, a penthouse complying with IBC Section 1509.2 is typically required. In instances where the roof is not occupied and does not have a penthouse containing elevator equipment, access can be provided either by an alternating tread device or a roof hatch of given area. Either method offers better protection to firefighters when compared to a hose connection on a roof because it allows them to deploy attack hose lines inside the building before ascending to the roof to commence manual firefighting operations.

CHANGE TYPE: Modification

CHANGE SUMMARY: Portable fire extinguishers are no longer required in many public and common areas of Group R-2 occupancies provided a complying extinguisher is provided within each individual dwelling unit.

2012 CODE: **906.1 Where Required.** Portable fire extinguishers shall be installed in the following locations.

> 1. In ~~new and existing~~ Group A, B, E, F, H, I, M, R-1, R-2, R-4, and S occupancies.
>
> **Exception:** ~~In new and existing Group A, B and E occupancies equipped throughout with quick-response sprinklers, portable fire extinguishers shall be required only in locations specified in Items 2 through 6.~~ <u>In Group R-2 occupancies, portable fire extinguishers shall be required only in locations specified in Items 2 through 6 where each dwelling unit is provided with a portable fire extinguisher having a minimum rating of 1-A:10-B:C.</u>

Items 2 through 6 remain unchanged.

CHANGE SIGNIFICANCE: The installation of portable fire extinguishers (PFEs) in low-hazard areas of new and existing Group A, B, and E occupancies where the occupancies are equipped with an automatic sprinkler system utilizing quick-response automatic sprinklers is now required. The removal of the exception reflects a reluctance to place complete reliance on automatic sprinkler systems for the protection of assembly, business, and educational occupancies.

Another issue expressed by code officials was the retrofitting of an automatic sprinkler system into existing buildings. In several cases, these retrofits resulted in the removal of PFEs. The removal of PFEs is widely considered as a reduction in the level of protection in the building. Given that the exception to Section 906.1 only included Groups A, B, and E occupancies, its deletion was viewed as appropriate.

The new exception to item 1 permits smaller PFEs in dwelling units of Group R-2 occupancies. Under the revised exception, the installation of 1-A:10-B:C PFEs within individual dwelling units now allows apartment owners to eliminate their installation in common areas such as corridors, laundry rooms, and swimming pool areas. PFEs in these areas are susceptible to vandalism or theft. Another issue is larger PFEs are more difficult for the infirmed and elderly to safely deploy and operate.

It is more logical to place PFEs inside dwelling units versus common areas because the extinguisher is located in an area that statistically has been shown to be where most fires occur. If the occupant cannot control the fire using the PFE, he or she can escape and allow the automatic sprinkler system to operate and control the fire. The safety of Group R-2 residents should be enhanced because they will not be required to leave a dwelling involved in a fire, find a PFE, and then return to the fire-involved dwelling unit to attempt incipient fire attack.

Including this requirement in the building code alerts designers and building officials that the extinguishers are required. This will allow designers to plan for recessed cabinets that may be used or to design locations where the extinguishers will not project into or obstruct the egress or circulation path.

906.1

Portable Fire Extinguishers in Group R-2 Occupancies

Portable fire extinguisher provided within dwelling unit in lieu of common areas

907.2.1

Fire Alarms Systems in Group A Occupancies

CHANGE TYPE: Modification

CHANGE SUMMARY: Requirements for a fire alarm system in a building housing two or more Group A occupancies are now based on whether or not the occupancies are in separate fire areas.

2012 CODE: 907.2.1 Group A.   A manual fire alarm system that activates the occupant notification system in accordance with Section 907.5 shall be installed in Group A occupancies ~~having an~~ where the occupant load ~~of~~ due to the assembly occupancy is 300 or more. Group A occupancies not separated from one another in accordance with Section 707.3.10 shall be considered as a single occupancy for the purposes of applying this section. Portions of Group E occupancies occupied for assembly purposes shall be provided with a fire alarm system as required for the Group E occupancy.

> **Exception:** Manual fire alarm boxes are not required where the building is equipped throughout with an automatic sprinkler system installed in accordance with Section 903.3.1.1 and the occupant notification appliances will activate throughout the notification zones upon sprinkler water flow.

CHANGE SIGNIFICANCE: Clarification has been made regarding what occupant load is appropriate to use when determining the fire alarm requirements for assembly occupancies. The revised provisions address three separate situations regarding the application of the alarm requirements for assembly areas: (1) where an assembly occupancy and another occupancy are involved, (2) where multiple assembly areas exist in

Assembly occupancy

International Code Council®

a building, and (3) where the assembly use occurs in and is a part of a Group E occupancy.

In situations where an assembly area and another occupancy are involved, the code will now specify that it is the occupant load "due to the assembly occupancy" that would need to be 300 or more before the manual fire alarm system is required. For example, if the building contains a Group A occupancy, such as a restaurant, with an occupant load of 250 and an adjacent Group B office area with an occupant load of 100, the assembly space would not require an alarm system because the occupant load "due to the assembly occupancy" is less than 300.

In buildings that contain multiple assembly areas, the second portion of the code text requires that the aggregate occupant load of the assembly areas be used unless the spaces are separated as required in the fire area requirements of Section 707.3.10. Consider two examples to address this portion of the requirements. In a multi-theater complex, the auditoriums are generally not separated from each other by the 2-hour fire-resistance rating that Table 707.3.10 would require. Therefore, the aggregate occupant load of all of the assembly spaces would be combined to determine if the occupant load was 300 or more. If it were, then the manual alarm would be required in all of the assembly spaces. As another example, consider a strip mall shopping center with a restaurant at one end of the building with an occupant load of 200 and a different restaurant with an occupant load of 150 at the other end of the building. Even though these are two completely separate establishments and have an amount of retail occupancy between them, the occupant load of the assembly areas does exceed 300. Therefore, unless a 2-hour fire-resistance-rated separation complying with Section 707.3.10 is provided somewhere between the two restaurants to separate them into different fire areas, a manual fire alarm would be required in the Group A occupancies. If a complying separation is provided at some point in the building, then each assembly space can be reviewed independently and would not require the installation of the fire alarm system. It should be noted that the separation of assembly spaces or the need to aggregate the occupant loads from them could occur not only on the same story within a building but also to assembly spaces located on different stories.

Code users should pay special attention to the requirements of Section 907.2.1.1. Even though that section was not modified, the revised text within Section 907.2.1 would presumably affect when an emergency voice/communications alarm system in assembly occupancies is required. Section 907.2.1.1 requires a system in Group A occupancies with an occupant load of 1000 or more. An emergency voice/communications alarm system would appear to be required if a building has several Group A occupancies that are not separated in accordance with Section 707.3.10. Where the assembly uses are properly separated, the occupant load of 1000 or more would be based on each individual assembly area and not on an aggregate total.

907.2.1.2

Emergency Voice/ Alarm Communication Captions

Emergency voice/alarm communication required in large Group A occupancies

CHANGE TYPE: Addition

CHANGE SUMMARY: Mass notification fire alarm signals in large stadiums, arenas, and grandstands now require captioned messages.

2012 CODE: **907.2.1.2 Emergency Voice/Alarm Communication Captions.**   Stadiums, arenas, and grandstands required to caption audible public announcements shall be in accordance with Section 907.5.2.2.4.

907.5.2.2.4 Emergency Voice/Alarm Communication Captions. Where stadiums, arenas, and grandstands are required to caption audible public announcements in accordance with Section 1108.2.7.3, the emergency/voice alarm communication system shall also be captioned. Prerecorded or live emergency captions shall be from an approved location constantly attended by personnel trained to respond to an emergency.

CHANGE SIGNIFICANCE: The court ruled in a 2008 U.S. federal court case that persons with hearing impairments who attend events at stadiums, grandstands, and arenas require a means of equivalent communications in lieu of the public address system. Providing occupant notification in these structures is challenging because of the building area and the number and diversity of occupants. Provisions were added in the code to require captioned messages in these buildings and grandstands when public address (PA) systems are prescribed by the accessibility requirements.

IBC Section 1108.2.7.3 sets forth requirements for audible PA systems in stadiums, arenas, and grandstands. It requires that equivalent text information be provided to the audience and that the delivery time for these messages be the same as those broadcasted from the PA system. These requirements apply to prerecorded and real-time messages. The captioning

Control elements for mass notification system *(Courtesy of Cooper Notifications, Long Branch, NJ)*

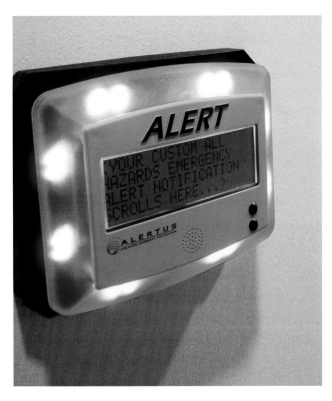

Example of EV/ACS captioning *(Courtesy of the Alertus Technologies LLC, Beltsville MD)*

of messages is mandated in stadiums, arenas, and grandstands that have more than 15,000 fixed seats.

Because messages being broadcasted can include instructions to building or site occupants explaining the actions they need to take in the event of an emergency, the requirements of NFPA 72, *National Fire Alarm and Signaling Code,* are applicable for alarm captioning systems. Such a system falls within the scope of NFPA 72's Chapter 24, "Emergency Communication Systems." NFPA 72 defines an emergency communications system (ECS) as a system designed for life safety that indicates the existence of an emergency and communicates the appropriate response and action. The ECS is required to be classified as either a one-way or two-way path system. Emergency responder radio coverage systems specified in IFC Section 510 are a part of the NFPA 72 ECS requirements. The messages that will be broadcast are based on an emergency response plan developed during a risk analysis by the project stakeholders and is approved by the fire code official.

Further information can be found in *Significant Changes to the International Fire Code,* 2012 Edition, authored by Scott Stookey.

907.2.3

Group E Fire Alarm Systems

CHANGE TYPE: Modification

CHANGE SUMMARY: An emergency voice/alarm communications system is now required in Group E occupancies with an occupant load of 30 or more.

2012 CODE: 907.2.3 Group E. A manual fire alarm system that ~~activates~~ initiates the occupant notification signal utilizing an emergency voice/alarm communication system meeting the requirements of Section 907.5.2.2 and installed in accordance with Section 907.6 shall be installed in Group E occupancies. When automatic sprinkler systems or smoke detectors are installed, such systems or detectors shall be connected to the building fire alarm system.

Exceptions:

1. A manual fire alarm system is not required in Group E occupancies with an occupant load of ~~less than 50~~ 30 or less.
2. Manual fire alarm boxes are not required in Group E occupancies where all of the following apply:
   2.1. Interior corridors are protected by smoke detectors.
   2.2. Auditoriums, cafeterias, gymnasiums, and similar areas are protected by heat detectors or other approved detection devices.
   2.3. Shops and laboratories involving dusts or vapors are protected by heat detectors or other approved detection devices.
   2.4. ~~The capability to activate the evacuation signal from a central point is provided.~~

Group E elementary school

International Code Council®

~~**2.5.** In buildings where normally occupied spaces are provided with a two-way communication system between such spaces and a constantly attended receiving station from where a general evacuation alarm can be sounded, except in locations specifically designated by the fire code official.~~

3. Manual fire alarm boxes shall not be required in Group E occupancies where the building is equipped throughout with an approved automatic sprinkler system installed in accordance with Section 903.3.1.1; the ~~notification appliances~~ emergency voice/ alarm communication system will activate on sprinkler water flow, and manual activation is provided from a normally occupied location.

CHANGE SIGNIFICANCE: Section 404.3.3 was a new requirement in the 2009 IFC that addressed the development and implementation of lockdown plans. These requirements were developed to ensure that the level of life safety inside of the building is not reduced or compromised during a lockdown. In order for a building to safely function in a lockdown condition, a means of communication is required between the established central location and each secured area. IFC Section 404.3.3.1 does not prescribe the means of communication, which could include the use of text messages to cell phones/mobile devices, e-mail messages, or the use of preestablished audio or visual signals. The provisions in IFC Section 404.3.3 are not specific to Group E occupancies—they are applicable to all occupancies that develop and implement lockdown plans.

Because of concerns of school campus safety serving kindergarten through 12th-grade students, specific requirements have been established for enhanced communication among the school administrators, teachers, and students when a lockdown plan is activated in Group E occupancies. As a result, emergency voice/alarm communication systems (EV/ACSs) are now prescribed in Group E occupancies with an occupant load of 30 or more. Previously, a manual fire alarm system was permitted to use audible and visible alarm notification appliances and would not have required the added capabilities that an EV/ACS provides.

Section 907.2.3 sets forth the requirements for automatic fire alarm and detection system requirements in Group E occupancies. This section was revised by reducing the occupant threshold for a fire alarm system from 50 to 30. In addition, Section 907.2.3 now prescribes the installation of an EV/ACS as opposed to a traditional horn/strobe occupant notification system. Because of the reduced occupant threshold, the installation of such a system could be required in portable classrooms.

The revisions within items 2 and 3 of Section 907.2.3 are related to the addition of the requirement for an EV/ACS. The previous items 2.4 and 2.5 have been deleted because these requirements are a part of the capabilities of the EV/ACS alarm notification system.

To gain a better understanding of the impact of this change and the requirements for an EV/ACS, please see the *Significant Changes to International Fire Code, 2012 Edition,* and the discussion within that document related to IFC Section 907.2.3.

907.2.9.3

Smoke Detection in Group R-2 College Buildings

CHANGE TYPE: Addition

CHANGE SUMMARY: A smoke detection system, tied into the occupant notification system, is now required in certain public and common spaces of Group R-2 college and university buildings, and the required smoke alarms within individual dwelling and sleeping units must be interconnected with the building's fire alarm and detection system.

2012 CODE: 907.2.9 Group R-2. Fire alarm systems and smoke alarms shall be installed in Group R-2 occupancies as required in Section 907.2.9.1 ~~and 907.2.9.2~~ through 907.2.9.3.

907.2.9.3 Group R-2 College and University Buildings. An automatic smoke detection system that activates the occupant notification system in accordance with Section 907.5 shall be installed in Group R-2 college and university buildings in the following locations:

1. Common spaces outside of dwelling units and sleeping units.
2. Laundry rooms, mechanical equipment rooms, and storage rooms.
3. All interior corridors serving sleeping units or dwelling units.

Required smoke alarms in dwelling units and sleeping units in Group R-2 college and university buildings shall be interconnected with the fire alarm system in accordance with NFPA 72.

Exception: An automatic smoke detection system is not required in buildings that do not have interior corridors serving sleeping units or dwelling units and where each sleeping unit or dwelling unit either

Dormitory at a university

International Code Council®

has a means of egress door opening directly to an exterior exit access that leads directly to an exit or a means of egress door opening directly to an exit.

CHANGE SIGNIFICANCE: The fire alarm provisions for college and university buildings now differ somewhat from those required for other Group R-2 buildings. While a Group R-2 occupancy will generally require only a manual fire alarm system, these new requirements will require the connection of smoke detection systems in public areas to smoke alarms that are within the dwelling and sleeping units, creating an automatic alarm and detection system.

These requirements would seem to apply differently to buildings that are owned by a college or university versus those that are privately owned but may be used as housing for college students. Because this is somewhat of a continuation of requirements that were added into Chapter 4 of the 2006 IFC for emergency preparedness and planning, and those requirements were intended to deal with buildings that were owned by a college or university, it seems reasonable to interpret that this new requirement is also limited to the buildings that are owned by the college or university and does not apply to other privately owned facilities.

Two items differ between these college and university buildings and most other Group R-2 occupancies. The differences are:

- A smoke detection system will be required in public and common spaces and will need to activate the building's fire alarm and detection system. In most other Group R-2 occupancies, smoke alarms are only required to be located within the individual dwelling or sleeping units. The only instance where the IBC requires smoke detection in the corridors of a Group R-2 occupancy is when the corridors serve the sleeping units. Section 907.2.9.3 is more restrictive for college and university Group R-2 buildings in comparison to other Group R-2 uses because it requires automatic smoke detection in corridors serving dwelling and sleeping units as well as common areas.

- Smoke alarms that are installed within the dwelling unit and sleeping units must be interconnected with the building's fire alarm system. Typically, the alarms within the units are only used to notify the occupants of that unit and are not connected to the building's fire alarm control unit.

Previously, the installation of single- or multiple-station smoke alarms were mandated on all levels "within" the dwelling unit, in the sleeping rooms, and in the immediate vicinity of the bedrooms. The new provision will expand the requirement and require a smoke detection system in laundry rooms, mechanical equipment rooms, storage rooms, and into common spaces and interior corridors that serve the units. The requirement for smoke detection in "laundry rooms, mechanical equipment rooms, and storage rooms" is intended to apply to those types of communal spaces that are located in public and common areas outside of the individual units and not to any rooms within the units that are used for those purposes. Because smoke alarms within the units provide coverage for these types of uses within the unit, it seems reasonable to limit the ap-

907.2.9.3 continues

907.2.9.3 continued plication of item 2 to the communal laundry, mechanical, and storage areas that are outside of the individual units.

Although smoke detectors are specifically required in the indicated areas of Group R-2 college and university buildings, building officials should be cognizant that the installation of smoke detectors in certain areas may ultimately reduce fire safety by serving as a source of nuisance alarm activations. In common areas where cooking is allowed, experience has found that smoke detection is not appropriate because of the potential for burnt food causing an accidental alarm activation. In communal laundry rooms, accumulations of lint and flocking from dryers that are not well maintained may initiate a false alarm signal. In these areas, the use of listed heat detectors should be considered, and is allowed by Section 907.4.3.

The smoke detection system serving the public and common spaces as well as the smoke alarms within the units are to be connected to the alarm system and notify the occupants when a problem occurs within the building. Connecting the smoke detection systems in the public and common spaces outside of the units with the alarm system will help provide automatic notification if a problem should develop in those areas. Another substantial change is the requirement that the smoke alarms within the dwelling and sleeping units be interconnected with the fire alarm system. Previously, the activation of these smoke alarms provided an alarm notification within the unit but did not activate the building's fire alarm system. This requirement for interconnection of the unit smoke alarms and the building's fire alarm system will be a significant difference between these college and university buildings and any other Group R-2 occupancy such as a typical apartment building.

The exception allows for the elimination of the smoke detection system in a specific situation. The use of the wording "automatic smoke detection system" is important in the exception because it helps distinguish between the system that is required in the public and common areas by the first paragraph (along with its three numbered items) and the smoke alarms that are required within the units. Within the units, Sections 907.2.9.2 and 907.2.11 will require smoke alarms at specific locations. The second paragraph of Section 907.2.9.3 will require these smoke alarms to be interconnected with the fire alarm system. The exception does not eliminate the requirement for these smoke alarms within the units or the requirement that they be interconnected with the fire alarm system. The exception is intended to only apply to the "automatic smoke detection system" of the first paragraph and not to the "smoke alarm" requirement of the second paragraph.

CHANGE TYPE: Modification

CHANGE SUMMARY: The smoke alarm interconnection requirements are now applicable to Group I-1 occupancies and include allowances for use of wireless alarms.

2012 CODE: 907.2.11.3 Interconnection. Where more than one smoke alarm is required to be installed within an individual dwelling unit or sleeping unit in Group ~~R-1, R-2, R-3 or R-4~~ R or I-1 occupancies, the smoke alarms shall be interconnected in such a manner that the activation of one alarm will activate all of the alarms in the individual unit. Physical interconnection of smoke alarms shall not be required where listed wireless alarms are installed and all alarms sound upon activation of one alarm. The alarm shall be clearly audible in all bedrooms over background noise levels with all intervening doors closed.

CHANGE SIGNIFICANCE: The addition of the Group I-1 occupancy classification helps coordinate the IBC provisions with requirements that have previously existed within both the IFC and the *International Existing Building Code* (IEBC). In addition, the provisions now recognize the use of listed wireless smoke alarms. Listed wireless alarms are now permitted to substitute for wired interconnection of the smoke alarms in both new and existing construction. Some building officials have previously accepted these wireless alarms as acceptable alternatives because they met the intent of ensuring that all of the alarms would sound if one detector were activated. It is now clear that listed wireless smoke alarms do comply with the code.

In addition to this change in the IBC, similar requirements can now be found in the IRC, IFC, and IEBC for both new and existing buildings to allow the installation of wireless smoke alarms. All wireless smoke alarms are listed to UL 217, *Single and Multiple Station Smoke Alarms,* and are classified by NFPA 72 as low-power systems.

All of the devices available in the marketplace utilize a single smoke alarm that serves as the "host" device that is wirelessly connected to the "guest" smoke alarms in the dwelling and sleeping spaces. The master smoke alarm may be wired to a 120-volt AC branch circuit, or it may be battery powered. In IBC-regulated new construction, the "host" device is required to be wired into a branch circuit receiving electrical energy from a commercial source (see Section 907.2.11.4). Wiring is no longer required to interconnect the additional "guest" smoke alarms. The "guest" smoke alarms are battery powered.

NFPA 72 Section 23.18 requires the "host" device supervise all the "guest" smoke alarms. Required supervisory signals include loss or depletion of battery power in the "guest" smoke alarms and the integrity of the signal frequency and path interconnecting all of the devices. Before a battery reaches a power level that can render a smoke alarm inoperable, or in the event of a failure of the communications path, NFPA 72 requires that an audible and visual supervisory signal be transmitted and annunciated so it can be identified and repaired. NFPA 72 specifies a maximum polling frequency of 60 seconds between the "master" and "guest" devices. "Host" and "guest" smoke alarms are specifically listed as to their

907.2.11.3 continues

907.2.11.3
Wireless Interconnection of Smoke Alarms

Wireless smoke alarm *(Courtesy of BRK Electronics, Aurora IL)*

907.2.11.3 continued

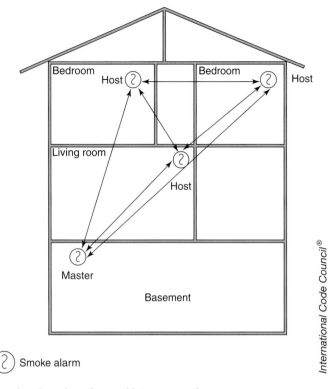

Smoke alarm location and interconnection

function in accordance with NFPA 72. The response time between a smoke alarm's activation and the transmission of a signal that causes the interconnected alarms to activate cannot exceed 20 seconds.

NFPA 72 requires that wireless low-power smoke alarms be capable of reliably communicating at a distance of 100 feet inside dwellings. The particular test method specified in NFPA 72 is based on the system's ability to attenuate (transmit and receive) a wireless radio-frequency signal inside of a Type V building with four walls and two floors constructed of wood, gypsum wallboard, and plywood and a floor covered with tile. In systems where smoke and carbon monoxide alarms are wirelessly interconnected, NFPA 72 requires the fire alarm signal take precedence over other alarm signals. The wireless smoke alarm that initiates an alarm signal must be manually reset to silence the audible alarm signal. Finally, a failure of any "guest" alarm cannot cause the loss of signal to other transceivers on the wirelessly monitored circuit. Depending on the design and listing, a single "host" smoke alarm may be capable of serving 12 to 18 "guest" smoke or CO alarms.

908.7
Carbon Monoxide Alarms

CHANGE TYPE: Addition

CHANGE SUMMARY: In new and existing buildings, carbon monoxide (CO) alarms are now required in Group R and I occupancies with fuel-burning appliances or attached garages.

2012 CODE: **908.7 Carbon Monoxide Alarms.**   Group I or R occupancies located in a building containing a fuel-burning appliance or a building which has an attached garage shall be equipped with single-station carbon monoxide alarms. The carbon monoxide alarms shall be listed as complying with UL 2034 and be installed and maintained in accordance with NFPA 720 and the manufacturer's instructions. An open parking garage, as defined in Chapter 2, or enclosed parking garage ventilated in accordance with Section 404 of the *International Mechanical Code* shall not be considered an attached garage.

> **Exception:**   Sleeping units or dwelling units which do not themselves contain a fuel-burning appliance or have an attached garage, but which are located in a building with a fuel-burning appliance or an attached garage, need not be equipped with single-station carbon monoxide alarms provided that:
>
> 1. The sleeping unit or dwelling unit is located more than one story above or below any story that contains a fuel-burning appliance or an attached garage.
>
> 2. The sleeping unit or dwelling unit is not connected by ductwork or ventilation shafts to any room containing a fuel-burning appliance or to an attached garage.
>
> 3. The building is equipped with a common-area carbon monoxide alarm system.

908.7.1 Carbon Monoxide Detection Systems.   Carbon monoxide detection systems, that include carbon monoxide detectors and audible notification appliances—installed and maintained in accordance with this section for carbon monoxide alarms and NFPA 720 shall be permitted. The carbon monoxide detectors shall be listed as complying with UL 2075.

908.7 continues

Carbon monoxide alarm *(Courtesy of UTC Fire and Home Security, Mebane NC)*

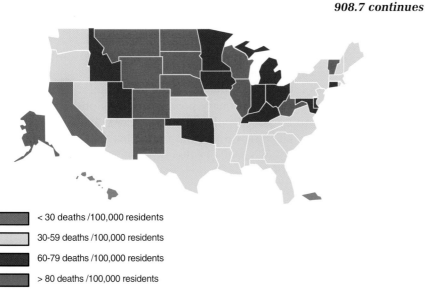

▨	< 30 deaths /100,000 residents
▨	30-59 deaths /100,000 residents
▨	60-79 deaths /100,000 residents
▨	> 80 deaths /100,000 residents

Unintentional carbon monoxide deaths by state, 1979–1988
(JAMA, August 7, 1991, Vol. 266, No. 5, p. 661)

908.7 continued

Chapter 35

NFPA.   720-2005, *Standard for the Installation of Carbon Monoxide (CO) Warning Equipment in Dwelling Units*

UL.   2034-2008, *Standard for Single and Multiple Station Carbon Monoxide Alarms*

CHANGE SIGNIFICANCE: Section 908.7 contains new requirements for carbon monoxide detectors in all residential (Group R) and institutional (Group I) occupancies. These provisions apply to new construction, and a similar requirement was added into the IFC to deal with existing buildings.

Carbon monoxide (CO) detectors were first required by the 2009 IRC for all one- and two-family dwellings. Technical data in a 1998 article published by the *Journal of the American Medical Association (JAMA)* was the basis of the decision to first mandate CO detectors. This particular paper stated that approximately 2100 deaths occur annually as a result of CO poisoning. That annual number is based on the findings of a paper prepared by the U.S. Department of Health, Centers for Disease Control and Prevention (CDC).[1] The referenced paper documented epidemiological research by two CDC physicians who examined 56,133 death certificates over a 10-year period. When the researchers excluded suicides, homicides, structure fires, and deaths resulting from CO poisoning in motor vehicles, the death rate steadily decreased for the sample period, from a value of 1513 people in 1979 to 878 in 1988.

Section 908.7 now requires the installation of a CO alarm in any new Group I or R occupancy when it contains a fuel-burning appliance or it has an attached garage. CO alarms are not required in open or enclosed parking garages as defined by the IBC. The exception indicates a single-station CO alarm is not required in each sleeping or dwelling unit where they are located more than one story above or below the floor or level housing the fuel-burning appliance or an attached garage and where there are no ducts or ventilation shafts that connect between the unit and the fuel-burning appliance or attached garage. However, in such a building, a common-area CO detection system is required. Such a system would be required to comply with the requirements of NFPA 72 and NFPA 720, *Standard for the Installation of Carbon Monoxide (CO) Warning Equipment in Dwelling Units,* including the installation of listed detectors and occupant notification devices.

CO alarms installed in accordance with the IBC are listed in accordance with UL 2034, *Standard for Single and Multiple Station Carbon Monoxide Alarms.* They are designed to initiate an audible alarm when the level of CO is below that which can cause a loss of the ability to react to the dangers of CO exposure.

Unless listed as low-power wireless, CO alarms require a primary and secondary power supply. The primary power supply is utility power, and secondary power supply is typically a battery. NFPA 720 requires a CO alarm outside of each sleeping unit in the immediate vicinity of the bedroom and on every occupiable level of a dwelling, including basements. CO alarms are not required in attics or crawl spaces. When a combination CO/ smoke alarm is provided, the fire alarm signal takes precedence over any other alarm signals. NFPA 720 requires the CO alarm be capable of transmitting a distinct audible signal that is different than the smoke alarm signal.

In jurisdictions adopting the 2012 IFC, retroactive provisions in IFC Secion 1103.9 are applicable to existing buildings classified as Group I or R.

---

[1]Cobb, Nathaniel, and Etzel, Ruth A., "Unintentional Carbon Monoxide-Related Deaths in the United States, 1979 through 1988," *Journal of the American Medical Association,* August 7, 1991, Vol. 266, No. 5, pp. 659–663.

PART 4

Means of Egress

Chapter 10

■ **Chapter 10 Means of Egress**

The criteria set forth in Chapter 10 regulating the design of the means of egress are established as the primary method for protection of people in buildings. Both prescriptive and performance language is utilized in the chapter to provide for a basic approach in the determination of a safe exiting system for all occupancies. It addresses all portions of the egress system and includes design requirements as well as provisions regulating individual components. A zonal approach to egress provides a general basis for the chapter's format through regulation of the exit access, exit, and exit discharge portions of the means of egress. ■

153

CHANGE TYPE: Addition

CHANGE SUMMARY: A reference is now provided to the IFC provisions addressing emergency planning, procedures, and training programs in order to have consistent requirements for the development of evacuation plans.

2012 CODE: 1001.4 Fire Safety and Evacuation Plans.   Fire safety and evacuation plans shall be provided for all occupancies and buildings where required by the *International Fire Code.* Such fire safety and evacuation plans shall comply with the applicable provisions of Sections 401.2 and 404 of the *International Fire Code.*

CHANGE SIGNIFICANCE: Required compliance with the IFC regarding fire safety and evacuation plans is another new provision based on the recommendations of the National Institute of Standards and Technology (NIST) report addressing the collapse of the World Trade Center towers. Reference to the IFC provisions provides consistent requirements for

1001.4 continues

1001.4
Fire Safety and Evacuation Plans

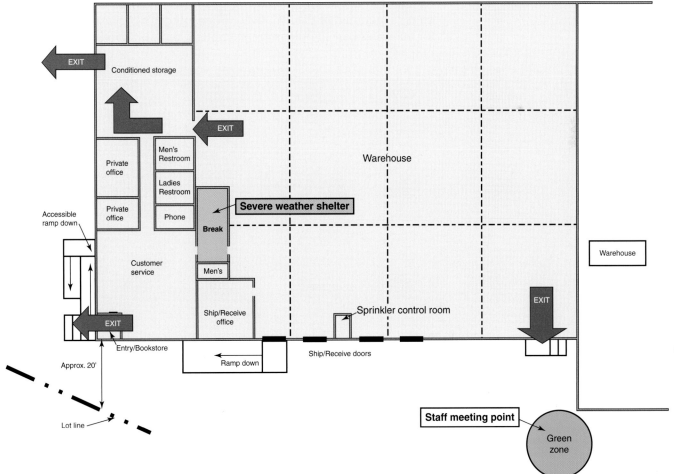

Illustration of evacuation plan

International Code Council®

1001.4 continued jurisdictions regarding the need for, and requirements of, fire safety and evacuation plans.

Section 404 of the IFC requires fire safety and evacuation plans in certain Group A, B, E, F, H, I, M, and R occupancies, in high-rise buildings and underground buildings as well as in specific covered mall buildings and buildings with an atrium. These plans are required to include or address a number of different types of issues that may affect the egress of occupants from the building. Along with other items, these include the identification of potential hazards, exits, primary and secondary egress routes, and occupant assembly points as well as establishing procedures for assisted rescue for people who are unable to use the general means of egress unassisted.

The inclusion of this requirement within the IBC should ensure that all means of egress issues in the IFC and IBC are addressed before the building department issues the certificate of occupancy. Making the requirement for safety and evacuation plans apparent in the IBC increases the likelihood that plans will be developed and submitted and will assist the fire department when they perform means of egress maintenance reviews.

By providing the direct reference to the IFC provisions of Sections 401.2 and 404, these two provisions of the fire code are essentially "adopted by reference" even if the jurisdiction has not adopted the IFC. Section 401.2 of the IFC will place the enforcement and approval burden for these plans on the fire code official. However, because this requirement is included in the IBC, the building official is ultimately responsible for ensuring that the plans are submitted and approved by the fire code official.

CHANGE TYPE: Modification

CHANGE SUMMARY: An occupant load factor for museums and exhibit galleries has been established at 30 square feet per occupant.

2012 CODE: ~~1004.1.1~~ **1004.1.2 Areas without Fixed Seating.**   The number of occupants shall be computed at the rate of one occupant per unit of area as prescribed in Table ~~1004.1.1~~ 1004.1.2. For areas without fixed seating, the occupant load shall not be less than that number determined by dividing the floor area under consideration by the occupant <u>load</u> ~~per unit of area~~ factor assigned to the ~~occupancy~~ <u>function of the space</u> as set forth in Table ~~1004.1.1~~ 1004.1.2. Where an intended <u>function</u> ~~use~~ is not listed in Table ~~1004.1.1~~ 1004.1.2, the building official shall establish a <u>function</u> ~~use~~ based on a listed <u>function</u> ~~use~~ that most nearly resembles the intended <u>function</u> ~~use~~.

> **Exception:**   Where approved by the building official, the actual number of occupants for whom each occupied space, floor, or building is designed, although less than those determined by calculation, shall be permitted to be used in the determination of the design occupant load.

TABLE ~~1004.1.1~~ <u>1004.1.2</u>   Maximum Floor Area Allowances per Occupant

Function of Space	Occupant Load Factor[a] ~~Floor Area In Sq. Ft. Per Occupant~~
Assembly	
Gaming floors (keno, slots, etc.)	11 gross
<u>Exhibit gallery and museum</u>	<u>30 net</u>
<u>Mall buildings—covered and open</u>	See Section 402.4.1

For SI: 1 square foot = 0.0929 m^2.
a. <u>Floor area in square feet per occupant.</u>

Note: (no changes to remainder of table)

CHANGE SIGNIFICANCE: A 30-square-foot per person occupant load factor has been added for museums and exhibit galleries in the assembly use entry. The manner in which these spaces function is different than the way most assembly uses are used and the new factor recognizes this difference.

Museums and exhibit areas are typically used in ways that are not typical of other assembly spaces. What must be taken into consideration is the way an exhibit is viewed. Using even an "unconcentrated" occupant load factor of 15 square feet per person is typically not appropriate because the display could not be seen by the vast majority of the people in the room at that density of people. Very few displays are actually viewed from close proximity. In fact, most artworks are best viewed from distances, and most people are not within 10 to 15 feet of the object being viewed. People do make close inspections, but only after viewing the object from a distance and, when approaching a display, most people

1004.1.2, Table 1004.1.2

Design Occupant Load—Areas without Fixed Seating

Museum exhibit gallery

International Code Council®

1004.1.2, Table 1004.1.2 continues

1004.1.2, Table 1004.1.2 continued

would be courteous and would not step in front of or near the object until other viewers have left the area or completed their distant inspection. Consequently, a museum gallery would not be filled to a high-density design capacity simply because of how the spaces are used.

While museums and galleries do have need for high-occupancy rooms, for gala openings or other special events, most facilities have dedicated spaces for such purposes or would only use a space that was not set up for an exhibit. It is the actual gallery/exhibit spaces that are used at this lower density, and it is only those spaces that this change addresses. When determining the anticipated occupant load of any space listed in the table the intent is to consider how the space will function and be used. To determine the occupant load of these spaces it is appropriate to consider how exhibits are viewed.

It is important to remember that Section 1004.3 continues to require posting of occupant loads within assembly uses. In addition, Section 302.1 addresses the occupant load for spaces where owners want to use the space for more than one use such as parties or lectures. If the intended use is for other than as a gallery or museum, that should be taken into consideration in the design, with an appropriate occupant load for the function being established by the building official.

Several more limited changes can also be found within these provisions. Both Section 1004.1.2 and the second column of Table 1004.1.2 have been modified to use the term "occupant load factor" versus the previous language. The fact that these occupant load factors are based on the floor area per occupant is found by the new footnote a to the table. In addition, an entry for "mall buildings" has been added and will direct the code users to Chapter 4, where the method for establishing an occupant load for a mall building is addressed. Lastly, a somewhat editorial change has been made to Section 1004.1.2 and will relate the title of the first column of the table to the code text by using the word "function" versus looking at the "use or occupancy" of the space.

CHANGE TYPE: Modification

CHANGE SUMMARY: Reduced exit width factors have been established for sprinklered buildings provided with an emergency voice/alarm communication system, and the exit width/capacity requirements are now presented in a more logical and organized layout.

2012 CODE: ~~**1004.4 Exiting From Multiple Levels.** Where exits serve more than one floor, only the occupant load of each floor considered individually shall be used in computing the required capacity of the exits at that floor, provided that the exit capacity shall not decrease in the direction of egress travel.~~

~~**1004.5 Egress Convergence.** Where means of egress from floors above and below converge at an intermediate level, the capacity of the means of egress from the point of convergence shall not be less than the sum of the two floors.~~

~~**1005.1 Minimum Required Egress Width.** The means of egress width shall not be less than required by this section. The total width of means of egress in inches (mm) shall not be less than the total occupant load served by the means of egress multiplied by 0.3 inch (7.62 mm) per~~

1005 continues

1005
Means of Egress Capacity Determination

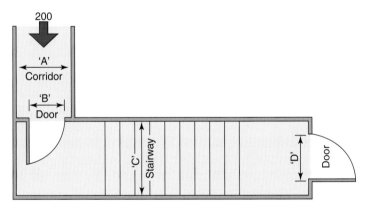

Example : Assuming exit is serving 200 people

Component	Min width based on component (1005.2)	Min width based on occupant load (1005.3)	
		General[1]	Sprinklered building with EV/ACS[2]
Corridor 'A'	44"	40"	30"
Door 'B'	32"	40"	30"
Stairway 'C'	44"	60"	40"
Door 'D'	32"	40"	30"

1. Building without sprinkler system or EV/ACS; (also includes Group H and I-2 occupancies)

2. Other than Group H and I-2 occupancies

International Code Council®

Means of egress sizing

1005 continued ~~occupant for stairways and by 0.2 inch (5.08 mm) per occupant for other egress components. The width shall not be less than specified elsewhere in this code. Multiple means of egress shall be sized such that the loss of any one means of egress shall not reduce the available capacity to less than 50 percent of the required capacity. The maximum capacity required from any story of a building shall be maintained to the termination of the means of egress.~~

> **Exception:** ~~Means of egress complying with Section 1028.~~

1005.1 General. All portions of the means of egress system shall be sized in accordance with this section.

> **Exception:** Means of egress complying with Section 1028.

1005.2 Minimum Width Based on Component. The minimum width, in inches, of any means of egress components shall not be less than that specified for such component elsewhere in this code.

1005.3 Required Capacity Based on Occupant Load. The required capacity, in inches, of the means of egress for any room, area, space, or story shall not be less than that determined in accordance with the following:

1005.3.1 Stairways. The capacity, in inches, of means of egress stairways shall be calculated by multiplying the occupant load served by such stairway by a means of egress capacity factor of 0.3 inches (7.62 mm) per occupant. Where stairways serve more than one story, only the occupant load of each story considered individually shall be used in calculating the required capacity of the stairways serving that story.

> **Exception:** For other than Group H and I-2 occupancies, the capacity, in inches, of means of egress stairways shall be calculated by multiplying the occupant load served by such stairway by a means of egress capacity factor of 0.2 inches (5.1 mm) per occupant in buildings equipped throughout with an automatic sprinkler system installed in accordance with Section 903.3.1.1 or 903.3.1.2 and an emergency voice/alarm communication system in accordance with Section 907.5.2.2.

1005.3.2 Other Egress Components. The capacity, in inches, of means of egress components other than stairways shall be calculated by multiplying the occupant load served by such component by a means of egress capacity factor of 0.2 inches (5.08 mm) per occupant.

> **Exception:** For other than Group H and I-2 occupancies, the capacity, in inches, of means of egress components other than stairways shall be calculated by multiplying the occupant load served by such component by a means of egress capacity factor of 0.15 inches (3.8 mm) per occupant in buildings equipped throughout with an automatic sprinkler system installed in accordance with Section 903.3.1.1 or 903.3.1.2 and an emergency voice/alarm communication system in accordance with Section 907.5.2.2.

1005.4 Continuity.   The capacity of the means of egress required from any story of a building shall not be reduced along the path of egress travel until arrival at the public way.

1005.5. Distribution of Egress Capacity.   Where more than one exit, or access to more than one exit, is required, the means of egress shall be configured such that the loss of any one exit, or access to one exit, shall not reduce the available capacity to less than 50 percent of the required capacity.

~~1004.5~~ 1005.6 Egress Convergence.   Where the means of egress from stories above and below converge at an intermediate level, the capacity of the means of egress from the point of convergence shall not be less than the sum of the required capacities for the two adjacent stories.

   Provisions in 2009 IBC Sections 1005.2 and 1005.3 regulating permissable encroachment of doors also have been reformatted as new Section 1005.7.

CHANGE SIGNIFICANCE: The multiple requirements related to egress width that were previously contained in a single paragraph in Section 1005.1 have been reorganized and clarified, and the related provisions from Section 1004.4 and 1004.5 have been relocated to a more logical location with the other egress width/capacity provisions.

   In addition, the reduced egress width factors for sprinklered buildings that had been in the 2000 through 2006 IBC but were removed in the 2009 edition have been reintroduced. The exceptions allow for use of reduced width factors for sprinklered buildings but only where an emergency voice/communications alarm system (EV/ACS) is provided for the building.

   The EV/ACS system provides the ability to communicate instructions to the occupants that could facilitate evacuation or relocation during a fire or other emergency. This additional information and direction could lead to more efficient use of the egress system. Studies have shown that most people do not react to an initial alarm; therefore, requiring a voice alarm will increase safety by providing occupants with additional information about the emergency and evacuation.

   The following list will help guide code users in finding the new location of the previous requirements and illustrate the editorial nature of this revision:

- Section 1005.1 provides a new charging paragraph and clarifies that it applies to all portions of the egress system.
- Section 1005.2 replaces the second sentence of the previous code's Section 1005.1 and notes that minimum width requirements for means of egress components may be specified in other locations in the code.
- Section 1005.3 provides the egress width factors in subsections that deal with the various types of components. Note the new exceptions in Sections 1005.3.1 and 1005.3.2 for sprinklered buildings that allow for a reduction in the minimum required calculated width.
- The provisions of the former Section 1004.4 have been incorporated as the last sentence of Section 1005.3.1.

1005 continues

1005 continued

- Section 1005.4 replaces the last sentence of the previous code's Section 1005.1, and notes that once a minimum capacity is required along a means of egress, it must be provided along the entire path of egress travel.

- Section 1005.5 is consistent with the fourth sentence of the previous code's Section 1005.1.

- The "egress convergence" provisions from Section 1004.5 can now be found in Section 1005.6. This is basically an issue of egress capacity/width and is more appropriately located here, instead of within the code section regulating occupant load.

- Revisions have also been made in Sections 3404 and 3412 related to reduced egress width factors.

CHANGE TYPE: Modification

CHANGE SUMMARY: Exterior areas for assisted rescue can now be provided on stories above the level of exit discharge. In addition, open interior exit access stairways are now recognized as accessible means of egress components.

1007
Accessible Means of Egress

1007 continues

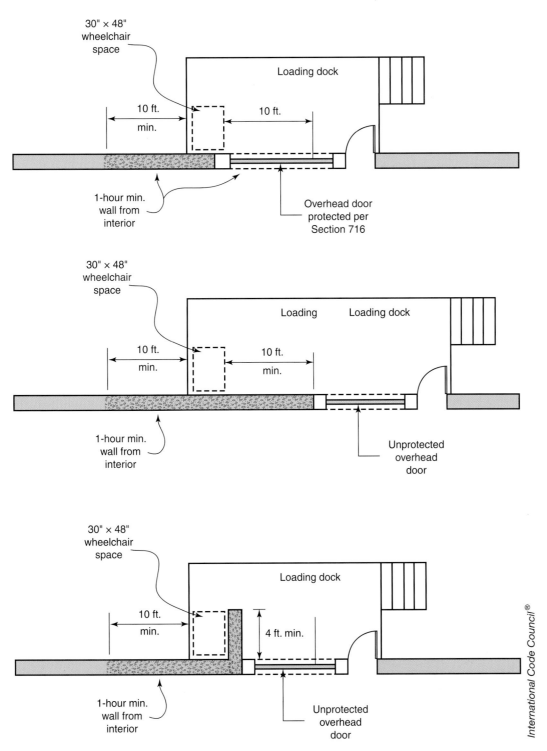

Protection options for exterior area for assisted rescue

International Code Council®

1007 continued   **2012 CODE: 1007.2 Continuity and Components.** Each required accessible means of egress shall be continuous to a public way and shall consist of one or more of the following components:

1. Accessible routes complying with Section 1104.
2. Interior exit stairways complying with Sections 1007.3 and 1022.
3. Interior exit access stairways complying with Section 1007.3 and Section 1009.3.
3.4. Exterior exit stairways complying with Sections 1007.3 and 1026 and serving levels other than the level of exit discharge.
4.5. Elevators complying with Section 1007.4.
5.6. Platform lifts complying with Section 1007.5.
6.7. Horizontal exits complying with Section 1025.
7.8. Ramps complying with Section 1010.
8.9. Areas of refuge complying with Section 1007.6.
10. Exterior area for assisted rescue complying with Section 1007.7.

Exceptions:

1. ~~Where the exit discharge is not accessible, an exterior area for assisted rescue shall be provided in accordance with Section 1007.7.~~
2. ~~Where the exit stairway is open to the exterior, the accessible means of egress shall include either an area of refuge in accordance with Section 1007.6 or an exterior area for assisted rescue in accordance with Section 1007.7.~~

1007.7 Exterior Area for Assisted Rescue.   Exterior areas for assisted rescue shall be accessed by an accessible route from the area served. Exterior areas for assisted rescue shall be permitted in accordance with Section 1007.7.1 or 1007.7.2.

1007.7.1 Level of Exit Discharge.   Where the exit discharge does not include an accessible route from an exit located on a level of exit discharge to a public way, an exterior area of assisted rescue shall be provided on the exterior landing in accordance with Sections 1007.7.3 through 1007.7.6.

1007.7.2 Outdoor Facilities.   Where exit access from the area serving outdoor facilities is essentially open to the outside, an exterior area of assisted rescue is permitted as an alternative to an area of refuge. Every required exterior area of assisted rescue shall have direct access to an interior exit stairway, exterior stairway, or elevator serving as an accessible means of egress component. The exterior area of assisted rescue shall comply with Sections 1007.7.3 through 1007.7.6 and shall be provided with a two-way communication system complying with Sections 1007.8.1 and 1007.8.2.

1007.7.3 Size.   Each exterior area for assisted rescue shall be sized to accommodate wheelchair spaces in accordance with Section 1007.6.1. ~~The exterior area for assisted rescue must be open to the outside air and meet the requirements of Section 1007.6.1.~~

1007.7.4 Separation. ~~Separation walls shall comply with the requirements of Section 705 for exterior walls. Where walls or openings are between the area for assisted rescue and the interior of the building, the building exterior walls within 10 feet (3048 mm) horizontally of a nonrated wall or unprotected opening shall have a fire-resistance rating of not less than 1 hour.~~ Exterior walls separating the exterior area of assisted rescue from the interior of the building shall have a minimum fire resistance rating of 1 hour, rated for exposure to fire from the inside. The fire-resistance-rated exterior wall construction shall extend horizontally 10 feet (3048 mm) beyond the landing on either side of the landing, or equivalent fire-resistance-rated construction is permitted to extend out perpendicular to the exterior wall 4 feet (1220 mm) minimum on the side of the landing. The fire-resistance-rated construction shall extend vertically from the ground to a point 10 feet (3048 mm) above the floor level of the area for assisted rescue or to the roof line, whichever is lower. Openings within such fire-resistance-rated exterior walls shall be protected in accordance with Section 716. ~~By opening protectives having a fire protection rating of not less than 3/4 hour. This construction shall extend vertically from the ground to a point 10 feet (3048 mm) above the floor level of the area for assisted rescue or to the roof line, whichever is lower.~~

~~1007.7.1~~ 1007.7.5 Openness. The exterior area for assisted rescue shall be open to the outside air. The sides other than the separation walls shall be at least 50 percent open, and the open area ~~above the guards~~ shall be ~~so~~ distributed so as to minimize the accumulation of smoke or toxic gases.

~~1007.7.2~~ 1007.7.6 ~~Exterior Exit~~ Stairway. ~~Exterior exit~~ Stairways that are part of the means of egress for the exterior area for assisted rescue shall provide a clear width of 48 inches (1219 mm) between handrails.

> **Exception:** The clear width of 48 inches (1219 mm) between handrails is not required at stairways serving buildings equipped throughout with an automatic sprinkler system installed in accordance with Section 903.3.1.1 or 903.3.1.2.

CHANGE SIGNIFICANCE: The provisions addressing exterior areas for assisted rescue have been clarified to focus on where such accessible means of egress components are intended to be utilized. The addition of item 10 in Section 1007.2, along with the elimination of the two previous exceptions, reinforces the intent that an exterior area of assisted rescue is a component of an accessible means of egress and a viable portion of the egress system. Because the list previously included areas of refuge, it was appropriate to add exterior areas of rescue assistance and eliminate the exceptions.

A new item 3 now specifically lists interior exit access stairs as an acceptable component. Including these stairways, which are generally open stairways between adjacent levels, in this list will help avoid any confusion or conflict with the provisions of Section 1007.3. The change in item 4 of Section 1007.2 helps clarify that the exterior "exit" stairways are for stairs serving a story other than the level of exit discharge and that those provisions do not apply to exit discharge elements such as a couple of steps serving a landing or other steps between the exit and the public way.

1007 continues

1007 continued

The revisions in Section 1007.7 for the exterior area for assisted rescue will split the provisions into smaller sections addressing limited aspects of the requirements, plus the modifications include several technical changes. Section 1007.7.2 has been added to clearly state that exterior areas for rescue assistance can be located at other than the level of exit discharge. Although primarily intended to be used for open-air assembly occupancies such as a Group A-5, the scope only limits it to areas serving outdoor facilities that are essentially open to the exterior, including open parking garages, rooftops, and large open balconies. Users should also note that when the exterior area of assisted rescue is located at other than the level of exit discharge, a two-way communications system is required.

The separation requirements of Section 1007.7.4 have been modified and affect the construction and protection of exterior areas for assisted rescue. The exterior wall adjacent to an exterior area for assisted rescue is to have a fire-resistance rating based on requiring protection from fire exposure from the inside of the building. This is similar to the provision for exterior walls in Section 705.5. Protecting from interior fire exposure makes sense because that is where the anticipated problem will occur, and this allows the wall protecting the exterior area for assisted rescue to be consistent with the general exterior wall construction requirements versus having to be rated from both the interior and exterior. The opening protection requirements for the wall between the building and the exterior area for assisted rescue are now based on the provisions of Section 716 versus specifying a ¾-hour fire-protection rating. While Section 716 does require a ¾-hour rating for openings in a 1-hour exterior wall, the previous requirements could have caused a conflict with Section 716 if the exterior wall had a higher level of protection based on some other provision. Perhaps of more significance is the inclusion of an option to turn the protection perpendicular for 4 feet as an alternative to rating the exterior wall for a distance of 10 feet. The 4-foot requirement was selected because it is consistent with the distance used for several fire wall provisions in Sections 706.5 and 706.5.1. In addition, this 4-foot requirement would be adequate to completely shield the 30-inch by 48-inch wheelchair space behind the wall. Designers could find this option of protection useful as an alternative to protecting the exterior wall and any adjacent opening in the exterior wall. This perpendicular wall protection would conceptually match what has been a long-held code concept: that fires do not turn corners, and therefore, unprotected openings are permitted if they are located perpendicular to the exposure hazard.

The addition of the exception to Section 1007.7.6 now eliminates the requirement for at least 48 inches between stair handrails where the building is sprinklered. Previously, no reductions were permitted for an exterior area of assisted rescue serving a sprinklered building, even though an interior area of refuge could take advantage of several allowable reductions.

As previously mentioned, the separation and protection of exterior areas for assisted rescue may be a bit confusing or different when it occurs in an area that is "essentially open to the outside" and is on a story above the level of exit discharge. Note that the provisions of Sections 1007.7.4 and 1007.7.5 are applicable to these areas. The first bit of potential confusion is the construction of a wall to separate the area from the "interior of the building" when it is essentially an open area; reference the parking deck or roof deck example mentioned previously. Secondly, it would be important to note that the separation is a wall, and there is no mention of a requirement for protecting the exterior space from areas beneath it and there is no reference to the various wall continuity/support provisions of Chapter 7.

CHANGE TYPE: Clarification

CHANGE SUMMARY: The occupant load used to determine the door swing requirement is not to be based on an assigned or distributed occupant load, but on the entire occupant load of the space served by the door.

2012 CODE: 1008.1.2 Door Swing. Egress doors shall be of the pivoted or side-hinged swinging.

> **Exceptions:** (no changes to exceptions)

Doors shall swing in the direction of egress travel where serving <u>a room or area containing</u> an occupant load of 50 or more persons or a Group H occupancy.

CHANGE SIGNIFICANCE: The provision addressing door swing has been clarified to recognize that the total occupant load of the space is to be considered in the regulation of door swing direction. If the occupant load of the room or area is 50 or more persons, the egress doors must swing in the direction of travel. The previous language had occasionally been viewed as allowing a distributed or tributary occupant load to be used for determining the door swing. The additional text clarifies that it is not the code's intent to allow a distributed occupant load to be used for the determination of these basic minimum requirements.

To illustrate the difference in application, consider a space with an occupant load of 60 people and two egress doors serving the area. The total occupant load of 60 should be considered when deciding that the doors serve an occupant load of 50 or more and need to swing in the direction of travel. It is not intended that the 60 occupants be distributed to the two doors so that each door is viewed as only serving 30 people and therefore able to swing against the direction of egress travel.

Most code users will see this as a clarification and not a technical modification. This new language reaffirms the long-standing intent and practice for this door swing requirement.

1008.1.2
Door Swing

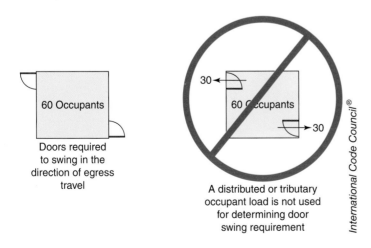

Door swing

1008.1.9.9

Electromagnetically Locked Egress Doors

CHANGE TYPE: Modification

CHANGE SUMMARY: Electromagnetically locked egress doors may now be used at locations that require panic hardware provided the operation of the hardware releases the magnetic lock by interrupting the power to the electromagnet.

2012 CODE: <u>1008.1.9.9</u> ~~1008.1.9.8~~ **Electromagnetically Locked Egress Doors.** Doors in the means of egress ~~that are not otherwise required to have panic hardware~~ in buildings with an occupancy in Group A, B, E, M, R-1, or R-2 and doors to tenant spaces in Group A, B, E, M, R-1, or R-2 shall be permitted to be electromagnetically locked if equipped with listed hardware that incorporates a built-in switch and meet the requirements below:

1. The listed hardware that is affixed to the door leaf has an obvious method of operation that is readily operated under all lighting conditions.
2. The listed hardware is capable of being operated with one hand.
3. Operation of the listed hardware <u>directly</u> ~~releases~~ <u>interrupts the power</u> to the electromagnetic lock and unlocks the door immediately.
4. Loss of power to the listed hardware automatically unlocks the door.
5. <u>Where panic or fire exit hardware is required by Section 1008.1.10, operation of the listed panic or fire exit hardware also releases the electromagnetic lock.</u>

Electromagnet used to lock door

International Code Council®

CHANGE SIGNIFICANCE: The use of electromagnetic locks on egress doors required to have panic hardware was prohibited under the provisions of the 2009 IBC. The use of such locking devices is now acceptable because there are several "panic" bars that are tested and listed for use in the release of the electromagnetic lock. While these are "listed" for magnetic locking devices, they are generally installed for security or normal locking arrangements and are not specifically intended for panic hardware purposes.

The previous prohibition related to panic hardware has been deleted, while the changes in item 3 and the addition of item 5 ensure a direct connection that will interrupt the power to, and release, the electromagnetic lock. When power is removed from a listed electromagnetic lock, it typically will release in less than ½ second. It is these changes in items 3 and 5 that clarify the release of the lock must be automatic with the operation of the push bar and that these magnetic locks are tested and listed devices.

Directly connecting the hardware to stop the power flow to the magnetic lock ensures that the hardware complies with the single operation required by Section 1008.1.9.5 and that the hardware can open as panic hardware when required by simply pressing against the bar. Panic or fire exit hardware must still be listed in accordance with the standards and requirements found in Section 1008.1.10.

Although the text of Section 1008.1.9.9 appears to clearly allow these magnetic locks on doors with panic hardware when the hardware and magnet are interconnected, the provisions of Sections 1008.1.10 and 1008.1.10.1 have not been completely coordinated with these new provisions. Although Section 1008.1.9.9 permits the magnetic locks, panic and fire exit hardware must still meet the requirements found in Section 1008.1.10. The primary issue is the language of Section 1008.1.10 that indicates the doors "shall not be provided with a latch or lock unless it is panic hardware or fire exit hardware." This generally is interpreted to limit the use of any type of lock or restraint that is not a part of the panic hardware device. It appears that the intended application of Section 1008.1.9.9 establishes it as a specific requirement that would override this general provision if the magnetic locks are arranged to have the hardware "directly interrupt the power" and that the "operation of the listed panic or fire exit hardware also releases the electromagnetic lock." The inconsistency is that panic and fire exit hardware are required to comply with UL 305 (see Section 1008.1.10.1) and that standard views the hardware as a mechanical device. It does not address or evaluate devices with electromagnetic locks included in them. Magnetic locking devices simply are not covered by the currently specified standard. Because of this, it appears that the revisions of Section 1008.1.9.9 will at the least create confusion if not simply conflict with the requirements of Section 1008.1.10 and its subsection.

It is difficult to see how the device can be "listed" as indicated in both Sections 1008.1.9.9 and 1008.10.1 when the applicable standard for the panic hardware does not include the option of the magnetic lock or release for it. It appears that the "listing" for electromagnetically locked devices and the separate listing for panic hardware were intended to be combined—even though they are separate listings and the magnetic locks have not and are not tested as a part of the UL 305 listing for panic and fire exit hardware.

Electromagnetic lock shown at top right of door leaf

1009, 1010, 202

Interior Stairways and Ramps

CHANGE TYPE: Clarification

CHANGE SUMMARY: Revisions have been made throughout the code to coordinate the provisions for unenclosed interior stairways and ramps that can be used as a portion of the means of egress.

2012 CODE:

202 Definitions.

EXIT. That portion of a means of egress system ~~which is separated from other interior spaces of a building or structure by fire-resistance-rated construction and opening protectives as required to provide a protected path of egress travel~~ between the exit access and the exit discharge <u>or public way</u>. <u>Exits</u> <u>components</u> include exterior exit doors at the level of exit discharge, ~~vertical exit enclosures~~ <u>interior exit stairways, interior exit ramps</u>, exit passageways, exterior exit stairways<u>, and</u> exterior exit ramps and horizontal exits.

<u>EXIT ACCESS RAMP.</u> <u>An interior ramp that is not a required interior exit ramp.</u>

<u>EXIT ACCESS STAIRWAY.</u> <u>An interior stairway that is not a required interior exit stairway.</u>

EXIT ENCLOSURE. ~~An exit component that is separated from other interior spaces of a building or structure by fire-resistance-rated construction and opening protectives, and provides for a protected path of egress travel in a vertical or horizontal direction to the exit discharge or the public way.~~

<u>INTERIOR EXIT RAMP.</u> <u>An exit component that serves to meet one or more means of egress design requirements, such as required number of exits or exit access travel distance, and provides for a protected path of egress travel to the exit discharge or public way.</u>

Interior exit stairway

Exit access stairway

International Code Council®

INTERIOR EXIT STAIRWAY.   An exit component that serves to meet one or more means of egress design requirements, such as required number of exits or exit access travel distance, and provides for a protected path of egress travel to the exit discharge or public way.

1009.1 General.   Stairways serving occupied portions of a building shall comply with the requirements of this section.

1009.2 Interior Exit Stairways.   Interior exit stairways shall lead directly to the exterior of the building or shall be extended to the exterior of the building with an exit passageway conforming to the requirements of Section 1023, except as permitted in Section 1027.1.

1009.2.1 Where Required.   Interior exit stairways shall be included, as necessary, to meet one or more means of egress design requirements, such as required number of exits or exit access travel distance.

1009.2.2 Enclosure.   All interior exit stairways shall be enclosed in accordance with the provisions of Section 1022.

1009.3 Exit Access Stairways.   Floor openings between stories created by exit access stairways shall be enclosed.

Exceptions:

1. In other than Group I-2 and I-3 occupancies, exit access stairways that serve, or atmospherically communicate between, only two stories are not required to be enclosed.

2. Exit access stairways serving and contained within a single residential dwelling unit or sleeping unit in Group R-1, R-2, or R-3 occupancies are not required to be enclosed.

3. In buildings with only Group B or M occupancies, exit access stairway openings are not required to be enclosed provided that the building is equipped throughout with an automatic sprinkler system in accordance with Section 903.3.1.1, the area of the floor opening between stories does not exceed twice the horizontal projected area of the exit access stairway, and the opening is protected by a draft curtain and closely spaced sprinklers in accordance with NFPA 13.

4. In other than Groups B and M occupancies, exit access stairway openings are not required to be enclosed provided that the building is equipped throughout with an automatic sprinkler system in accordance with Section 903.3.1.1, the floor opening does not connect more than four stories, the area of the floor opening between stories does not exceed twice the horizontal projected area of the exit access stairway, and the opening is protected by a draft curtain and closely spaced sprinklers in accordance with NFPA 13.

5. Exit access stairways within an atrium complying with the provisions of Section 404 are not required to be enclosed.

1009, 1010, 202 continues

1009, 1010, 202 continued

6. Exit access stairways and ramps in open parking garages that serve only the parking garage are not required to be enclosed.

7. Stairways serving outdoor facilities where all portions of the means of egress are essentially open to the outside are not required to be enclosed.

8. Exit access stairways serving stages, platforms, and technical production areas in accordance with Sections 410.6.2 and 410.6.3 are not required to be enclosed.

9. Stairways are permitted to be open between the balcony, gallery, or press box and the main assembly floor in occupancies such as theaters, places of religious worship, auditoriums, and sports facilities.

10. In Group I-3 occupancies, exit access stairways constructed in accordance with Section 408.5 are not required to be enclosed.

1010.1 Scope. The provisions of this section shall apply to ramps used as a component of a means of egress.

Exceptions:

1. Other than ramps that are part of the accessible routes providing access in accordance with Sections 1108.2 through 1108.2.4 and 1108.2.6, ramped aisles within assembly rooms or spaces shall conform with the provisions in Section 1028.11.

2. Curb ramps shall comply with ICC A117.1.

3. Vehicle ramps in parking garages for pedestrian exit access shall not be required to comply with Sections ~~1010.3~~ 1010.4 through ~~1010.9~~ 1010.10 when they are not an accessible route serving accessible parking spaces, other required accessible elements, or part of an accessible means of egress.

1010.2 Enclosure. All interior exit ramps shall be enclosed in accordance with the applicable provisions of Section 1022. Exit access ramps shall be enclosed in accordance with the provisions of Section 1009.3 for enclosure of stairways.

CHANGE SIGNIFICANCE: Although generally considered as a clarification of existing requirements, the multiple changes regarding interior stairways and ramps will provide for consistent application of the code requirements. Because so many code sections are affected by this change, including the revision of some of the basic means of egress terminology, it is important that code users are aware of the revisions even if they do not result in major technical changes.

Historically, the IBC has allowed the limited use of unenclosed exit stairs in a manner that has resulted in inconsistent interpretations. During previous code development cycles, numerous code changes were submitted, with some incorporated into the code, in order to clarify the intent and application of specific provisions. This new revision is considered as a comprehensive change that addresses the entire egress system and how unenclosed stairs affect issues such as exit versus exit access, travel distance measurements, contribution to the minimum number of required exits, etc.

To illustrate the need for a comprehensive revision, consider a two-story building that has one enclosed exit stairway and one open (unenclosed) stairway serving the second floor, which is required to have at least two exits. Because the open stairway did not meet the definition for an "exit," technically only one "exit" is provided from the second story even though the second stairway is permitted to be unenclosed. In the same example, the correct means of measuring exit access travel distance was possibly confusing depending on whether or not the open stairway was considered as an "exit" stairway or an "exit access" stairway from the story.

Code users should be aware of these changes because they will affect means of egress terminology. In addition, modifications result in a number of substantial revisions to Sections 1009, 1010, 1016, 1021, and 1022 as well as sections in Chapters 4, 7, and 8, the IFC and IMC. It should be noted that these revisions are primarily a clarification and are intended to provide consistency throughout the code. The new and revised definitions and those sections that were revised within the code are based on the following concepts:

- All stairs within a building are elements of the means of egress system and must comply with Chapter 10.
- Unenclosed stairways are not considered as an *exit.*
- All exit stairways, to qualify as *exits,* must be enclosed with a fire-resistance-rated enclosure consisting of exit stair shafts and passageways based on the previous exit enclosure provisions.
- All stairways that are permitted to be open, or are not required stairways for egress purposes, are *exit access stairways.*
- *Exit access stairways* must be enclosed with fire-resistance-rated enclosures based on shaft provisions or may be open in accordance with exceptions based on the previous code exceptions.
- Exit access travel distance is measured from an entrance to an *exit.*
- Exit access travel distance includes the travel distance on an *exit access stairway.*
- Entrances to exits on each story are not mandatory and access to exits on other stories is permissible within certain limitations.

1009.1

Application of Stairway Provisions

Example of stairway that should be regulated by the code

International Code Council®

CHANGE TYPE: Clarification

CHANGE SUMMARY: Section 1009.1 has been clarified to apply to any stairway serving occupied portions of a building, including "convenience" stairways that are not a portion of a required means of egress or required means of egress stairways.

2012 CODE: <u>**1009.1 General.**   Stairways serving occupied portions of a building shall comply with the requirements of this section.</u>

CHANGE SIGNIFICANCE: With the inclusion of a new scoping provision, it has been clarified that the requirements of Section 1009 apply to any stairway serving an occupied portion of the building. This charging language eliminates the potential for inappropriate interpretations that view stairways not required as a part of the means of egress system as not regulated by Chapter 10 or the provisions of Section 1009.

Whether stairways are serving as a required portion of the egress system or simply installed in additional numbers beyond the code minimum, it is appropriate for the stairways to meet the minimum safeguards that the code intends. Without the broad scope establish by Section 1009.1, it could be debated that items such as the rise and run, width, dimensional uniformity, handrails, etc., could all be considered as unregulated if the stairway was not a required means of egress stairway. This obviously was never the intent because steeper risers or inconsistent rise/run provisions would make the stairway unsafe regardless of whether it was used for convenience or for egress purposes.

1011.2

Floor-Level Exit Signs in Group R-1

CHANGE TYPE: Addition

CHANGE SUMMARY: Where general-use exit signs are required in Group R-1 occupancies, low-level exit signs must also be provided in the means of egress serving the guest rooms.

2012 CODE: **1011.2 Floor-Level Exit Signs in Group R-1.**   Where exit signs are required in Group R-1 occupancies by Section 1011.1, additional low-level exit signs shall be provided in all areas serving guest rooms in Group R-1 occupancies and shall comply with Section 1011.5.

<u>The bottom of the sign shall be not less than 10 inches (254 mm) nor more than 12 inches (305 mm) above the floor level. The sign shall be flush mounted to the door or wall. Where mounted on the wall, the edge of the sign shall be within 4 inches (102 mm) of the door frame on the latch side.</u>

CHANGE SIGNIFICANCE: To help guide building occupants to the exits during emergency conditions, the potential for recognition of exit signs has been increased by requiring additional exit signs within the egress system serving the guest rooms of a Group R-1 occupancy. These additional signs are to be located very close to the floor level near or on the exit doors.

Limiting the application to only those egress systems serving guest rooms of Group R-1 occupancies recognizes that the occupants of such facilities are transient and not familiar with their surroundings. If a corridor or other egress component serving guest rooms were to fill with smoke, the general exit signs that are located higher in the space could quickly be obscured by the rising smoke. As the space fills with smoke the evacuees are forced to crawl on the floor to reach the nearest exit. They will be confronted with many doors, all looking the same and will not know which door is the exit. The installation of these low-level exit signs will assist these persons in safely exiting the building when exit signs at the higher levels are obscured as the smoke layer develops at the ceiling.

Low-level exit signs will also serve to increase firefighter safety while on the fire scene. In their efforts to evacuate the occupants the firefighters will be in the building after the smoke has developed. Although they rely on several other techniques, fire service personnel may also become dependent upon this low-level signage while trying to locate the doors to the stair tower and safely egress the fire floor.

It should be noted that by the reference to Section 1011.5, internally illuminated signs are mandated for these low-level locations. These signs must be listed in accordance with UL 924, and therefore, the graphics and power requirements of Section 1011.6 for externally illuminated signs are not applicable because the standard will address these issues.

Traditional and low-level exit signs at an interior exit stairway

International Code Council®

1012.2
Handrail Height

CHANGE TYPE: Modification

CHANGE SUMMARY: Transition pieces of a continuous handrail are now permitted to exceed the maximum permitted handrail height.

2012 CODE: 1012.2 Height.   Handrail height, measured above stair tread nosings or finish surface of ramp slope, shall be uniform, not less than 34 inches (864 mm) and not more than 38 inches (965 mm). Handrail height of alternating tread devices and ship ladders, measured above tread nosings, shall be uniform, not less than 30 inches (762 mm) and not more than 34 inches (864 mm).

Exceptions:

1. When handrail fittings or bendings are used to provide continuous transition between flights the fittings or bendings shall be permitted to exceed the maximum height.

2. In Group R-3 occupancies, within dwelling units in Group R-2 occupancies, and in Group U occupancies that are associated with a Group R-3 occupancy or associated with individual dwelling units in Group R-2 occupancies, when handrail fittings or bendings are used to provide continuous

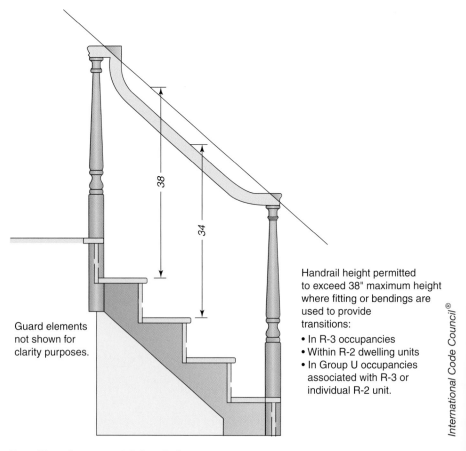

Guard elements not shown for clarity purposes.

38

34

Handrail height permitted to exceed 38" maximum height where fitting or bendings are used to provide transitions:
- In R-3 occupancies
- Within R-2 dwelling units
- In Group U occupancies associated with R-3 or individual R-2 unit.

International Code Council®

Transition pieces on stair handrail

transition between flights, transition at winder treads, transition from handrail to guard, or when used at the start of a flight, the handrail height at the fittings or bendings shall be permitted to exceed the maximum height.

Transition from handrail to guard

CHANGE SIGNIFICANCE: Fittings such as easings and gooseneck risers are commonly used features intended to provide rail continuity at locations where a straight transition is not possible. Incorporating such features is one means of complying with the provisions of Section 1012.4 (Continuity) and is now fairly common architectural and construction practice, especially within residential occupancies. The inclusion of the two new exceptions allows for a more stylized handrail design and permits the handrail heights on a flight of stairs to vary and exceed the height maximums at these transitions. The code previously has always required a "uniform" height for the handrail.

Depending on the proposed handrail height and guard height, the application of the permitted variation will probably be more common in residential occupancies at the locations mentioned in Exception 2. This is due to the fact that guard height on the open side of a stairway can be reduced down to handrail height. Therefore, if the combined handrail/guard on the stair were built at a 34-inch height, a transition would be needed to match up with a 36-inch minimum height guard on the landing. However, it should also be pointed out that if the combination handrail/guard on a stairway was built to a height of 36 inches or 38 inches, it could match up with a guard of a similar height at a landing, and there would be no need for the transition or the use of the exception. With the inclusion of a new Exception 1 in Section 1013.3, the lower guard height for specific residential occupancies will reduce the number of situations where these transitions may be needed. Please see the discussion addressing Section 1013.3 for additional commentary related to this change.

The need for these transition pieces is less of an issue for railings in commercial occupancies because the IBC requires a minimum 42-inch guard height on the open side of a stair as well as a 34 to 38-inch handrail height. Because separate handrails and guards are being provided, the need to transition from one height to another is less of an issue. A 42-inch minimum guard on a stair will easily match up with a similar height guard on a floor or landing without any transition being needed for the handrail.

These transitions are especially common where the handrail transitions from one flight of stairs to another at a dog-leg or switch-back stair landing. Although handrails are not typically required at the landing, the mandate for handrail extensions or for handrail continuity often creates the need for some type of transition, especially at turns. Previously, the height of a continuous handrail at these landings was not regulated. Therefore, the height of the handrail could transition rather abruptly as it transitioned from one stair flight to the next at the landing. The use of the new exceptions will permit a more gradual variation in the height even though it will allow for portions of the handrail to exceed the normal 38-inch maximum height—the belief being that a "continuous" handrail is more important than staying within the height limitation.

Exception 2 will differ from Exception 1 because it is limited to the specified occupancies and is permitted at additional locations. While Exception 1 is only permitted at the "continuous transition between flights," Exception 2 can be used for transitions at winder treads, from handrail to guard, or at the start of a flight.

1012.3.1, 1012.8

Handrail Graspability and Projections

CHANGE TYPE: Modification

CHANGE SUMMARY: A minimum cross-section dimension has now been established for the graspability of noncircular Type I handrails.

2012 CODE: **1012.3 Handrail Graspability.**   All required handrails shall comply with Section 1012.3.1 or shall provide equivalent graspability.

> **Exception:**   In Group R-3 occupancies, within dwelling units in Group R-2 occupancies, and in Group U occupancies that are accessory to a Group R-3 occupancy or accessory to individual dwelling units in Group R-2 occupancies, handrails shall be Type I in accordance with Section 1012.3.1, Type II in accordance with Section 1012.3.2, or shall provide equivalent graspability.

1012.3.1 Type I.   Handrails with a circular cross section shall have an outside diameter of at least 1¼ inches (32 mm) and not greater than 2 inches (51 mm). When the handrail is not circular, it shall have a perimeter dimension of at least 4 inches (102 mm) and not greater than 6¼ inches (160 mm) with a maximum cross-section dimension of 2¼ inches (57 mm) and minimum cross-section dimension of 1 inch (25 mm). Edges shall have a minimum radius of 0.01 inch (0.25 mm).

1012.8 Projections.   On ramps, the clear width between handrails shall be 36 inches (914 mm) minimum. Projections into the required width of stairways and ramps at each ~~handrail~~ side shall not exceed 4½ inches

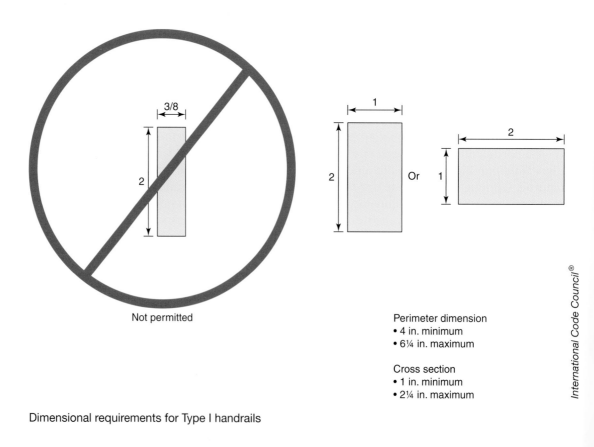

Not permitted

Perimeter dimension
• 4 in. minimum
• 6¼ in. maximum

Cross section
• 1 in. minimum
• 2¼ in. maximum

Dimensional requirements for Type I handrails

International Code Council®

(114 mm) at or below the handrail height. Projections into the required width shall not be limited above the minimum headroom height required in Section ~~1009.2~~ 1009.5. <u>Projections due to intermediate handrails shall not constitute a reduction in the egress width.</u>

CHANGE SIGNIFICANCE: A minimum cross-section dimension has previously not been specified for Type I handrails that were not circular. A circular cross-section has historically been limited to a 1¼-inch minimum, but a minimum dimension has not been required of other handrail shapes, resulting in the acceptance of rails that may not allow a secure grip. The human hand gets its most secure grip on handrail cross sections that allow the hand to fit comfortably around the rail and do not require a pinching grip. While a handrail shape such as a ⅜-inch by 2-inch tube will fall within the code's previously specified dimensional requirements, the ability to grip the rail would be severely limited if the 2-inch dimension was oriented vertically.

While the same shaped section turned horizontally would be more comfortable and accommodating to the hand and grip of most users, there was no requirement that restricted the orientation to the horizontal position. In addition, the limited depth of the member would affect many users who tried to grasp it.

This requirement for a 1-inch minimum cross section when combined with the maximum dimension and the specified perimeter range will provide a shape that is more comfortable and accommodating to the hand's natural grasping shape. The 1-inch dimension was selected because it will allow the use of the maximum 2-inch cross section in one direction combined with the 1-inch dimension on the perpendicular axis and not exceed the maximum allowed 6¼-inch perimeter limitation.

The revisions to the projection provisions are intended to clarify the code's application and provide for more consistent enforcement. One change coordinates with the code language limiting the projection depth "at and below the handrail height." Because the item that protrudes the farthest may be a handrail, baluster, stringer, or an element of trim that is below the handrail, the provision now applies the limit to the "side" of the stair or ramp and is not limited to the handrail itself.

In addition, it has been clarified that an intermediate handrail on a stair or in an aisle is to be considered as a permitted projection and not as a reduction in the required egress width. For example the 48-inch aisle stairway required by Section 1028.9.1 that has a 2-inch-wide intermediate handrail would be viewed as providing 48 inches of egress width even though the aisle is arranged to provide 23 inches of clear width between the handrail and seating on both sides with the other 2 inches occupied by the handrail.

1013.1, 1013.8

Guards at Operable Windows

CHANGE TYPE: Modification

CHANGE SUMMARY: The guard requirements for operable windows having a sill height more than 72 inches above the finished grade have been relocated from Chapter 14 to the general guard provisions of Chapter 10 and the minimum window sill height at which a guard is not required has been increased from 24 inches to 36 inches.

2012 CODE: **1013.1 General.** <u>Guards shall comply with the provisions of Sections 1013.2 through 1013.7. Operable windows with sills located more than 72 inches (1.83 m) above finished grade or other surface below shall comply with Section 1013.8.</u>

1013.8 ~~1405.13.2~~ **Window Sills.** In Occupancy Groups R-2 and R-3, one- and two-family and multiple-family dwellings, where the opening of the sill portion of an operable window is located more than 72 inches (1829 mm) above the finished grade or other surface below, the lowest part of the clear opening of the window shall be at a height not less than ~~24 inches (610 mm)~~ <u>36 inches (915 mm)</u> above the finished floor surface of the room in which the window is located. ~~Glazing between the floor and a height of 24 inches (610 mm) shall be fixed or have openings through which a 4-inch (102 mm) diameter sphere cannot pass.~~ <u>Operable sections of windows shall not permit openings that allow passage of a 4-inch (102-mm) diameter sphere where such openings are located within 36 inches (915 mm) of the finished floor.</u>

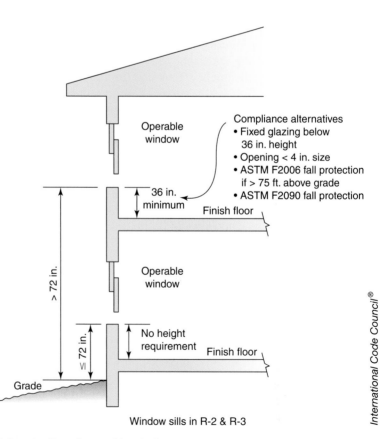

Fall protection at operable windows

Exceptions:

1. ~~Openings that are~~ Operable windows where the sill portion of the opening is located more than 75 feet (22.86 m) above the finished grade or other surface below and that are provided with window ~~guards~~ fall prevention devices that comply with ASTM F 2006 ~~or F 2090~~.

2. Windows whose openings will not allow a 4-inch (102-mm) diameter sphere to pass through the opening when the window is in its largest opened position.

3. Openings that are provided with window fall prevention devices that comply with ASTM F2090.

4. Windows that are provided with window opening control devices that comply with Section 1013.8.

1013.8.1 Window Opening Control Devices. Window opening control devices shall comply with ASTM F 2090. The window opening control device, after operation to release the control device allowing the window to fully open, shall not reduce the minimum net clear opening area of the window unit to less than the area required by Section 1029.2.

CHANGE SIGNIFICANCE: The fall protection requirements related to low-height window sills have been relocated from Chapter 14 to Section 1013. In addition, the minimum height of the window sill at which a guard is not required has been revised from 24 inches to 36 inches. The 36-inch sill height was chosen to reduce the ability of a child to climb onto the sill and thus enabling them to fall through the opening. While the 24-inch height was above the center of gravity for most children under 4½ years of age, the lower height was easily climbed by most standing children.

The modified Exception 1 makes two changes that better coordinate the code with the scope of the standard addressing window fall prevention devices. Most notable will be the fact that the exception is now limited to only those operable windows that are located more than 75 feet above grade. This revision is coordinated with the scoping provisions found within the ASTM F2006 standard itself. Section 1.2 of the standard states, "This safety specification applies only to window fall prevention devices that are to be used on windows that are not intended for escape (egress) and rescue (ingress)." Further, Section 1.3 states, "This safety specification applies only to devices intended to be applied to windows installed at heights of more than 75 above ground level in multiple-family dwelling buildings. This safety specification is not intended to apply to windows below 75 feet because all windows below 75 feet that are operable could be used as a possible secondary means of escape."

Users will also notice that the ASTM F2090 standard that was previously referenced has been deleted and is now addressed in a new Exception 3. With the revised height limitation in Exception 1 and the fact that emergency escape and rescue openings are not required above 75 feet, the ASTM F2090 standard is no longer applicable. ASTM F 2090 includes window fall prevention devices (the new Exception 3) and window opening control devices (Exception 4 and the new Section 1013.8.1). The standard is specifically written for window openings within 75 feet

1013.1, 1013.8 continues

1013.1 1013.8 continued

of grade and specifically allows for windows to be used for emergency escape and rescue. Opening control devices allow for normal operation to result in a 4 inch maximum opening, thus meeting the requirements of the last sentence in the base paragraph, but can be released to allow the window to be fully opened in order to comply with the emergency escape provisions of Section 1029.2. The window control devices and their operation are regulated by the new Section 1013.8.1 to ensure they can serve both the fall protection concerns as well as the escape and rescue opening functions.

The 4-inch opening size limitation specified in Exception 2 is consistent with the guard provisions of Section 1013.4. Although not stated directly within the exception, the requirements of Exception 2 are limited to windows or portions of windows where the opening is located between the floor surface and 36 inches in height above the floor surface. Due to the height limitations within the base paragraph of Section 1013.8, any opening that above the 36-inch height would not be regulated by the 4-inch limitation.

1013.3
Guard Height

CHANGE TYPE: Modification

CHANGE SUMMARY: The minimum required height for guards in Group R-3 occupancies and within individual Group R-2 dwelling units has been decreased from 42 inches to 36 inches.

2012 CODE: 1013.3 ~~1013.2~~ Height.   Required guards shall not be less than 42 inches (1067 mm) high, measured vertically ~~above the~~ as follows:

1. From the adjacent walking surfaces. ~~adjacent fixed seating or~~
2. On stairs, from the line connecting the leading edges of the tread ~~treads~~ nosings, and
3. On ramps, from the ramp surface at the guard.

Exceptions:

1. For occupancies in Group R-3 not more than three stories above grade in height and within individual dwelling units in occupancies in Group R-2 not more than three stories above grade in height with separate means of egress, required guards shall not be less than 36 inches (914 mm) in height measured vertically above the adjacent walking surfaces or adjacent fixed seating.

2.~~1.~~ For occupancies in Group R-3, and within individual dwelling units in occupancies in Group R-2, guards on the open sides of stairs shall have a height not less than 34 inches (864 mm) measured vertically from a line connecting the leading edges of the treads.

1013.3 continues

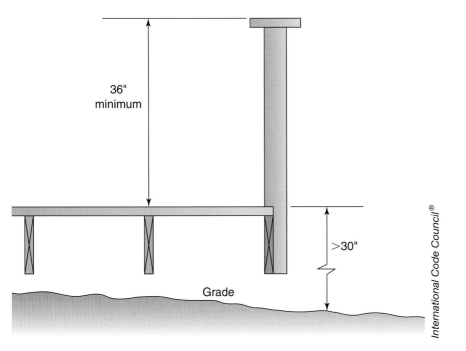

Minimum residential guard height

1013.3 continued

Guard height for R-3 and within R-2 dwelling units

International Code Council®

3.2. For occupancies in Group R-3, and within individual dwelling units in occupancies in Group R-2, where the top of the guard also serves as a handrail on the open sides of stairs, the top of the guard shall not be less than 34 inches (864 mm) and not more than 38 inches (965 mm) measured vertically from a line connecting the leading edges of the treads.

4.3. The guard height in assembly seating areas shall comply ~~be in accordance~~ with Section 1028.14.

5.4. Along alternating tread devices and ship ladders, guards whose top rail also serves as a handrail shall have height not less than 30 inches (762 mm) and not more than 34 inches (864 mm), measured vertically from the leading edge of the device tread nosing.

CHANGE SIGNIFICANCE: The minimum required guard height for certain residential occupancies has been reduced from 42 inches to 36 inches in height in order to coordinate with existing provisions in the IRC. The exception is limited only to those listed residential occupancies located no more than three stories above grade. Previously the IBC required a 42-inch guard height for all occupancies. The 42-inch height was selected because the center of gravity for the 95th percentile male population (and about 97 percent of the total population) is below that height, and therefore, it was deemed unlikely that an accidental fall would occur from simply leaning over the rail. Although the lower guard height will address a lower percentage of the population, the reduced height has historically been recognized as acceptable under the IRC.

The reduction in required guard height for the floors and landings in residential units will help reduce the situations where a handrail fitting or bending addressed in the new exceptions to Section 1012.2 are applied. See the discussion of Section 1012.2 for more information on transitions between handrails or between a handrail and a guard.

1021.2
Exits from Stories

CHANGE TYPE: Modification

CHANGE SUMMARY: Exits are now permitted to be arranged where they serve a portion of a story instead of requiring that all of the required exits from the story be accessible to all of the occupants.

2012 CODE: ~~1021.1~~ **1021.2 Exits from Stories.** Two exits, or exit access stairways or ramps providing access to exits, from any story or occupied roof shall be provided where one of the following conditions exists:

(Items 1–3 not shown for clarity)

Exceptions:

(Exceptions 1 through 6 not shown for clarity)

7. Exits serving specific spaces or areas need not be accessed by the remainder of the story when all of the following are met:
    7.1. The number of exits from the entire story complies with Section 1021.2.4;
    7.2. The access to exits from each individual space in the story complies with Section 1015.1; and
    7.3. All spaces within each portion of a story shall have access to the minimum number of approved independent exits based on the occupant load of that portion of the story, but not less than two exits.

Note: Because Section 1021 was substantially reformatted, the entire 2009 code text is not shown. See 2009 IBC Section 1021.1 for comparison.

CHANGE SIGNIFICANCE: Section 1021.2 is essentially a revised version of Section 1021.1 from the 2009 edition. There are a number of changes in this new Section 1021.2 that are not shown here but were related to the change discussed previously with Section 1009 and the definitions

1021.2 continues

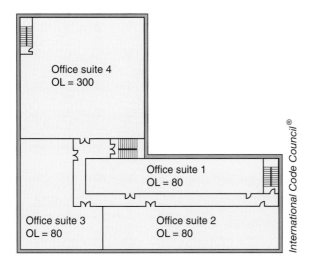

Exits serving specific spaces or areas

1021.2 continued for various types of stairways. The intent of the added Exception 7 is to clearly state that not all occupants of a story are required to have access to all of the exits provided from that story. This exception recognizes that in some building arrangements an exit may serve a specific area, such as being within an individual tenant space and may not be useable by other occupants on this story. Having these isolated exits is not a problem provided the three established requirements are met. These requirements ensure that (1) the entire story is provided with the proper number of exits, (2) that each space on the story has access to the required number of exits that space needs, and (3) that all portions of the story have access to the minimum number of required exits. It is important to note that occupants above the first floor generally need access to at least two exits.

The previous requirement that "all spaces shall have access to the minimum number of approved independent exits as specified in Table 1021.1" often resulted in designs where access was provided to every exit required from the story even though the exiting layout was such that certain exits served distinct portions of the story and those portions may have only required access to one or two exits based on the occupant load of that area. An exception now allows exits serving a specific area or portion of a building to not be accessible from other portions of the building provided (1) the overall number of exits are provided from the story as required by Section 1021.2, (2) access to exits from individual spaces comply with Section 1015.1, and (3) all spaces within each portion of a story have access to the minimum number of approved independent exits as specified in Section 1021.2 "based on the occupant load of that portion of the story." With the added phrasing "but not less than two exits" this third item also ensures occupants above the first floor have access to multiple exits from that story.

CHANGE TYPE: Clarification

CHANGE SUMMARY: A ratio equation is now to be used to determine if a single exit is allowed to serve the combined occupant load from different occupancies.

2012 CODE: 1021.2.1 ~~Single Exits.~~ Mixed Occupancies. ~~Only one exit shall be required from Group R-3 occupancy buildings or from stories of other buildings as indicated in Table 1021.2. Occupancies shall be permitted to have a single exit in buildings otherwise required to have more than one exit if the areas served by the single exit do not exceed the limitations of Table 1021.2.~~ Where one exit, or exit access stairway or ramp providing access to exits at other stories, is permitted to serve individual stories, mixed occupancies shall be permitted to be served by single exits provided each individual occupancy complies with the applicable requirements of Table 1021.2(1) or Table 1021.2(2) for that occupancy. Where applicable, cumulative occupant loads from adjacent occupancies shall be considered in accordance with the provisions of Section 1004.1. ~~Basements with a single exit shall not be located more than one story below grade plane.~~

In each story of a mixed occupancy building, the maximum number of occupants served by a single exit shall be such that the sum of the ratios of the calculated number of occupants of the space divided by the allowable number of occupants for each occupancy does not exceed one.

CHANGE SIGNIFICANCE: In the evaluation of a mixed occupancy building that is served by a single exit, it has been clarified as to how the occupant load is to be evaluated. Previously, a single exit was permitted provided each individual occupancy was in compliance. The new provisions address mixed occupancy buildings in a ratio manner similar to that used for allowable floor area limitations in separated occupancy buildings.

This approach is also applicable to a single-story multiple tenant building where each tenant space has its own exterior exit. If a designer chooses to have two tenant spaces of different occupancies share the same exit, then this provision could be applied. As an option, individual exits can be provided from each space. Consistency is also provided with the new Exception 7 and Item 7.3 in Section 1021.2.

1021.2.1
Exits from Mixed Occupancy Buildings

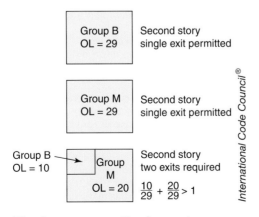

Group B OL = 29	Second story single exit permitted
Group M OL = 29	Second story single exit permitted
Group B OL = 10 → Group M OL = 20	Second story two exits required $\frac{10}{29} + \frac{20}{29} > 1$

International Code Council®

Mixed occupancy exiting from a story

1021.2.3, Table 1021.2(1)

Exits from Dwelling Units

CHANGE TYPE: Modification

CHANGE SUMMARY: A new section clarifies when a single exit is permitted within or from an individual dwelling unit. Changes to Section 1021.2 and the tables will also provide a second option for compliance.

2012 CODE: ~~1021.1~~ **1021.2 Exits from Stories.** ~~All spaces within each story shall have access to the minimum number of approved independent exits as specified in Table 1021.1 based on the occupant load of the story. For the purposes of this chapter, occupied roofs shall be provided with exits as required for stories.~~

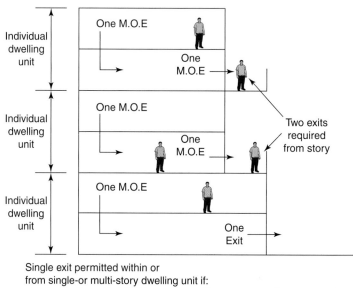

Single exit permitted within or
from single-or multi-story dwelling unit if:
• Unit complies with 1015.1 as space with one means of egress, and
• Discharges directly to exterior at level of exit discharge, or
• Exit access outside unit provides access to not less than two exits

Egress from multistory dwelling units

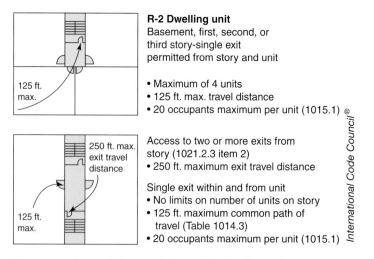

R-2 Dwelling unit
Basement, first, second, or third story-single exit permitted from story and unit

• Maximum of 4 units
• 125 ft. max. travel distance
• 20 occupants maximum per unit (1015.1)

Access to two or more exits from story (1021.2.3 item 2)
• 250 ft. maximum exit travel distance

Single exit within and from unit
• No limits on number of units on story
• 125 ft. maximum common path of travel (Table 1014.3)
• 20 occupants maximum per unit (1015.1)

Egress requirements from individual R-2 dwelling units

Exceptions:

4. ~~In Group R-2 and R-3 occupancies, one means of egress is permitted within and from individual dwelling units with a maximum occupant load of 20 where the dwelling unit is equipped throughout with an automatic sprinkler system in accordance with Sections 903.3.1.1 or 903.3.1.2.~~

Two exits, or exit access stairways or ramps providing access to exits, from any story or occupied roof shall be provided where one of the following conditions exists:

1. The occupant load or number of dwelling units exceeds one of the values in Table 1021.2(1) or 1021.2(2).

2. The exit access travel distance exceeds that specified in Table 1021.2(1) or 1021.2(2) as determined in accordance with the provisions of Section 1016.1.

3. Helistop landing areas located on buildings or structures shall be provided with two exits, or exit access stairways or ramps providing access to exits.

Exceptions:

1. Rooms, areas and spaces complying with Section 1015.1 with exits that discharge directly to the exterior at the level of exit discharge, are permitted to have one exit.

2. Exception not shown for clarity

3. Exception not shown for clarity

4. Exception not shown for clarity

5. Individual dwelling units in compliance with Section 1021.2.3.

6. Exception not shown for clarity

7. Exception not shown for clarity

1021.2.3 Single-Story or Multi-Story Dwelling Units. Individual single-story or multi-story dwelling units shall be permitted to have a single exit within and from the dwelling unit provided that all of the following criteria are met:

1. The dwelling unit complies with Section 1015.1 as a space with one means of egress and

2. Either the exit from the dwelling unit discharges directly to the exterior at the level of exit discharge, or the exit access outside the dwelling unit's entrance door provides access to not less than two approved independent exits.

Because the code was substantially reformatted in Section 1021, only a portion of the 2009 code text is shown. See 2009 IBC Section 1021.1 for comparison.

CHANGE SIGNIFICANCE: The requirements for residential dwelling units include a number of changes that affect both the egress requirements from the unit and those from the story of the building where the unit is located. Some of these provisions, such as the new Section 1021.2.3, apply to any

1021.2.3, Table 1021.2(1) continued

TABLE 1021.2(1)   **Stories with One Exit or Access to One Exit for R-2 Occupancies**

Story	Occupancy	Maximum Number of Dwelling Units	Maximum Exit Access Travel Distance
Basement, first, second or third story	R-2[a, b]	4 dwelling units	125 feet
Fourth story and above	NP	NA	NA

For SI: 1 foot = 3048 mm.
NP – Not Permitted
NA – Not Applicable
a. Buildings classified as Group R-2 equipped throughout with an automatic sprinkler system in accordance with Section 903.3.1.1 or 903.3.1.2 and provided with emergency escape and rescue openings in accordance with Section 1029.
b. This Table is used for R-2 occupancies consisting of dwelling units. For R-2 occupancies consisting of sleeping units, use Table 1021.2(2).

TABLE 1021.2(2)   **Stories with One Exit or Access to One Exit for Other Occupancies**

Story	Occupancy	Maximum Occupants Story (Or Dwelling Units) Per Floor And Travel Distance	Maximum Exit Access Travel Distance
First story or basement	A, B[bd], E[e], F[bd], M, U, S[bd]	49 occupants and 75 feet travel distance	75 feet
	H-2, H-3	3 occupants and 25 feet travel distance	25 feet
	H-4, H-5, I, R, R-1, R-2[a,c] [c,f], R-4	10 occupants and 75 feet travel distance	75 feet
	S[a]	29 occupants and 100 feet travel distance	100 feet
Second story	B[b], F, M, S[a]	29 occupants and 75 feet travel distance	75 feet
	R-2	4 dwelling units and 50 feet travel distance	
Third story	R-2[c]	4 dwelling units and 50 feet travel distance	
Third story and above	NP	NA	NA

For SI: 1 foot = 304.8 mm.
NP – Not Permitted
NA – Not Applicable
a. For the required number of exits for parking structures, see Section 1021.1.2.
b. For the required number of exits for air traffic control towers, see Section 412.1.
a. c. Buildings classified as Group R-2 equipped throughout with an automatic sprinkler system in accordance with Section 903.3.1.1 or 903.3.1.2 and provided with emergency escape and rescue openings in accordance with Section 1029.
b. d. Group B, F, and S occupancies in buildings equipped throughout with an automatic sprinkler system in accordance with Section 903.3.1.1 shall have a maximum travel distance of 100 feet.
e. Day care occupancies shall have a maximum occupant load of 10.
c. This Table is used for R-2 occupancies consisting of sleeping units. For R-2 occupancies consisting of dwelling units, use Table 1021.2(1).

residential dwelling unit while others, such as those in Table 1021.2(1), apply only to Group R-2 dwelling units (not to Group R-2 sleeping units). One important aspect to note is that there are now two separate means of egress compliance options in these residential occupancies.

New Section 1021.2.3 has replaced the previous Exception 4 in order to clarify the single means of egress requirements for a dwelling unit and organize all of the provisions in a single location. The base paragraph refers to both single-story and multi-story dwelling units so it is clear that the provision can be applied to a multi-story unit even if it has an unenclosed exit access stairway as permitted by Section 1009.3, Exception 2.

Item 1 of Section 1021.2.3 and its reference to Section 1015.1 directs users to the two provisions for spaces with one exit or exit access including, (1) the 125 foot common path of travel limit from Section 1014.3 (referenced from Section 1015.1, Item 2) and (2) the occupant load limitation of 20 that is found in Exception 1 to Section 1015.1, Item 1. The second item addresses the code requirements for the means of egress after the occupant has left the individual dwelling unit. The two possible situations, (a) the occupant discharges directly to the exterior at the level of exit discharge, or (b) the occupant enters a common exit access which leads to at least two exits, are also addressed elsewhere in the code. If the person leaves the building, they are in the exit discharge and are considered safe because they are outside the building at ground level and have access to the public way. If the occupant has left the dwelling unit and is not on the level of exit discharge, then the occupant is now continuing through the exit access portion of the building, and will generally require access to at least two exits from the point that the occupant traveled out of the dwelling unit. However, Section 1021.2 and Table 1021.2(1) will provide another option where up to 4 dwelling units may be located on a story with access to only a single exit from the basement, first, second, or third story.

As mentioned previously, Section 1021.2 provides two options for exiting from the individual dwelling unit and the story of the building. These options are most clearly seen by comparing the requirements of Section 1021.2.3 (primarily item 2) with the general egress requirements that apply when using Sections 1015.1 and Table 1021.2(1). Table 1021.2(1) will allow a single exit from the basement, first, second, or third story if there are a maximum of 4 dwelling units with a limited travel distance; while Section 1021.2.3 can be used for any story (including the basement, first, second, or third) without a limitation on the number of units on the floor and a more generous travel distance limit. These two options are located in Section 1021.2 and the general egress requirements or in Exception 7 and its reference to Section 1021.2.3.

A second table to specifically address Group R-2 dwelling units continues the effort to clarify the application of the requirements for a single exit from a building or story (Section 1021) versus the requirements for egress from a space (Section 1015). Dividing the previous table into two tables addresses code modifications that have occurred over the past two editions that dealt with the common path of travel for sprinklered Group R-2 dwelling units. When the IBC was initially developed, not all residential occupancies were required to be sprinklered and as such a shorter common path of travel was imposed on the Group R-2 dwelling units. Splitting the tables allows for Table 1021.2(1) to address the requirements for Group R-2 dwelling units based on the number of units on the story while Table 1021.2(2) will regulate Group R-1, R-2 sleeping units and R-4 occupancies based on the number of occupants. Listing the individual residential occupancies also helps to clarify that Group R-3 occupancy buildings are always permitted to have a single exit (see Section 1021.2, Exception 2).

1022.5

Enclosure Penetrations of Interior Exit Stairways

CHANGE TYPE: Modification

CHANGE SUMMARY: Penetrations of the outside membrane of a fire barrier utilized to enclose an interior exit stair or ramp are now permitted provided the penetration is properly protected.

2012 CODE: ~~1022.4~~ <u>1022.5</u> **Penetrations.** Penetrations into and openings through ~~an exit enclosure~~ <u>interior exit stairways and ramps</u> are prohibited except for required exit doors, equipment, and ductwork necessary for independent ventilation or pressurization, sprinkler piping, standpipes, electrical raceway for fire department communication systems, and electrical raceway serving the ~~exit enclosure~~ <u>interior exit stairway and ramp</u> and terminating at a steel box not exceeding 16 square inches (0.010 m^2). Such penetrations shall be protected in accordance with Section 714. There shall be no penetrations or communication openings, whether protected or not, between adjacent ~~exit enclosures~~ <u>interior exit stairways and ramps</u>.

> **Exception:** <u>Membrane penetrations shall be permitted on the outside of the interior exit stairway and ramp. Such penetrations shall be protected in accordance with Section 714.3.2.</u>

CHANGE SIGNIFICANCE: Unless specifically permitted, penetrations have historically been prohibited at the fire-resistance-rated enclosure around an interior exit stairway or ramp. The strict limitations were

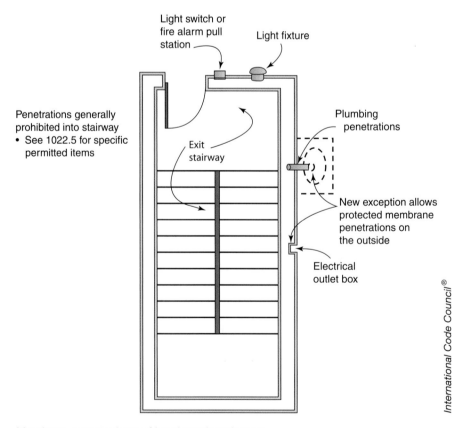

Membrane penetrations of interior exit stairways

International Code Council®

deemed necessary to ensure that the fire-resistive integrity of the exit system was not compromised by penetrations of the protective enclosure. Penetrations of the exterior membrane of the fire-resistance-rated assembly are now permitted provided they are in compliance with the membrane penetration provisions of Section 714.3.2.

Virtually all penetrations have been prohibited in the past, regardless of purpose with very limited exceptions. The prohibition applied to an alarm pull station next to a door into the stair enclosure, fire hose cabinets, fire extinguisher cabinets, alarm notification appliances, electrical wiring for exit signs, electrical outlets, and other items. The new exception will not limit the type of or purpose for the penetration but will simply limit the location to the exterior membrane and require the proper protection.

1028.1.1.1

Separation of Spaces under Grandstands and Bleachers

CHANGE TYPE: Addition

CHANGE SUMMARY: Spaces beneath grandstands and bleachers are now required to be adequately separated to protect the assembly seating area from any potential hazards.

2012 CODE: <u>**1028.1.1.1 Spaces under Grandstands and Bleachers.**</u> <u>When spaces under grandstands or bleachers are used for purposes other than ticket booths less than 100 square feet (9.29 m^2) and toilet rooms, such spaces shall be separated by fire barriers complying with Section 707 and horizontal assemblies complying with Section 711 with not less than 1-hour fire-resistance-rated construction.</u>

CHANGE SIGNIFICANCE: In order to protect the assembly seating area of bleachers and grandstands from an exposure to hazards, a minimum separation of 1-hour fire-resistance-rated construction is now required to protect the seating from the spaces below. This separation requirement applies where the space beneath the seating is used for any purpose other than restrooms of any size or limited size ticket booths, even if the spaces are protected by an automatic sprinkler system in accordance with Section 903.2.1.5. Conceptually this requirement for protection on the underside of the seating is similar to the required protection of enclosed usable space beneath a stairway (see Section 1009.9.3).

The location of the new requirements was chosen for its proximity to the reference to the ICC 300 standard. Placing the provision directly after that reference section will help ensure the protection requirement is not overlooked.

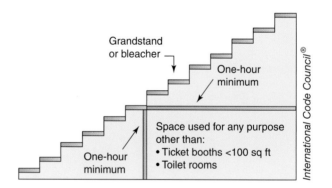

Spaces under grandstands and bleachers

PART 5

Accessibility

Chapter 11

■ **Chapter 11** Accessibility

Chapter 11 is intended to address the accessibility and usability of buildings and their elements to persons having physical disabilities. The provisions within the chapter are generally considered as scoping requirements that state what and where accessibility is required or how many accessible features or elements must be provided. The technical requirements, addressing how accessibility is to be accomplished, are found in ICC A117.1, as referenced by Chapter 11. The concept of the code is to initially mandate that all buildings and building elements be accessible and then to reduce the level of required accessibility only to that which is logical and reasonable. ■

1104.3.1
Employee Work Areas

1107.6.1
Accessible Units in R-1 Occupancies

1108.2.7.3
Captioning of Public Address Announcements

1109.2, 1109.5
Accessible Children's Facilities

1109.6
Accessible Saunas and Steam Rooms

1110.4
Variable Message Signs

1104.3.1

Employee Work Areas

Employee work area in office

CHANGE TYPE: Modification

CHANGE SUMMARY: Where an employee work area is less than 1000 square feet in floor area, the common use circulation path need not meet the accessible route requirements.

2012 CODE: 1104.3.1 Employee Work Areas.   Common-use circulation paths within employee work areas shall be accessible routes.

Exceptions:

1. Common use circulation paths, located within employee work areas that are less than ~~300~~ 1000 square feet (~~27.9~~ 93 m^2) in size and defined by permanently installed partitions, counters, casework, or furnishings, shall not be required to be accessible routes.

2. Common-use circulation paths, located within employee work areas, that are an integral component of equipment, shall not be required to be accessible routes.

3. Common-use circulation paths, located within exterior employee work areas that are fully exposed to the weather, shall not be required to be accessible routes.

CHANGE SIGNIFICANCE: In general, employee work areas are exempt from the accessibility requirements of Chapter 11 based on the "general exception" found in Section 1103.2.3. However, where the circulation path through an employee work area is for the shared use of two or more employees, then the circulation path is required to comply with the accessible route requirements of both the IBC and the A117 standard. Previously, the code allowed the common-use circulation path to be exempt from this requirement if the size of the space was limited by some type of permanent boundary and was less than 300 square feet in floor area.

Employee work area < 1,000 sq. ft.

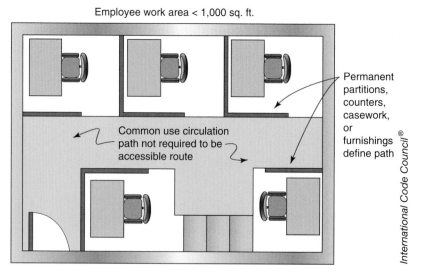

Accessible route for common use circulation path

The size limitation has now been increased to 1000 square feet versus the previous 300 square feet. Increasing the size will help match the requirements of the IBC with those of the Americans with Disabilities Act (ADA). Because the newly released ADA and Architectural Barriers Act (ABA) Accessibility Guidelines accept this larger space, having the IBC area limit increased will help in the coordination efforts and reduce confusion regarding the application of the requirements.

1107.6.1

Accessible Units in R-1 Occupancies

CHANGE TYPE: Modification

CHANGE SUMMARY: A reduced number or percentage of the facilities in Accessible units are now required to be accessible.

2012 CODE: 1107.6.1 Group R-1. Accessible units and Type B units shall be provided in Group R-1 occupancies in accordance with Sections 1107.6.1.1 and 1107.6.1.2.

1107.6.1.1 Accessible Units. ~~In Group R-1 occupancies,~~ Accessible dwelling units and sleeping units shall be provided in accordance with Table 1107.6.1.1. All ~~R-1~~ <u>dwelling units and sleeping</u> units on a site shall be considered to determine the total number of Accessible units. Accessible units shall be dispersed among the various classes of units. Roll-in showers provided in Accessible units shall include a permanently mounted folding shower seat.

<u>1107.6.1.1.1 Accessible Unit Facilities.</u> <u>All interior and exterior spaces provided as part of or serving an Accessible dwelling unit or sleeping unit shall be accessible and be located on an accessible route.</u>

<u>**Exceptions:**</u>

1. <u>Where multiple bathrooms are provided within an Accessible unit, at least one full bathroom shall be accessible.</u>

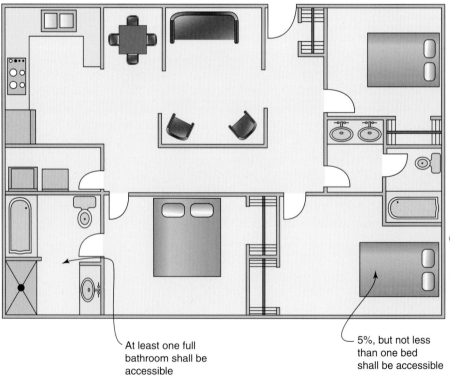

At least one full bathroom shall be accessible

5%, but not less than one bed shall be accessible

Accessible unit facilities

2. Where multiple family or assisted bathrooms serve an Accessible unit, at least 50 percent but not less than one room for each use at each cluster shall be accessible.

3. Five percent, but not less than one bed shall be accessible.

CHANGE SIGNIFICANCE: New exceptions to Section 1107.6.1.1.1 modify the requirement that every element within or serving an Accessible unit has to be accessible. The first exception will clearly limit the accessible bathroom requirements to only one bathroom within the unit. For hotel rooms or suites containing two or more bathrooms, the code previously would have required all of the bathrooms to be fully accessible. It is not only reasonable to limit this requirement to a single bathroom within the unit, but it will also coordinate this provision with a similar limitation that is found within the new A117.1-2009 edition.

The second exception will be more limited in its application. It's intent is to address the rare situations where bathrooms are not provided within the unit but the units are instead served by multiple single-user bathrooms at some location within the facility. Where these multiple single-user bathrooms are provided and serve the Accessible units, then only 50 percent of the bathrooms must be accessible. This reduced limit on accessible bathrooms is similar to the reduced percentage allowed by Section 1109.2, Exception 3, which applies to this type of single-user facility in assembly and mercantile occupancies. Although this type of arrangement is not generally common in most hotels and motels, it does exist, especially in some budget locations or in older structures. Without this exception, then each of the bathrooms would be required to be accessible based on the requirements of both Section 1109.2 and the base paragraph of Section 1107.6.1.1.1.

The third exception that addresses beds is an item that is not typically covered by the building code but is found within the Appendix E requirements that help to coordinate the IBC and A117.1 with the ADA. A clear floor space be provided around the bed to allow access to it. While the "not less than one bed" requirement will generally be adequate for most Accessible units, if a large barracks- or dormitory-type arrangement of beds was used in an Accessible unit, this "5 percent" requirement would require that additional beds provide the access space where there are 20 or more beds within the unit.

Users should also be aware that A117.1-2009 contains a new section within the Accessible unit provisions that is applicable to the beds. The change in the A117.1 standard uses the same scoping ("5 percent but not less than one") and requires that at least one bed within the Accessible unit have an open frame beneath it. This requirement allows a bed lift such as a Hoyer lift to be used for people to transfer from a wheelchair into the bed. For additional discussion on the A117.1 requirement, please see the publication *Significant Changes to the A117.1 Accessibility Standard, 2009 edition* authored by Jay Woodward.

1108.2.7.3

Captioning of Public Address Announcements

CHANGE TYPE: Modification

CHANGE SUMMARY: The captioning of audible public announcements is now only required for assembly spaces having a public address system and 15,000 or more seats.

2012 CODE: ~~1108.2.7.2~~ **1108.2.7.3 Public Address Systems.**   Where stadiums, arenas, and grandstands <u>have 15,000 fixed seats or more and</u> provide audible public announcements, they shall also provide ~~equivalent text information regarding events and facilities in compliance with Sections 1108.2.7.2.1 and 1108.2.7.2.2~~ <u>prerecorded or real-time captions of those audible public announcements.</u>

~~**1108.2.7.2.1 Prerecorded Text Messages.**   Where electronic signs are provided and have the capability to display prerecorded text messages containing information that is the same, or substantially equivalent to information that is provided audibly, signs shall display text that is equivalent to audible announcements.~~

> ~~**Exception:**   Announcements that cannot be prerecorded in advance of the event shall not be required to be displayed.~~

~~**1108.2.7.2.2 Real-time Messages.**   Where electronic signs are provided and have the capability to display real-time messages containing information that is the same, or substantially equivalent, to information that is provided audibly, signs shall display text that is equivalent to audible announcements.~~

International Code Council®

Stadium with less than 15,000 seats

CHANGE SIGNIFICANCE: The requirement for providing captions of audible announcements is now applicable only when the stadium, arena, or grandstand has 15,000 or more fixed seats. Previously, any grandstand with a public address system was required to provide equivalent text—that is, captioning—of the information announced over a public address system. Therefore, under the 2009 IBC, a little league ballpark where the seating meets the definition of a grandstand required captioning if there was a microphone with a loud speaker that would address the public. There are a number of similar facilities in most communities that fell into this category without any practical way of complying.

Captioning requires a certain level of expertise and equipment that is not typically available in a small facility. These smaller facilities are often staffed by volunteers rather than the trained staff found in the more sophisticated facilities. Solving the technological challenges alone does not ensure effective captioning. Because of the difficulties and the expense of the equipment and captioning, this requirement now only applies to those stadiums, arenas, and grandstands that would have adequate infrastructure to adequately caption announcements.

Although the previous text dealing with prerecorded and real-time messages has been deleted, the two options for captioning of audible announcements have been maintained by virtue of the last portion of Section 1108.2.7.3. The decision regarding which type of message, prerecorded text, or real-time captioning is done at the discretion of the facility. However, it should be noted that the code previously contained an exception that eliminated the requirement for captioning of announcements that could not be prerecorded. Because this exception did not get carried over into the new text, announcements that cannot be prerecorded must be provided as a real-time caption.

The revision from the 2009 edition's use of the term "electronic signs" to the new code's use of "captions" is essentially an editorial revision and a clarification that distinguishes the purpose of the system versus a type of electronic advertising billboard. "Captions" of audible information is nationally recognized as the term for providing text information that matches audible announcements. In a court decision related to the ADA, captioning of information that is announced over the PA system was ruled as being needed for equivalent communication with persons having a hearing impairment.

Users should also be aware of a new requirement in the A117.1-2009 standard. The standard added specific technical requirements related to variable message signs (VMS), and although not required by this provision or by the new Section 1110.4, the requirements could be used for the captioning system. There are several exceptions within the VMS provisions of the standard that modify the requirements for the size of the sign lettering when used in assembly occupancies with large viewing distances to the sign. For additional discussion on the A117.1 requirements, see the publication *Significant Changes to the A117.1 Accessibility Standard, 2009 Edition.*

1109.2, 1109.5

Accessible Children's Facilities

CHANGE TYPE: Modification

CHANGE SUMMARY: Toilet facilities and drinking fountains that are "primarily for children's use" may now be installed at a lower height than generally permitted for accessible elements and considered as the required accessible elements.

2012 CODE: **1109.2 Toilet and Bathing Facilities.**   Each toilet room and bathing room shall be accessible. Where a floor level is not required to be connected by an accessible route, the only toilet rooms or bathing rooms provided within the facility shall not be located on the inaccessible floor. At least one of each type of fixture, element, control, or dispenser in each accessible toilet room and bathing room shall be accessible.

> **Exceptions:**   Exceptions 1–5 are unchanged.
>
> **6.** Where toilet facilities are primarily for children's use, required accessible water closets, toilet compartments, and lavatories shall be permitted to comply with children's provision of ICC A117.1.

1109.5 Drinking Fountains.   Where drinking fountains are provided on an exterior site, on a floor, or within a secured area, the drinking fountains shall be provided in accordance with Sections 1109.5.1 and 1109.5.2.

1109.5.1 Minimum Number.   No fewer than two drinking fountains shall be provided. One drinking fountain shall comply with the requirements for people who use a wheelchair and one drinking fountain shall comply with the requirements for standing persons.

Toilet facilities sized for children's use (*Courtesy Blue Valley School District*)

Exception Exceptions:

1. A single drinking fountain that complies with the requirements for people who use a wheelchair and standing persons shall be permitted to be substituted for two separate drinking fountains.

2. Where drinking fountains are primarily for children's use, drinking fountains for people using wheelchairs shall be permitted to comply with the children's provisions in ICC A117.1, and drinking fountains for standing children shall be permitted to provide the spout at 30 inches (762 mm) minimum above the floor.

1109.5.2 More Than the Minimum Number.   Where more than the minimum number of drinking fountains specified in Section 1109.5.1 are provided, 50 percent of the total number of drinking fountains provided shall comply with the requirements for persons who use a wheelchair, and 50 percent of the total number of drinking fountains provided shall comply with the requirements for standing persons.

Exception Exceptions:

1. Where 50 percent of the drinking fountains yields a fraction, 50 percent shall be permitted to be rounded up or down, provided that the total number of drinking fountains complying with this section equals 100 percent of the drinking fountains.

2. Where drinking fountains are primarily for children's use, drinking fountains for people using wheelchairs shall be permitted to comply with the children's provisions in ICC A117.1, and drinking fountains for standing children shall be permitted to provide the spout at 30 inches (762 mm) minimum above the floor.

CHANGE SIGNIFICANCE: Three added exceptions allow for toilet facilities and drinking fountains to be designed using the children's size provisions of the A117.1 standard and still be considered as being accessible. In facilities or portions of buildings that are primarily designed for children's use, the general accessibility requirements may result in the elements not being usable by the major portion of the occupants. For example, the current height of 38 to 43 inches specified in the A117.1 for standing height drinking fountains is too high for small children. Therefore, the exception in Section 1109.5.2 will override the standard and will allow the height to be reduced to 30 inches.

These added exceptions provide the "scoping" requirement that will allow designers to use the children's height requirements of the A117.1 standard when determining the accessibility requirements for areas that are primarily for children's use. Chapter 6 of the A117.1 standard contains a number of provisions such as those for drinking fountains, water closets, compartments, lavatories, and grabs bars for children's size elements. However, because these provisions were not previously scoped or accepted under the IBC, designers were unclear whether or not they needed to provide additional accessible fixtures for children beyond those required for adults, how many additional fixtures were required, or whether or not they were permitted to substitute for the adult fixtures.

1109.2, 1109.5 continues

1109.2, 1109.5 continued

These new exceptions will clarify that if the area is "primarily for children's use," the accessible children's height elements can replace the required adult size accessible elements. Therefore, no additional fixtures are required beyond the general occupancy requirement.

The A117.1 standard defines *children's use* as "spaces and elements specifically designed for use primarily by people 12 years old and younger." Examples of where the adults are in the minority and the space is "primarily for children's use" include most areas of an elementary school, preschool, or kindergarten; a children's library; and a children's museum. Adult-dimensioned fixtures should be provided in other areas or spaces where there is a mix of all ages or where the space serves staff, parents, older students, or the general public. Therefore, if a restroom is provided in the staff area of an elementary school, that toilet room should meet the adult requirements and not attempt to use the reduced-size children's provisions in that location. Even though the code now permits children's-sized accessible elements to substitute for accessible adult-sized provisions, the intent is to clearly limit the application to the areas that the children occupy and the fixtures that they would be using.

It should be noted that the children's sizes are typically provided within the A117.1 standard. The standard does not, however, list a height for a children's-sized standing drinking fountain. The exceptions in Section 1109.5 will specify a minimum 30-inch height for the children's drinking fountain. This minimum height and the 43-inch maximum height that is specified in the general A117.1 requirements will provide designers with a range from which they may select the appropriate height based upon the height of the typical users. Therefore, a drinking fountain in a kindergarten or preschool may be lower than what would be installed in other areas of an elementary school. It is possible that this 30-inch minimum height may be relocated to the A117.1 standard at some future point in time, but it is not included in the newly referenced 2009 edition of the standard.

1109.6

Accessible Saunas and Steam Rooms

CHANGE TYPE: Addition

CHANGE SUMMARY: Saunas and steam rooms are now specifically identified as features and facilities that must be accessible.

2012 CODE: **1109.6 Saunas and Steam Rooms.** Where provided, saunas and steam rooms shall be accessible.

> **Exception:** Where saunas or steam rooms are clustered at a single location, at least 5 percent of the saunas and steam rooms but not less than one of each type in each cluster shall be accessible.

CHANGE SIGNIFICANCE: Because the new edition of the A117.1 standard has included technical requirements for saunas and steam rooms, the IBC added this scoping section to ensure they are properly constructed and accessible. The previously applicable 2003 edition of the standard did not provide technical requirements for saunas or steam rooms. The addition of the IBC's Section 1109.6 scoping requirements and the standard's new technical requirements help to coordinate the building code with the ADA and ABA Accessibility Guidelines.

The requirements within the standard are not especially onerous but will simply provide references to the bench requirements of A117.1 Section 903, limit the potential for a door to swing into the required clear floor space serving the bench, and also require that a turning space complying with A117.1 Section 304 is provided within the room. The bench requirements of A117.1 Section 903 regulate the need for a clear floor space, address the size and height of the bench, require back support, regulate the structural strength, and state the need for slip-resistant surfaces on the seat. For additional discussion on the revised A117.1 requirements for the benches, please see Section 903 in the *Significant Changes to the A117.1 Accessibility Standard, 2009 Edition.*

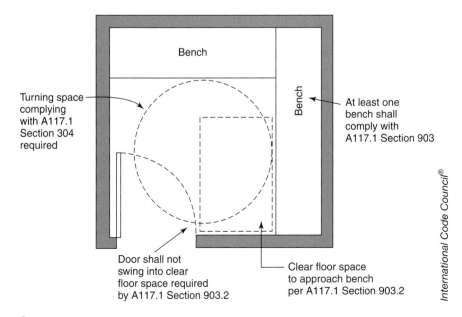

Sauna or steam room

1110.4

Variable Message Signs

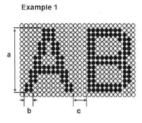

Example 1

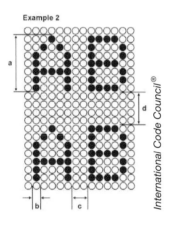

Example 2

International Code Council®

	Property	Example 1	Example 2
a	Character Height	14 Pixels	7 Pixels
b	Stroke Width	2 Pixels	1 Pixel
c	Character Spacing	3 Pixels	2 Pixels
d	Line Spacing		4 Pixels

Low resolution VMS signage characters

CHANGE TYPE: Addition

CHANGE SUMMARY: Variable message signs in transportation facilities and emergency shelters are now required to comply with the provisions of the A117.1-2009 standard.

2012 CODE: 1110.4 Variable Message Signs.   Where provided in the locations in Sections 1110.4.1 and 1110.4.2, Variable Message Signs (VMS) shall comply with the VMS requirements of ICC A117.1.

1110.4.1 Transportation Facilities.   Where provided in transportation facilities, variable message signs conveying transportation-related information shall comply with Section 1110.4.

1110.4.2 Emergency Shelters.   Where provided in buildings that are designated as emergency shelters, variable message signs conveying emergency-related information shall comply with Section 1110.4.

> **Exception:**   Where equivalent information is provided in an audible manner, VMS signs are not required to comply with ICC A117.1.

CHANGE SIGNIFICANCE: Variable message sign (VMS) requirements have been developed by the ICC A117.1 committee and added into the 2009 edition of the standard. This new text within the IBC provides scoping provisions to ensure that VMS signage in both transportation facilities and emergency shelters is usable by most of the population.

The scoping provisions of each section are important to properly apply the requirements. First of all, Section 1110.4 is similar to many accessibility requirements in the fact that the provisions do not "require" VMS signage to be installed in any facility but instead require that "where provided" within transportation facilities and emergency shelters, the VMS will comply with the standard and be usable by most of the population. The transportation section then limits the application to VMS "conveying transportation-related information," while the shelter provisions regulate them when "conveying emergency-related information." The phrasing "conveying transportation-related information" and "conveying emergency-related information" was included to limit the application of the A117.1 VMS requirements to signs that are necessary for the most effective use of transportation information and information in emergency shelters. A sign that presents information that is not necessary for transportation or emergency use of facilities would be exempt. For example, a VMS that advertises what dining options are available in an airport is not required to comply with the standard's VMS provisions, nor is a television set within the waiting area that has closed captioning of a news program. However, any VMS that indicates what flight is departing from what gate or the time of the flight is regulated. In the case of an emergency shelter located at a school, only the signage related to the shelter and the actual emergency is covered. The signage dealing with the school and possibly listing what school activities are occurring that week is not expected to comply with the VMS requirements, nor is shelter information that is not emergency related. For example, an announcement that

VMS sign at subway boarding platform

the mayor will be visiting the shelter later that day or that specific insurance companies will be there in a few days are not "emergency related," even though they may be related to the event that necessitated the opening of the shelter.

Selecting and limiting the application of the VMS provisions to transportation and emergency shelters was done for several reasons. Transportation facilities are included because riders with reduced vision are especially dependent on public transportation for travel and are required by the ADA to have information provided that is equivalent to that provided to riders having unimpaired vision. In addition, VMS signs are often used within transportation facilities (see Sections E109.2.2.2, E109.2.2.3, E109.2.7, and E110.3) as a primary means of conveying information to riders.

Although any building could ultimately be pressed into service as an emergency shelter in some circumstances, the intent of Section 1110.4.2 is to apply the requirement only to those facilities that are designated in advance or during the planning stage as an intended shelter. Emergency shelters are typically identified by a jurisdiction when they are studying emergency planning or working with the Federal Emergency Management Agency (FEMA) to develop community response plans. In many communities, this may include certain schools, civic administration buildings, or even large convention facilities that a community designates as intended emergency shelters.

The emergency shelter provisions contain an exception that eliminates the requirement to meet the VMS provisions of the standard. Where audible information is conveyed that is either the same or equivalent to the information provided by the sign, compliance with ICC A117.1 VMS requirements is not needed. The VMS provides the information in a visual method, and the A117.1 provisions ensure that the information is clearly visible and legible to both the general population and to people with some

1110.4 continues

1110.4 continued

level of visual impairment. Providing the equivalent information in an audible manner makes the information accessible to people with severe visual impairment or who are blind.

The exception for equivalent audible information is not included within the section dealing with transportation facilities. This is done not only to coordinate with the ADA but is based upon the fact that, in many transportation facilities, the audible information is simply not easily understood. Anyone who has stood on a subway or train platform and tried to understand an audible message at the same time that the train is pulling up to the platform will understand why the exception is not included for transportation facilities. Because of the problems with hearing messages in many transportation facilities, the audible message delivery is simply not considered as being adequate to replace or eliminate the clear visual sign information. The A117.1 standard modifies requirements for VMS signs primarily by requiring increased character height and spacing for low-resolution VMS signs. Even for users with unimpaired vision, there is strong research evidence that character height in low-resolution VMS (such as the LED signs that are common in the transportation environment and elsewhere) must be approximately 30 percent greater than for equivalent print signs or VMS with high resolution. To better understand the unique differences between VMS and the general sign requirements, users should look at the technical requirements of the A117.1 standard. For additional discussion on the A117.1 requirements, please see the *Significant Changes to the A117.1 Accessibility Standard, 2009 Edition.*

6

Building Envelope, Structural Systems, and Construction Materials

Chapters 12 through 26

The interior environment provisions of Chapter 12 include requirements for lighting, ventilation, and sound transmission. Chapter 13 provides a reference to the *International Energy Conservation Code* for provisions governing energy efficiency. Regulations governing the building envelope are located in Chapters 14 and 15, addressing exterior wall coverings and roof coverings, respectively. Structural systems are regulated through the structural design provisions of Chapter 16, whereas structural testing and special inspections are addressed in Chapter 17. The provisions of Chapter 18 apply to soils and foundation systems.

The requirements for materials of construction, both structural and nonstructural, are located in Chapters 19 through 26. Structural materials regulated by the code include concrete, lightweight metals, masonry, steel, and wood. Glass and glazing, gypsum board, plaster, and plastics are included as regulated nonstructural materials. ■

1203.1
Mechanical Ventilation Required

1203.2
Ventilation of Attic Spaces

1208.3
Minimum Kitchen Floor Area

1210
Toilet and Bathroom Requirements

1403.5
Flame Propagation at Exterior Walls

1404.12, 1405.18, 202
Polypropylene Siding

1405.6
Anchored Masonry Veneer

1503.4
Roof Drainage Systems

continues

1203.1

Mechanical Ventilation Required

CHANGE TYPE: Addition

CHANGE SUMMARY: The option of natural ventilation rather than mechanical ventilation is now unavailable when a dwelling unit is tested using a blower door test and it is determined that an adequate number of air changes are not provided.

2012 CODE: 1203.1 General.   Buildings shall be provided with natural ventilation in accordance with Section 1203.4, or mechanical ventilation in accordance with the *International Mechanical Code.*

Where the air infiltration rate in a dwelling unit is less than 5 air changes per hour when tested with a blower door at a pressure 0.2 inch w.c. (50 Pa) in accordance with Section R402.4.1.2 of the *International Energy Conservation Code*, the dwelling unit shall be ventilated by mechanical means in accordance with Section 403 of the *International Mechanical Code.*

CHANGE SIGNIFICANCE: As the building's thermal envelope gets tighter resulting in less outdoor air leaking into the building's interior, mechanical ventilation may be necessary to maintain the indoor air quality. For dwelling units, a measured cutoff point is provided and a reference is made to the applicable provisions of the *International Mechanical Code* for providing the necessary ventilation. It should be noted that the new text in the IBC does not require that a blower door test be conducted, but rather, acts on the results of any such test that is conducted. However, code users should be aware that Section R402.4.1.2 of the 2012 *International Energy Conservation Code* (IECC) does require that residential buildings conduct a blower door test to determine the amount of air leakage. Therefore, in jurisdictions that have adopted the IECC those blower door test results will be used to determine if the IBC requirement for a mechanical ventilation system is to be imposed. Because the IECC requires a maximum air leakage rate of 5 air changes per hour, any building other than those in Climate Zones 1 and 2 that have exactly 5 air changes per hour would need a mechanical ventilation system based on IBC Section 1203.1.

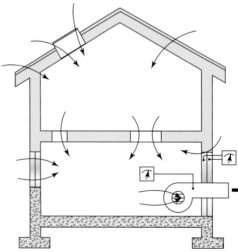

Infiltration of dwelling unit tested with blower door at pressure of 0.2 in. w.c.
- <5 air changes per hour (ACH)—requires mechanical ventilation
- ≥5 ACH—natural or mechanical ventilation permitted

Blower door test determines air infiltration rate

International Code Council®

Generally, the designer has a choice of either providing natural ventilation or mechanical ventilation. The option of using natural ventilation instead of mechanical ventilation is no longer permitted when the dwelling unit is tested using the blower door test and the infiltration level is below 5 air changes per hour (ACH). As building construction practices have improved, buildings have become tighter; as buildings become tighter, mechanical ventilation must be introduced to provide sufficient levels of ventilation to ensure indoor air quality. This 5-air-change requirement at the specified pressure provides a clear point to determine when infiltration is not adequate and must be supplemented by a mechanical system.

The 5 ACH limit was selected even though it is less than would generally be obtained by the IMC's requirement for 0.35 ACH of outdoor airflow for a dwelling using a mechanical ventilation system. By following the calculation procedures of the ASHRAE 136 standard, it can be shown that a natural infiltration rate of 0.35 ACH is equivalent to somewhere between 7 to 10 ACH at the specified pressure, depending on the local climatic conditions of the building. As fewer air changes are provided, the reliance on window operation is not sufficient and the concern for the effect on indoor air quality goes up. An additional source of justification for the 5 ACH limit is the National Association of Home Builder's (NAHB's) *National Green Building Standard* that requires whole-house mechanical ventilation when the infiltration rate falls below the 5 ACH value.

The actual effect of this change may be less than what it would first appear because most dwelling units do provide a mechanical ventilation system and do not rely on natural ventilation. If the dwelling unit is using a mechanical system, the IMC requires the necessary 0.35 ACH of outdoor airflow which is adequate for providing the proper ventilation.

1203.2

Ventilation of Attic Spaces

CHANGE TYPE: Modification

CHANGE SUMMARY: The minimum required ventilation area for attics has been clarified and exceptions are now provided that either allow a reduction in the vent area or eliminate the requirement completely.

2012 CODE: **1203.2 Attic Spaces.** Enclosed attics and enclosed rafter spaces formed where ceilings are applied directly to the underside of roof framing members shall have cross ventilation for each separate space by ventilation openings protected against the entrance of rain and snow. Blocking and bridging shall be arranged so as not to interfere with the movement of air. ~~A minimum of 1 inch (25 mm) of~~ <u>An</u> airspace <u>of not less than 1 inch (25 mm)</u> shall be provided between the insulation and the roof sheathing. The net free ventilating area shall not be less than ~~$^1/_{300}$~~ <u>$^1/_{150}$th</u> of the area of the space ventilated.

Exceptions:

1. <u>The net free cross-ventilation area shall be permitted to be reduced to $^1/_{300}$ provided that not less than</u> ~~with~~ <u>50 percent and not more than 80 percent</u> of the required ventilating area provided by ventilators located in the upper portion of the space to be ventilated at least 3 feet (914 mm) above eave or cornice vents with the balance of the required ventilation provided by eave or cornice vents.

2. <u>The net free cross-ventilation area shall be permitted to be reduced to $^1/_{300}$ where a Class I or II vapor barrier is installed on the warm-in-winter side of the ceiling.</u>

3. <u>Attic ventilation shall not be required when determined not necessary by the building official due to atmospheric or climatic conditions.</u>

CHANGE SIGNIFICANCE: A minimum ventilation rate of 1 square foot for every 150 square feet of attic space has been established, with a reduced ventilation rate of $^1/_{300}$ permitted when the previous Exception 1 or the new Exception 2 is applicable. Historically the ventilation rate of $^1/_{150}$ has been considered as the general requirement with a reduction allowed where the design helped promote airflow through the attic or

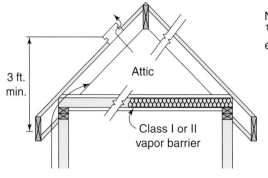

Attic ventilation requirements

where a means of limiting the transfer of moisture from the building into the attic was provided.

Some of the uncertainty with the 2009 provisions that this revised language clarifies is (1) addressing what the general ventilation rate is required to be, (2) providing guidance if the roof does not have at least 3 feet of vertical rise, and (3) limiting the maximum area of vents that can be located at the upper portion of the roof when using the reduced ventilation rate. Previously, the requirements were based on the assumption that it was possible to locate some of the vents a minimum of 3 feet above eave or cornice vents; therefore, the reduced ventilation rate was justified. However, no guidance had previously been provided as to the appropriate ventilation rate for low-slope or flat roofs where the vertical separation could not be obtained. In addition, a minimum amount of vent area was previously required to be located in the upper portion of the attic but a maximum amount was not established. Because the purpose of having vents in both the high and low portions is to encourage the natural flow of air, providing too little ventilation at the lower part of the attic reduces the amount and effectiveness of the natural air movement.

The new second exception recognizes that the installation of a vapor barrier between the conditioned space and the ventilated attic will help reduce the movement of moisture into the attic. When the amount of moisture moving into or condensing in the attic is reduced, it was considered reasonable to reduce the required vent area. This option—or variations of it—had previously been allowed in earlier editions of the IBC but was removed in the 2009 edition. Allowing for this reduced ventilation area reflects that these systems have performed adequately in the past, and it will also create consistency with Section R806.2 of the IRC, which allows for this option in climate zones 6, 7, and 8.

The addition of the third exception allows for the elimination of the ventilation requirement when the historic evidence of a community shows that the ventilation is not needed. Depending on the climate, occupancy, and the location of the structure, the building official is allowed to waive the ventilation requirement where experience or data justifies it. This allowance would typically have more application in dryer or warmer climates, where the buildup of moisture or condensation within the attic has not historically been an issue. Because this is an exception that is available only when approved by the building official, code users should not expect this allowance without prior consultation and approval. It is possible that some communities may grant a blanket approval, while others will review each structure on a case-by-case basis or will never allow the use of the exception.

1208.3

Minimum Kitchen Floor Area

Kitchens do not require a minimum floor area but must have minimum 3 foot clear passageways.

International Code Council®

CHANGE TYPE: Deletion

CHANGE SUMMARY: The minimum floor area requirement of 50 square feet for kitchens has been deleted.

2012 CODE: **1208.3 Room Area.** Every dwelling unit shall have ~~at least~~ no fewer than one room that shall have not less than 120 square feet (13.9 m^2) of net floor area. Other habitable rooms shall have a net floor area of not less than 70 square feet (6.5 m^2).

> **Exception:** ~~Every kitchen in a one- and two-family dwelling shall have not less than 50 square feet (4.64 m^2) of gross floor area.~~ Kitchens are not required to be of a minimum floor area.

CHANGE SIGNIFICANCE: In recognition that the size limitation for a kitchen is not related to either life safety or health concerns, kitchens are no longer required to be a minimum of 50 square feet in floor area. Other requirements—such as Section 1208.1 that requires at least 3 feet between counter fronts and appliances or walls—will generally be adequate to ensure that the kitchen has adequate maneuvering space and is usable.

The scope of the exception has also been expanded by eliminating the wording that restricted the exception to kitchens "in a one- and two-family dwelling." A kitchen in any dwelling unit, such as in a Group R-2 apartment or a Group R-4 congregate residence, is also no longer regulated for minimum floor area.

In addition to the limitations found within Section 1208.1, any dwelling unit that is required to be either an Accessible, Type A, or Type B dwelling unit would need to meet the provisions specified in ICC A117.1 for those particular units. Sections 1002.12, 1003.12, and 1004.12 of ICC A117.1 will, depending on the layout, impose a minimum requirement of 40 inches or 60 inches between the opposing cabinets, walls, and appliances. Other accessibility requirements help ensure that maneuvering clearances to the various appliances and work surfaces are provided.

1210
Toilet and Bathroom Requirements

CHANGE TYPE: Clarification

CHANGE SUMMARY: The water closet compartment and urinal partition requirements have been relocated from Chapter 29 to Section 1210.

2012 CODE:

SECTION 1210
~~SURROUNDING MATERIALS~~ TOILET AND BATHROOM REQUIREMENTS

1210.1 Required Fixtures.   The number and type of plumbing fixtures provided in any occupancy shall comply with Chapter 29.

1210.2 Finish Materials.   Walls, floors, and partitions in toilet and bathrooms shall comply with Sections 1210.2.1 through 1210.2.4.

~~1210.1~~ 1210.2.1 Floors and Wall ~~Base Finish Materials~~ Bases.   In other than dwelling units, toilet, bathing, and shower room floor finish materials shall have a smooth, hard, nonabsorbent surface. The intersections of such floors with walls shall have a smooth, hard, nonabsorbent vertical base that extends upward onto the walls ~~at least~~ not less than 4 inches (102 mm).

~~1210.2~~ 1210.2.2 Walls and Partitions.   Walls and partitions within 2 feet (610 mm) of service sinks, urinals, and water closets shall have a smooth, hard, nonabsorbent surface to a height of 4 feet (1219 mm) above

1210 continues

Urinal privacy partitions are required.

1210 continued   the floor, and except for structural elements, the materials used in such walls shall be of a type that is not adversely affected by moisture.

> **Exceptions:** This section does not apply to the following buildings and spaces:
>
> 1. Dwelling units and sleeping units.
> 2. Toilet rooms that are not accessible to the public and which have not more than one water closet.

Accessories such as grab bars, towel bars, paper dispensers, and soap dishes, provided on or within walls, shall be installed and sealed to protect structural elements from moisture. For walls and partitions, also see Section 2903.

~~1210.3~~ **1210.2.3 Showers.** Shower compartments and walls above bathtubs with installed shower heads shall be finished with a smooth, nonabsorbent surface to a height not less than 70 inches (1778 mm) above the drain inlet.

~~1210.4~~ **1210.2.4 Waterproof Joints.** Built-in tubs with showers shall have waterproof joints between the tub and adjacent wall.

1210.3 Privacy. Privacy at water closets and urinals shall be provided in accordance with Sections 1210.3.1 and 1210.3.2.

~~2903.1~~ **1210.3.1 Water Closet Compartment.** Each water closet utilized by the public or employees shall occupy a separate compartment, with walls or partitions and a door enclosing the fixtures to ensure privacy.

> **Exceptions:**
>
> 1. Water closet compartments shall not be required in a single-occupant toilet room with a lockable door.
> 2. Toilet rooms located in child day care facilities and containing two or more water closets shall be permitted to have one water closet without an enclosing compartment.
> 3. This provision is not applicable to toilet areas located within Group I-3 housing areas.

~~2903.2~~ **1210.3.2 Urinal Partitions.** Each urinal utilized by the public or employees shall occupy a separate area with walls or partitions to provide privacy. The walls or partitions shall begin at a height not more than 12 inches (305 mm) from and extend not less than 60 inches (1524 mm) above the finished floor surface. The walls or partitions shall extend from the wall surface at each side of the urinal not less than 18 inches (457 mm) or to a point not less than 6 inches (152 mm) beyond the outermost front lip of the urinal measured from the finished backwall surface, whichever is greater.

> **Exceptions:**
>
> 1. Urinal partitions shall not be required in a single occupant or family or assisted use toilet room with a lockable door.

2. Toilet rooms located in child day care facilities and containing two or more urinals shall be permitted to have one urinal without partitions.

~~1210.5~~ **1210.4 Toilet Room <u>Location</u>.** Toilet rooms shall not open directly into a room used for the preparation of food for service to the public.

~~**SECTION 2903**~~
~~**TOILET ROOM REQUIREMENTS**~~

CHANGE SIGNIFICANCE: Previously, both Sections 1210 and 2903 addressed design issues and finish materials for spaces that serve as toilet rooms or bathrooms. The primary distinction was that the items found within Chapter 29 were sections that had been reproduced from the *International Plumbing Code*, while those in Chapter 12 were IBC developed requirements. The provisions have now been consolidated into one location for ease of use. Sections 2903.1 and 2903.2 addressing the "water closet compartment" and "urinal partitions" have been relocated to Sections 1210.3.1 and 1210.3.2, respectively.

Sections 1210.1, 1210.2, and 1210.3 were also created to provide scoping for the various requirements. These new sections do not affect the application of the requirements but simply provide a means of arranging and referencing the subsections that follow.

The only technical change that has occurred is the addition of "service sinks" into the text of Section 1210.2.2 that is intended to protect the walls and floors adjacent to the various plumbing fixtures from the damage and decay that can be associated with moisture or cleaning products. Water splash or cleaning materials that are used with service sinks can lead to the degradation of materials that are not "smooth, hard, nonabsorbent" and therefore any finish material within 2 feet of a service sink needs to be "of a type that is not adversely affected by moisture."

1403.5

Flame Propagation at Exterior Walls

CHANGE TYPE: Addition

CHANGE SUMMARY: A flame-spread test of the wall assembly is now required where combustible water-resistive barriers are used in the exterior walls of Type I, II, III, and IV buildings that are greater than 40 feet in height.

2012 CODE: <u>**1403.5 Vertical and Lateral Flame Propagation.** Exterior walls on buildings of Type I, II, III, or IV construction that are greater than 40 feet (12 192 mm) in height above grade plane and contain a combustible water-resistive barrier shall be tested in accordance with and comply with the acceptance criteria of NFPA 285.</u>

CHANGE SIGNIFICANCE: Where combustible water-resistive barriers are installed in the exterior walls of buildings of Type I, II, III, or IV construction, such wall assemblies must now be tested using the NFPA 285 test standard. The application of this requirement to the higher types of construction that are mentioned is due to the fact that these types of construction typically allow either no or limited combustibles in the exterior walls.

Newer construction practices, such as the addition of combustible water-resistant barriers, can result in significant amounts of combustible materials (other than foam plastics) being installed within the exterior walls. Testing has shown that the addition of a combustible water-resistive barrier can cause an exterior wall system to fail the NFPA 285 test, even if the wall had successfully passed the test prior to the addition of the barrier. Small-scale testing has shown that these types of materials can provide significant amounts of combustible fuel loading to a wall assembly.

The code has historically only used the NFPA 285 testing for exterior walls that contain foam plastic insulation (see Section 2603.5.5) or include metal composite material (MCM) exterior veneers (see Section 1407.10.4). The use of this testing requirement is now also deemed appropriate for evaluating exterior walls that contain a combustible weather-resistive barrier in order to address the potential vertical and lateral flame spread that can occur either on or within exterior wall systems that contain the combustible barriers. Due to the built-in standoffs that are used in the

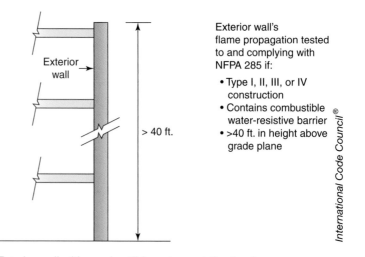

Exterior wall with combustible water-resistive barrier

installation of these systems, the barrier materials can exhibit significant vertical or lateral flame propagation.

The height limit of 40 feet was selected because that height is commonly used as the threshold for other combustible exterior wall materials. For example, the wood veneer provisions of Section 1405.5, Item 1; the wall covering provisions of Section 1406.2.2, Item 2; and the MCM provisions of Section 1407.11.1 are all based on a 40-foot height limit.

1404.12, 1405.18, 202

Polypropylene Siding

Pooling fire caused by burning polypropylene. (*Photo courtesy of National Institute of Standards and Technology*)

CHANGE TYPE: Addition

CHANGE SUMMARY: In order to address the hazards created by polypropylene siding, such materials are now regulated for flame-spread, testing requirements and fire-separation distance.

2012 CODE: **202 Definitions.**

POLYPROPYLENE SIDING. A shaped material, made principally from polypropylene homopolymer, or copolymer, which in some cases contain fillers or reinforcements, that is used to clad exterior walls of buildings.

1404.12 Polypropylene Siding. Polypropylene siding shall be certified and labeled as conforming to the requirements of ASTM D7254 and those of Section 1404.12.1 or 1404.12.2 by an approved quality control agency. Polypropylene siding shall be installed in accordance with the requirements of Section 1405.18 and in accordance with the manufacturer's installation instructions. Polypropylene siding shall be secured to the building so as to provide weather protection for the exterior walls of the building.

1404.12.1 Flame Spread Index. The certification of the flame spread index shall be accompanied by a test report stating that all portions of the test specimen ahead of the flame front remained in position during the test in accordance with ASTM E84 or UL 723.

1404.12.2 Fire Separation Distance. The fire separation distance between a building with polypropylene siding and the adjacent building shall be no less than 10 feet (3.05 m).

1405.18 Polypropylene Siding. Polypropylene siding conforming to the requirements of this section and complying with Section 1404.12 shall be limited to exterior walls of Type VB construction located in areas where the wind speed specified in Chapter 16 does not exceed 100 miles per hour (45 m/s) and the building height is less than or equal to 40 feet (12 192 mm) in Exposure C. Where construction is located in areas where the basic wind speed exceeds 100 miles per hour (45 m/s), or building heights are in excess of 40 feet (12 192 mm), tests or calculations indicating compliance with Chapter 16 shall be submitted. Polypropylene siding shall be installed in accordance with the manufacturer's installation instructions. Polypropylene siding shall be secured to the building so as to provide weather protection for the exterior walls of the building.

CHAPTER 35
ASTM D7254 *Standard Specification for Polypropylene (PP) Siding*

CHANGE SIGNIFICANCE: The regulation of polypropylene siding, although used in construction, is new to the IBC. Except for the fire testing, the new requirements are similar to the existing provisions addressing vinyl siding and will provide a way for the polypropylene siding to be used safely.

Vinyl siding is known to have adequate fire performance because the siding needs to be made of rigid (unplasticized) PVC in accordance with ASTM D3679. Unless polypropylene is manufactured with fire retarders in the material, the typical testing methods do not adequately evaluate the fire performance of the material. Therefore, the flame-spread certification must include an additional note stating the material stayed in place during the test. When polypropylene siding material (which does not have the appropriate fire performance) is tested using ASTM E84 (Steiner tunnel), the test specimen will often fall ahead of the arrival of the flame, giving incorrect results. It is these falling test samples that affect the continuity of the flame-spread along the top of the tunnel and result in flaming drips from the material that create a pool fire on the bottom of the test tunnel and skew the measured flame-spread index at the top of the tunnel. This additional requirement for the reporting of the test helps ensure the test specimen remains in place during the test and that flaming drips and falling test specimens do not occur.

Siding made of polypropylene can spread a fire by several means. Polypropylene that has not been appropriately fire retarded will release a greater amount of heat (about four times greater) than non-fire-retarded polypropylene or other combustible sidings that are permitted by the code such as wood siding or vinyl (PVC) siding. This greater heat release and resultant radiant heat can help spread the fire to other adjacent structures. In addition, the flaming droplets from the plastic can also contribute to the fire spreading by igniting grass, mulch, and debris found near the building. For comparison, in a PVC siding fire, the material may soften, char, and burn but will stay substantially intact. Because of the added and unique hazards that polypropylene creates when compared to other combustible sidings, a minimum fire separation distance must be provided. In addition, the use of polypropylene siding is limited to only those buildings of Type VB construction.

One item of confusion is the actual separation that Section 1404.12.2 will require. The new code text states that the 10 feet is measured "between a building with polypropylene siding and the adjacent building" and yet the section title and the first portion of this paragraph indicate that it is a "fire separation distance." A review of the code's definition of the term "fire separation distance" will show the distance is measured to the lot line, the centerline of a public way, or to an imaginary line between two buildings on the same lot.

1405.6

Anchored Masonry Veneer

Anchored masonry veneer

CHANGE TYPE: Deletion

CHANGE SUMMARY: Anchored veneer on buildings located in Seismic Design Category D no longer must comply with the joint reinforcing requirements applicable to Seismic Design Categories E and F.

2012 CODE: 1405.6.2 Seismic Requirements. Anchored masonry veneer located in Seismic Design Category C, D, E, or F shall conform to the requirements of Section 6.2.2.10 of TMS 402/ACI 530/ASCE 5. ~~Anchored masonry veneer located in Seismic Design Category D shall also conform to the requirements of Section 6.2.2.10.3.3 of TMS 402/ACI 530/ASCE 5.~~

CHANGE SIGNIFICANCE: Previously, anchored veneer on structures assigned to Seismic Design Category D was required to comply with the wire-reinforcing requirements for anchored veneer in structures assigned to Seismic Design Categories E and F. In Seismic Design Categories E and F, the Masonry Standards Joint Committee (MSJC) code requires continuous single wire joint reinforcement of wire size W1.7 (MW11) at a maximum spacing of 18 inches on center vertically and be mechanically attached with clips or hooks. Shake table testing conducted on full-scale structures with clay masonry veneer at the University of California–San Diego conclusively demonstrated that incorporating wire reinforcement in anchored veneer does not improve the seismic performance or behavior of the anchored masonry veneer when subjected to the maximum considered earthquake (MCE) corresponding to Seismic Design Category D. The testing was conducted separately on two full-scale specimens: a 20 ft $\times$ 20 ft prototypical wood-stud frame structure with anchored brick veneer and a 20 ft $\times$ 20 ft prototypical concrete masonry structure with anchored brick veneer. For each specimen, two of the four veneer elevations included joint reinforcement with the remaining two veneer elevations constructed without joint reinforcement. Both specimens were each subjected separately to a seismic load imposed by the MCE with a 2 percent probability of exceedance in 50 years for Seismic Design Category D. In each specimen, no differences in performance or behavior were observed in the veneers with and without joint reinforcement. Additionally, the absense of unnecessary reinforcement in the veneer will likely decrease the possibility of corrosion occurring in the veneer. With no required reinforcement, the extent of metal accessories present in the veneer is substantially lowered, thus reducing the potential for corrosion. It should be noted that Section 6.2.2.10.3 of the 2008 MSJC code does require the joint reinforcement in anchored masonry veneer in Seismic Design Category E and F.

International Code Council®

CHANGE TYPE: Clarification

CHANGE SUMMARY: The design and installation provisions of the *International Plumbing Code* (IPC) for roof drainage systems are now specifically referenced in the IBC.

2012 CODE: 1503.4 Roof Drainage. Design and installation of roof drainage systems shall comply with Section 1503 <u>of this code</u> and <u>Sections 1106 and 1108 as applicable of</u> the *International Plumbing Code.*

1503.4.1 ~~Secondary Drainage Required~~. **Secondary (Emergency Overflow) Drains or Scuppers.** <u>Where roof drains are required,</u> secondary (emergency <u>overflow</u>) roof drains or scuppers shall be provided where the roof perimeter construction extends above the roof in such a manner that water will be entrapped if the primary drains allow buildup for any reason. <u>The installation and sizing of secondary emergency overflow drains, leaders, and conductors shall comply with Sections 1106 and 1108 as applicable of the *International Plumbing Code*.</u>

CHANGE SIGNIFICANCE: The requirements for roof drains and the detailed provisions for secondary emergency overflow drains and their sizing, location, and quantity are all found within the IPC. These IPC provisions are now addressed in the IBC by providing a specific reference to the location where the provisions exist. By providing the reference to the IPC and listing the items regulated and required by that code, code users will be made aware of the requirements so they are not overlooked.

Sections 1106 and 1108 of the IPC are based upon the 100-year, 1-hour rainfall rates for the building's location and design features. Information

1503.4 continues

1503.4
Roof Drainage Systems

Roof drainage system with secondary emergency overflow drain

International Code Council®

1503.4 continued   is included regarding how to calculate the roof area where vertical walls may divert rainfall onto a roof, the capacity of drainage piping at various sizes and slopes, and requirements for separate primary and emergency drain systems, as well as sizing and placement of the drains and scuppers. Having this direct reference in the IBC directing users to the IPC should help to ensure these important features are properly designed and constructed.

CHANGE TYPE: Addition

CHANGE SUMMARY: New provisions for the installation of roof covering underlayment have been added for buildings located in high-wind areas where the nominal design wind speed is equal to or greater than 120 mph.

2012 CODE: 1507.2.8.1 High Wind Attachment. Underlayment applied in areas subject to high winds (V_{asd} greater than 110 mph ~~as determined~~ ~~in accordance with Section 1609.3.1~~ in accordance with Figure 1609) shall be applied with corrosion-resistant fasteners in accordance with manufacturer's installation instructions. Fasteners are to be applied along the overlap at a maximum spacing of 36 inches (914 mm) on center.

Underlayment installed where V_{asd}, in accordance with Section 1609.3.1, equals or exceeds 120 mph (54 m/s) shall comply with ASTM D226 Type II, ASTM D4869 Type IV, or ASTM D6757. The underlayment shall be attached in a grid pattern of 12 inches (305 mm) between side laps with a 6-inch (152-mm) spacing at the side laps. Underlayment shall be applied in accordance with Section 1507.2.8 except all laps shall be a minimum of 4 inches (102 mm). Underlayment shall be attached using metal or plastic cap nails with a head diameter of not less than 1 inch (25 mm) with a thickness of at least 32 gauge [0.0134 inch (0.34 mm)] sheet metal. The cap nail shank shall be a minimum of 12 gage [0.105 inch (2.67 mm)], with a length to penetrate through the roof sheathing or a minimum of ¾ inch (19.1 mm) into the roof sheathing.

1507.2.8.1 continues

1507.2.8.1

Roof Covering Underlayment in High Wind Areas

International Code Council®

Underlayment in high wind areas has additional attachment requirements.

1507.2.8.1 continued

Exception: <u>As an alternative, adhered underlayment complying</u> <u>with ASTM D1970 shall be permitted.</u>

Because this code change affects many roof covering types and revised substantial portions of Section 1507, the entire code change text is too extensive to be included here. Refer to Code Change S15-09/10 in the *2012 IBC Code Changes Resource Collection* for the complete text and history of the code change.

CHANGE SIGNIFICANCE: Research in Florida has demonstrated that some roof-covering underlayment materials perform very poorly when subjected to pressures from wind speeds greater than 110 mph. In the laboratory tests, specimens covered with ASTM D226 Type I and Type II underlayments performed dramatically differently. The Type I felt (15-pound) material completely blew off some portions of the specimen at winds exceeding 110 mph and pulled over the plastic caps on other parts of the test specimens. In contrast, the Type II (30-pound) material remained in place and showed very few signs of distress. Plastic caps deformed much more than the metal caps in several installations. Consequently, new requirements for the installation of roof covering underlayment have been added for high-wind areas where the nominal design wind speed, V_{asd}, is equal to or greater than 120 mph. Now included are specific requirements for the type of underlayment used (ASTM D226 Type II or ASTM D4869 Type IV), the minimum lap length, minimum cap nail shank gauge, and minimum penetration of the cap nails into the roof sheathing. The exception exempts self-adhered, polymer, modified-bitumen underlayment complying with ASTM D1970. See Section 1609 for a discussion of the changes in the wind-speed designation.

CHANGE TYPE: Addition

CHANGE SUMMARY: The IBC now provides a reference to new IFC provisions on roof gardens and landscaped roofs as a means of controlling the potential hazards these combustible materials on the roof could create.

2012 CODE: 1507.16 Roof Gardens and Landscaped Roofs. Roof gardens and landscaped roofs shall comply with the requirements of this chapter and Sections 1607.11.2.3 and 1607.11.2.3.1 and the *International Fire Code*.

1507.16.1 Structural Fire-Resistance. The structural frame and roof construction supporting the load imposed upon the roof by the roof gardens or landscaped roofs shall comply with the requirements of Table 601.

CHANGE SIGNIFICANCE: The addition of rooftop vegetation or landscaping can provide a number of benefits in building construction, such as reducing the heat-gain and cooling demands, helping to control storm water runoff, and simply providing pleasant areas or sites within a community. However, these landscaped roofs also have a potential to place combustible vegetation in an area where it is less accessible to the fire department in the event of an emergency or to increase the potential fire exposure of either the building itself or of buildings nearby. To offset some of these concerns, the IFC has added new provisions (IFC Section 317), and references have been placed in the IBC that will direct users to these new IFC provisions.

The added text in the base paragraph of Section 1507.16 is essentially a reference to the new requirements that were added into IFC Section 317. The fire code provisions have added requirements that limit the size of a single landscaped roof area, require a minimum separation and protection between adjacent areas, and address maintenance issues and separation from other combustible rooftop elements such as penthouses or mechanical equipment. Perhaps the most important IFC provisions are the size limitation on the vegetation and the Class A–rated roof covering requirement for the 6-foot-wide buffer area. These requirements do not exist within the IBC itself.

The clearance/buffer area of at least 6 feet is important because it requires a Class A roof-covering assembly. Table 1505.1 of the IBC only requires a Class A roof covering in certain areas under the provisions of the *International Wildland–Urban Interface Code* (IWUIC) or when the fire district requirements of Appendix D are used. In applying the clearance requirement, it is permissible to provide a Class A roof covering for the 6-foot minimum distance and then allow the appropriate roof covering classification for any distance in excess of the minimum dimension.

It is intended that new Section 1507.16 limit the use of Table 601, Footnote a, to situations that do not include a roof garden or landscaped roof based on the premise that the permitted reduction in the roof's fire-resistance rating is only applicable where the structural frame or bearing walls "are supporting a roof only" and that support of the loads imposed by the landscaping and vegetation on a roof is beyond what the original

1507.16 continues

1507.16
Roof Gardens and Landscaped Roofs

Landscaped roof

Vegetation on roof must comply with IFC requirements.

1507.16 continued   limitation intended. The "roof-only" requirement has historically been applied to exclude the footnote's use where the roof system may support portions of other stories as opposed to rooftop mechanical equipment or, in this case, a roof garden or vegetation on top of the roof. Although it may seem unlikely that a fire involving the rooftop vegetation would affect the building's roof structural supports, the fire-resistance rating is not permitted to be reduced based on Footnote a.

For additional information related to this issue, please see the publication *Significant Changes to the International Fire Code, 2012 Edition* and the information related to IFC Section 317.

Construction supporting landscaped roofs must comply with Table 601 fire-resistance.

CHANGE TYPE: Addition

CHANGE SUMMARY: Photovoltaic elements (modules/shingles or systems) must now meet the general code requirements for roofing materials and rooftop structures.

2012 CODE:

202 Definitions.

PHOTOVOLTAIC MODULES/SHINGLES. A roof covering composed of flat-plate photovoltaic modules fabricated in sheets that resemble three-tab composite shingles.

1505.8 Photovoltaic Systems.   Rooftop-installed photovoltaic systems that are adhered or attached to the roof covering or photovoltaic modules/shingles installed as roof coverings shall be labeled to identify their fire classification in accordance with the testing required in Section 1505.1.

1507.17 Photovoltaic Modules/Shingles.   The installation of photovoltaic modules/shingles shall comply with the provisions of this section.

1507.17.1 Material Standards.   Photovoltaic modules/shingles shall be listed and labeled in accordance with UL 1703.

1507.17.2 Attachment.   Photovoltaic modules/shingles shall be attached in accordance with the manufacturer's installation instructions.

1507.17.3 Wind Resistance.   Photovoltaic modules/shingles shall be tested in accordance with procedures and acceptance criteria in ASTM D3161. Photovoltaic modules/shingles shall comply with the classification requirements of Table 1507.2.7.1(2) for the appropriate maximum nominal design wind speed. Photovoltaic modules/shingle packaging shall bear a label to indicate compliance with the procedures in ASTM D3161 and the required classification from Table 1507.2.7.1(2).

1509.7 Photovoltaic Systems.   Rooftop-mounted photovoltaic systems shall be designed in accordance with this section.

1509.7.1 Wind Resistance.   Rooftop-mounted photovoltaic systems shall be designed for wind loads for component and cladding in accordance with Chapter 16 using an effective wind area based on the dimensions of a single unit frame.

1509.7.2 Fire Classification.   Rooftop-mounted photovoltaic systems shall have the same fire classification as the roof assembly required by Section 1505.

1509.7.3 Installation.   Rooftop-mounted photovoltaic systems shall be installed in accordance with the manufacturer's installation instructions.

1507.17, 3111, 202
Photovoltaic Systems

Photovoltaic panel on roof

1507.17, 3111, 202 continues

1507.17, 3111, 202 continued

Photovoltaic modules/shingles being installed. (*Photo courtesy of Atlantis Energy Systems, Inc.*)

1509.7.4 Photovoltaic Panels and Modules. Photovoltaic panels and modules mounted on top of a roof shall be listed and labeled in accordance with UL 1703 and shall be installed in accordance with the manufacturer's installation instructions.

SECTION 3111
SOLAR PHOTOVOLTAIC PANELS/MODULES

3111.1 General. Solar photovoltaic panels/modules shall comply with the requirements of this code and the *International Fire Code.*

CHANGE SIGNIFICANCE: Guidance is now provided related to how photovoltaic systems are to be regulated. Both integrated systems (see definition and Sections 1505.8 and 1507.17) and separate roof-mounted systems (see Sections 1505.8 and 1509.7) are addressed. In general, photovoltaic systems need to comply with the same minimum roof covering classification as the underlying roof assembly that the photovoltaic system is installed on and must meet the same structural requirements as any other roof-mounted equipment.

Section 1505.8 will require that both integrated systems (photovoltaic modules/shingles) and roof-mounted systems comply with the fire classification requirements of Section 1505.1. Depending on the building's type of construction and size, this will typically require that either a Class B or Class C roof covering/assembly be used. Therefore, these systems will be tested and listed so that the designer and building official will only need to verify that the systems have the appropriate classification and are installed in accordance with their listing.

The new definition for photovoltaic modules/shingles along with the requirements of Section 1507.17 provide the guidance needed for the installation of the integrated modules/shingles. These shingles are integrated with the building's roof covering and resemble a typical composite shingle. Because the shingles provide both the roof covering and a source of electrical power, testing using the appropriate UL 1703 standard is required, with the attachment in accordance with the listing and the manufacturer's installation instructions. It is important to realize that even though this type of shingle system may appear similar to an asphalt shingle system, the appropriate design slope and fastening requirements are different for each manufacturer's product. That is the reason why this type of shingle system is included in a separate section of Section 1507 and not included under the asphalt shingle requirements of Section 1507.2.

Section 1509 addresses rooftop-mounted systems (such as a panel array) versus the integrated systems that are covered by Section 1507.17. References are provided to the general code requirements regarding wind resistance, fire classification, and installation, ensuring that the rooftop-mounted systems comply with the same minimum requirements that the underlying roof assembly would need to meet.

The electrical portion of the photovoltaic systems are not addressed within the IBC but would be regulated by the applicable electrical code.

New Section 3111 provides a reference that intends to direct users to the provisions of IFC Section 605. The IFC now includes a number of requirements that affect the location and installation of these solar photovoltaic systems. For additional information, please reference *Significant Changes to the International Fire Code, 2012 Edition* by Scott Stookey.

CHANGE TYPE: Modification

CHANGE SUMMARY: In addition to several technical changes, the provisions addressing rooftop structures have been reformatted to better organize and clarify the requirements.

1509, 202
Rooftop Structures

2012 CODE:

202 Definitions.

MECHANICAL EQUIPMENT SCREEN.   A ~~partially enclosed~~ rooftop structure, not covered by a roof, used to aesthetically conceal ~~heating, ventilating and air conditioning (HVAC)~~ plumbing, electrical, or mechanical equipment from view.

PENTHOUSE.   An enclosed, unoccupied rooftop structure ~~above the roof of a building, other than a tank, tower, spire, dome, cupola or bulkhead.~~ used for sheltering mechanical and electrical equipment, tanks, elevators and related machinery, and vertical shaft openings.

ROOF DECK.   The flat or sloped surface constructed on top of the exterior walls of a building or other supports for the purpose of enclosing the story below, or sheltering an area, to protect it from the elements, not including its supporting members or vertical supports.

ROOFTOP STRUCTURE.   ~~An enclosed~~ A structure erected on ~~or above~~ top of the roof deck or on top of any part of a building.

1509.1 General.   The provisions of this section shall govern the construction of rooftop structures.

1509.2 Penthouses.   ~~A penthouse or~~ Penthouses in compliance with Sections 1509.2.1 through ~~1509.2.4~~ 1509.2.5 shall be considered as a portion of the story directly below the roof deck on which such penthouses are located. All other penthouses shall be considered as an additional story of the building.

1509, 202 continues

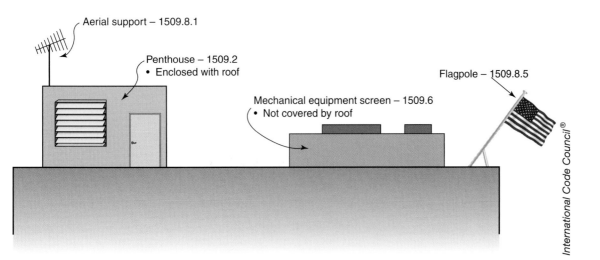

Rooftop structures

1509, 202 continued

1509.2.1 Height above Roof Deck.   A penthouse Penthouses or other projection above the roof in structures constructed on buildings of other than Type I construction shall not exceed 28 feet (8534 mm) above the roof where used as an enclosure for tanks or for elevators that run to the roof and in all other cases shall not exceed extend more than 18 feet (5486 mm) in height above the roof deck as measured to the average height of the roof of the penthouse.

Exceptions:

1. Where used to enclose tanks or elevators that travel to the roof level, penthouses shall be permitted to have a maximum height of 28 feet (8534 mm) above the roof deck.

2. Penthouses located on the roof of buildings of Type I construction shall not be limited in height.

1509.2.2 Area Limitation.   The aggregate area of penthouses and other enclosed rooftop structures shall not exceed one-third the area of the supporting roof deck. Such penthouses and other enclosed rooftop structures shall not be required to be included in determining contribute to either the building area or number of stories as regulated by Section 503.1. The area of the penthouse such penthouses shall not be included in determining the fire area defined specified in Section 901.7 902.

1509.2.3 Use Limitations.   A penthouse Penthouses bulkhead or any other similar projection above the roof shall not be used for purposes other than the shelter of mechanical or electrical equipment, tanks, or shelter of vertical shaft openings in the roof assembly.

1509.2.4 Weather Protection.   Provisions such as louvers, louver blades, or flashing shall be made to protect the mechanical and electrical equipment and the building interior from the elements. Penthouses or bulkheads used for purposes other than permitted by this section shall conform to the requirements of this code for an additional story. The restrictions of this section shall not prohibit the placing of wood flagpoles or similar structures on the roof of any building.

1509.2.4 1509.2.5 Type of Construction.   Penthouses shall be constructed with walls, floors, and roof as required for the type of construction of the building on which such penthouses are built.

Exceptions:

1. On buildings of Type I construction, the exterior walls and roofs of penthouses with a fire separation distance of more greater than 5 feet (1524 mm) and less than 20 feet (6096 mm) shall be permitted to have not less than a of at least 1-hour fire-resistance rating rated noncombustible construction. The exterior walls and roofs of penthouses with a fire separation distance of 20 feet (6096 mm) or greater shall be of noncombustible construction not be required to have a fire-resistance rating. Interior framing and walls shall be of noncombustible construction.

2. On buildings of Type I construction two stories or less in height above grade plane or less in height and of Type II construction, the exterior walls and roofs of penthouses with a fire separation distance of more greater than 5 feet

(1524 mm) and less than 20 feet (6096 mm) shall be permitted to have not less than a ~~of at least~~ 1-hour fire-resistance rating or a lesser fire-resistance rating as required by Table 602 ~~rated noncombustible or~~ and be constructed of fire-retardant-treated wood ~~construction~~. The exterior walls and roofs of penthouses with a fire separation distance of 20 feet (6096 mm) or greater shall be permitted to be constructed of ~~noncombustible or~~ fire-retardant-treated wood ~~construction~~ and shall not be required to have a fire-resistance rating. Interior framing and walls shall be permitted to be constructed of ~~noncombustible or~~ fire-retardant-treated wood.

3. On buildings of Type III, IV, or ~~and~~ V construction, the exterior walls of penthouses with a fire separation distance ~~of more~~ greater than 5 feet (1524 mm) and less than 20 feet (6096 mm) shall be permitted to have not less than a ~~at least~~ 1-hour fire-resistance rating or a lesser fire-resistance rating as required by Table 602 ~~rated construction~~. On buildings of Type III, IV, or VA construction, the exterior walls of penthouses with a fire separation distance of 20 feet (6096 mm) or greater ~~from a common property line~~ shall be permitted to be of Type IV ~~construction or~~ noncombustible construction or fire-retardant-treated wood ~~construction~~ and shall not be required to have a fire-resistance rating. ~~Roofs shall be constructed of materials and fire-resistance rated as required in Table 601 and Section 603, Item 25.3. Interior framing and walls shall be Type IV construction or noncombustible or fire-retardant-treated wood construction.~~

4. ~~On buildings of Type I construction, unprotected noncombustible enclosures housing only mechanical equipment and located with a minimum fire separation distance of 20 feet (6096 mm) shall be permitted.~~

5. ~~On buildings of Type I construction two stories or less above grade plane in height, or Type II, III, or IV and V construction, unprotected noncombustible or fire-retardant treated wood, enclosures housing only mechanical equipment, and located with a minimum fire separation distance of 20 feet (6096 mm) shall be permitted.~~

6. ~~On one-story buildings, combustible unroofed mechanical equipment screens, fences or similar enclosures are permitted where located with a fire separation distance of at least 20 feet (6096 mm) from adjacent property lines and where not exceeding 4 feet (1219 mm) in height above the roof surface.~~

7. ~~Dormers shall be of the same type of construction as the roof on which they are placed, or of the exterior walls of the building.~~

1509.3 Tanks. *(See IBC for changes made to Section 1509.3 and its subsections)*

1509.4 Cooling Towers. *(See IBC for changes made to Section 1509.4)*

1509.5 Towers, Spires, Domes, and Cupolas. *(See IBC for changes made to Section 1509.5 and its subsections)*

1509, 202 continues

1509, 202 continued

1509.6 Mechanical Equipment Screens. Mechanical equipment screens shall be constructed of the materials specified for the exterior walls in accordance with the type of construction of the building. Where the fire separation distance is greater than 5 feet (1524 mm), mechanical equipment screens shall not be required to comply with the fire-resistance-rating requirements.

1509.6.1 Height Limitations. Mechanical equipment screens shall not exceed 18 feet (5486 mm) in height above the roof deck, as measured to the highest point on the mechanical equipment screen.

> **Exception:** Where located on buildings of Type IA construction, the height of mechanical equipment screens shall not be limited.

1509.6.2 Types I, II, III, and IV Construction. Regardless of the requirements in Section 1509.6, mechanical equipment screens shall be permitted to be constructed of combustible materials where located on the roof decks of building of Type I, II, III, or IV construction in accordance with any one of the following limitations:

1. The fire separation distance shall not be less than 20 feet (6096 mm) and the height of the mechanical equipment screen above the roof deck shall not exceed 4 feet (1219 mm) as measured to the highest point on the mechanical equipment screen.

2. The fire separation distance shall not be less than 20 feet (6096 mm) and the mechanical equipment screen shall be constructed of fire-retardant-treated wood complying with Section 2303.2 for exterior installation.

3. Where exterior wall-covering panels are used, the panels shall have a flame spread index of 25 or less when tested in the minimum and maximum thicknesses intended for use with each face tested independently in accordance with ASTM E84 or UL 723. The panels shall be tested in the minimum and maximum thicknesses intended for use in accordance with, and shall comply with the acceptance criteria of, NFPA 285 and shall be installed as tested. Where the panels are tested as part of an exterior wall assembly in accordance with NFPA 285, the panels shall be installed on the face of the mechanical equipment screen supporting structure in the same manner as they were installed on the tested exterior wall assembly.

1509.6.3 Type V Construction. The height of mechanical equipment screens located on the roof decks of buildings of Type V construction, as measured from grade plane to the highest point on the mechanical equipment screen, shall be permitted to exceed the maximum building height allowed for the building by other provisions of this code where complying with any one of the following limitations, provided the fire separation distance is greater than 5 feet (1524 mm):

1. Where the fire separation distance is not less than 20 feet (6096 mm), the height above grade plane of the mechanical equipment screen shall not exceed 4 feet (1219 mm) more than the maximum building height allowed.

2. The mechanical equipment screen shall be constructed of noncombustible materials.

3. The mechanical equipment screen shall be constructed of fire-retardant-treated wood complying with Section 2303.2 for exterior installation.

4. Where fire separation distance is not less than 20 feet (6096 mm), the mechanical equipment screen shall be constructed of materials having a flame spread index of 25 or less when tested in the minimum and maximum thicknesses intended for use with each face tested independently in accordance with ASTM E84 or UL 723.

1509.7 Photovoltaic Systems. *(See code for changes made to Section 1509.7 and its subsections. These changes were discussed previously in this book with the page dealing with photovoltaic systems.)*

1509.8 Other Rooftop Structures. Rooftop structures not regulated by Sections 1509.2 through 1509.7 shall comply with Section 1509.8.1 through 1509.8.5 as applicable.

1509.8.1 Aerial Supports. Aerial supports shall be constructed of noncombustible materials.

> **Exception:** Aerial supports not greater than 12 feet (3658 mm) in height as measured from the roof deck to the highest point on the aerial supports shall be permitted to be constructed of combustible materials.

1509.8.2 Bulkheads. Bulkheads used for the shelter of mechanical or electrical equipment or vertical shaft openings in the roof assembly shall comply with Section 1509.2 as penthouses. Bulkheads used for any other purpose shall be considered as an additional story of the building.

1509.8.3 Dormers. Dormers shall be of the same type of construction as required for the roof in which such dormers are located or the exterior walls of the building.

1509.8.4 Fences. Fences and similar structures shall comply with Section 1509.6 as mechanical equipment screens.

1509.8.5 Flagpoles. Flagpoles and similar structures shall not be required to be constructed of noncombustible materials and shall not be limited in height or number.

CHANGE SIGNIFICANCE: The provisions of Section 1509 have been formatted so the requirements in Section 1509.2 are limited to penthouses while the provisions for mechanical equipment screens and other rooftop structures have been relocated in new Sections 1509.6 and 1509.8. Previously, Section 1509.2 also included items such as mechanical equipment screens, flagpoles, fences, and dormers. This formatting change, along with the revisions to some of the definitions, will help focus the application of Section 1509.2 on penthouses, while other rooftop structures that are not covered by a roof will be found either in the provisions

1509, 202 continues

1509, 202 continued

addressing mechanical equipment screens or those dealing with "other roof top structures." Previously, the provisions were somewhat disjointed and inconsistent due to an effort to address all types of rooftop structures within the one code section, even though there clearly are distinctions among the types and hazards of the structures.

While there are several important technical changes, the vast majority of the changes eliminate redundant language, relocate provisions to a more appropriate location, make the terminology more consistent, reformat the text to fit more effectively, and make it easier to distinguish among the various elements.

Section 1509.2 will only address unoccupied rooftop structures that are enclosed (covered by a roof) based on the definition for "penthouse" and the requirements within Section 1509.2. Where a penthouse is not covered by a roof, it will be regulated by either Sections 1509.3 through 1509.8 or it will be considered as an additional story of the building. The restrictions on the use of a penthouse were previously located in the use limitation provisions of Section 1509.2.3.

Section 1509.2.1 now clarifies that the height measurement is made from the roof "deck" and that it is measured to the average height of the penthouse. The code previously did not state how to measure the height if the penthouse had a sloped or gabled roof.

Section 1509.3 dealing with tanks has been editorially revised without any technical revisions.

Section 1509.4 has been revised to clarify that the requirements only apply to cooling towers located on the roof of a building and are not applicable to cooling towers that may be installed on the ground adjacent to a building. It also has been clarified as to how to measure the size of the equipment and where measurements are to be made.

The tower, spire, dome, and cupola construction requirements of Section 1509.5.1 contain technical changes that affect the application of the provisions. In addition, the provisions have been reformatted to distinguish between the requirements for the structures versus those for support of the structures. The 1½ -hour opening protection requirements between the building and any towers have been deleted, and compliance is now required with the general horizontal assembly opening protection requirements of Section 711. The means of protecting—and the type of protection needed for—these openings in the horizontal assembly have also been revised. Typically, either a shaft enclosure or a floor fire-door assembly with a fire-resistive rating is required.

Section 1509.6 is a new section that deals with mechanical equipment screens. The revised definition is important because it serves as a reminder that these screens are "not covered by a roof," whereas it was previously stated they were "partially enclosed" structures. A distinction is now made between roofed penthouses (Section 1509.2) and screens that are not covered by a roof (Section 1509.6). In general, the mechanical equipment screens will need to be constructed of materials consistent with the building's exterior wall type of construction requirements; however, the fire-resistance ratings would not be required. Because the screens are not roofed, they will present less of an exposure hazard than a penthouse. The location and combustibility of the screen itself are the key issues that are addressed.

Sections 1509.6.2 and 1509.6.3 essentially provide exceptions that modify the type of construction or general building height limitations. Item 1 in both of these sections is conceptually based on the previous

Exception 6 from Section 1509.2.4 (now Section 1509.2.5), however the one-story limitation has been deleted because the limited screen height was viewed as not creating a significant hazard. Both of the sections also contain an item to allow the use of fire-retardant-treated wood and will not impose the 4-foot height limit that is found within Item 1 of the provisions.

The last item in Section 1509.6.2 is a new, unique requirement. The limitations are based on a totally new concept where the combustible materials used to construct the mechanical equipment screen are limited to a maximum flame spread index of 25 (which is also required for fire-retardant-treated wood), and the materials are required to be successfully tested in accordance with NFPA 285. This is the same test method that is used to validate the use of foam plastic insulations in exterior walls of Types I, II, III, and IV construction, as well as for the use of metal composite materials (MCM) in accordance with Section 1407.10. Although the material would be tested as the outer face (or skin) of the exterior wall in the NFPA 285 test as part of an exterior wall assembly, the test clearly assesses the surface flame spread resistance of the materials constituting the outer face, as well as, to a certain degree, the inner face where it is exposed to any open cavities in the wall assembly. Because the NFPA 285 test is used to qualify combustible materials for use where noncombustible exterior walls are required, it seems reasonable to allow its use for this application for mechanical equipment screens without the need to have the entire wall assembly constructed as tested for the mechanical equipment screen, instead utilizing the materials tested on the exterior face of the wall system in accordance with NFPA 285.

As mentioned previously, the relationship and wording of the various requirements of Section 1509.6 need to be reviewed carefully. Section 1509.6.2 is essentially intended as an exception to the type of construction materials required by Section 1509.6 and that Section 1509.6.3 was an exception to the height limitations of Section 1509.6.1. Although not clearly stated, the relationship is somewhat apparent when considering how the provisions work together. Although Section 1509.6.3 does indicate that the height of the screens "shall be permitted to exceed the maximum building height allowed for the building by other provisions of this code" when one of the four items is used, the 18-foot height limitation above the roof deck in Section 1509.6.1 should still be applicable. This application would allow the mechanical screens to exceed the height limitations from Table 503, but would still limit the screen heights to 18 feet above the roof deck.

The requirements for photovoltaic systems in Section 1509.7 are also appropriate within Section 1509 because they address elements that are mounted on the roof. See the previous discussion in this publication related to photovoltaic systems for commentary related to the changes in Section 1509.7.

The provisions of Section 1509.8 regulate items that do not fall within the other categories of rooftop structures (penthouses, tanks, cooling towers, spires, or mechanical equipment screens). However, Sections 1509.8.2 and 1509.8.4 reference the penthouse and mechanical equipment screen requirements because bulkheads and fences closely resemble those elements from both an appearance and a hazard standpoint. The flagpole provisions of Section 1509.8.5 were relocated from the previous Section 1509.2.3 (see deleted text in 2012 Section 1509.2.4).

1510.3

Roof Covering Replacement

CHANGE TYPE: Modification

CHANGE SUMMARY: Existing ice dams are now permitted to remain in place when replacing a roof covering.

2012 CODE: 1510.3 Recovering versus Replacement. New roof coverings shall not be installed without first removing all existing layers of roof coverings down to the roof deck where any of the following conditions occur:

1. Where the existing roof or roof covering is water soaked or has deteriorated to the point that the existing roof or roof covering is not adequate as a base for additional roofing.
2. Where the existing roof covering is wood shake, slate, clay, cement, or asbestos-cement tile.
3. Where the existing roof has two or more applications of any type of roof covering.

Exceptions:

1. Complete and separate roofing systems—such as standing-seam metal roof systems—that are designed to transmit the roof loads directly to the building's structural system and that do not rely on existing roofs and roof coverings for support shall not require the removal of existing roof coverings.
2. Metal panel, metal shingle, and concrete and clay tile roof coverings shall be permitted to be installed over existing wood shake roofs when applied in accordance with Section 1510.4.

International Code Council®

Ice barrier membrane is permitted to remain in place when replacing roof covering.

3. The application of a new protective coating over an existing spray polyurethane foam roofing system shall be permitted without tear-off of existing roof coverings.

4. <u>Where the existing roof assembly includes an ice barrier membrane that is adhered to the roof deck, the existing ice barrier membrane shall be permitted to remain in place and be covered with an additional layer of ice barrier membrane in accordance with Section 1507.</u>

CHANGE SIGNIFICANCE: Section 1510.3 requires that, where certain conditions occur, the existing roof covering be removed "down to the deck" prior to the installation of a new roof covering. Where an adhered ice barrier membrane is present and there is either difficulty in removing it or its removal may create other problems, the ice barrier now does not need to be removed. In reroofing situations, it is oftentimes difficult, if not impossible, to remove the existing layer of adhered ice barrier membrane without damaging or replacing the roof deck or without leaving portions of the ice barrier membrane, which creates an uneven base surface.

An existing adhered ice barrier membrane is now permitted to remain in place during reroofing operations provided it is then covered with a new ice barrier membrane as a part of the reroofing process.

Table 1604.3

Deflection Limits

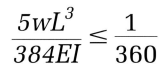

Beam deflection

CHANGE TYPE: Modification

CHANGE SUMMARY: Deflection limits for roof and wall members supporting plaster or stucco have been clarified. Footnote f was also modified to account for the new ultimate wind loads in the 2010 edition of ASCE/SEI 7 (ASCE 7-10), *Minimum Design Loads for Buildings and Other Structures.*

2012 CODE:

TABLE 1604.3 Deflection Limits[a, b, c, h, i]

Construction	L	S or W^{f}	$D + L^{\text{d, g}}$
Roof members:[e]			
Supporting plaster or stucco ceiling	l/360	l/360	l/240
Supporting nonplaster ceiling	l/240	l/240	l/180
Not supporting ceiling	l/180	l/180	l/120
Floor members	l/360	—	l/240
Exterior walls and interior partitions:			
With plaster or stucco finishes	—	l/360	—
With other brittle finishes	—	l/240	—
With flexible finishes	—	l/120	—
Farm buildings	—	—	l/180
Greenhouses	—	—	l/120

f. The wind load is permitted to be taken as ~~0.7~~ 0.42 times the "component and cladding" loads for the purpose of determining deflection limits herein.
(no changes to other footnotes)

CHANGE SIGNIFICANCE: Table 1604.3 now includes a line item for the deflection limit on roofs and walls with plaster or stucco finishes. The intent is to clarify the terminology and coordinate the language in the deflection limits table with the corresponding IRC table and ASTM C926-98a, *Standard Specification for Application of Portland Cement-Based Plaster.* In preparing the new wind maps for ASCE 7-10, the committee decided to use multiple ultimate event or strength design maps in conjunction with a wind load factor of 1.0 for strength design and 0.6 for allowable stress design. Footnote f of Table 1604.3 was modified to be 0.42 (0.7 × 0.6 = 0.42) because serviceability (deflection) calculations are done at an allowable stress design level.

CHANGE TYPE: Modification

CHANGE SUMMARY: The term "occupancy category" has been changed to "risk category" to better reflect the intended meaning and to coordinate with the terminology used in ASCE 7-10.

2012 CODE: 202 Definitions.

~~OCCUPANCY CATEGORY~~ ~~A category used to determine structural requirements based on occupancy.~~

<u>**RISK CATEGORY.** A categorization of buildings and other structures for determination of flood, wind, snow, ice, and earthquake loads based on the risk associated with unacceptable performance.</u>

1604.5 ~~Occupancy~~ <u>Risk</u> Category. Each building and structure shall be assigned ~~an occupancy~~ <u>a risk</u> category in accordance with Table 1604.5. <u>Where a referenced standard specifies an occupancy category, the risk category shall not be taken as lower than the occupancy category specified therein.</u>

1604.5.1 Multiple Occupancies. Where a building or structure is occupied by two or more occupancies not included in the same ~~occupancy~~ <u>risk</u> category, it shall be assigned the classification of the highest ~~occupancy~~ <u>risk</u> category corresponding to the various occupancies. Where buildings or structures have two or more portions that are structurally separated, each portion shall be separately classified. Where a separated portion of a building or structure provides required access to, required egress from, or shares life safety components with another portion having a higher ~~occupancy~~ <u>risk</u> category, both portions shall be assigned to the higher ~~occupancy~~ <u>risk</u> category.

1604.5, 202 continues

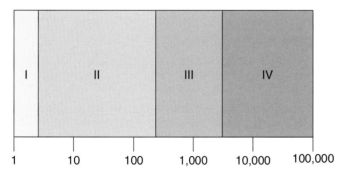

Approximate relationship between number of lives at risk by failure based on risk category *(ASCE 7-10 Commentary)*

1604.5, 202 continued

TABLE 1604.5 ~~Occupancy~~ <u>Risk</u> Category of Buildings and Other Structures

~~Occupancy~~ <u>Risk</u> Category	Nature of Occupancy
II	Buildings and other structures except those listed in ~~Occupancy~~ <u>Risk</u> Categories I, III, and IV.
III	Buildings and other structures that represent a substantial hazard to human life in the event of failure, including but not limited to: • Buildings and other structures whose primary occupancy is public assembly with an occupant load greater than 300 • Buildings and other structures containing elementary school, secondary school, or day care facilities with an occupant load greater than 250 • Buildings and other structures containing adult education facilities, such as colleges and universities, with an occupant load greater than 500 • Group I-2 occupancies with an occupant load of 50 or more resident ~~patients~~ <u>care recipients</u> but not having surgery or emergency treatment facilities • Group I-3 occupancies • Any other occupancy with an occupant load greater than 5000[a] • Power-generating stations, water treatment for potable water, wastewater treatment facilities, and other public utility facilities not included in ~~Occupancy~~ <u>Risk</u> Category IV • Buildings and other structures not included in ~~Occupancy~~ <u>Risk</u> Category IV containing ~~sufficient~~ quantities of toxic or explosive ~~substances~~ <u>materials that:</u> <u>Exceed maximum allowable quantities per control area as given in Table 307.1(1) or 307.1(2) or per outdoor control area in accordance with the *International Fire Code*; and are sufficient</u> to ~~be dangerous~~ <u>pose a threat</u> to the public if released[b]
IV	Buildings and other structures designated as essential facilities, including but not limited to: • Group I-2 occupancies having surgery or emergency treatment facilities • Fire, rescue, ambulance, and police stations and emergency vehicle garages • Designated earthquake, hurricane, or other emergency shelters • Designated emergency preparedness, communications, and operations centers and other facilities required for emergency response • Power-generating stations and other public utility facilities required as emergency backup facilities for ~~Occupancy~~ <u>Risk</u> Category IV structures • <u>Buildings and other structures containing quantities of</u> highly toxic materials ~~as defined by Section 307 where the quantity of the material~~ <u>that:</u> Exceed~~s the~~ maximum allowable quantities ~~of~~ <u>per control area as given in</u> Table 307.1(2) <u>or per outdoor control area in accordance with the International Fire Code; and are sufficient to pose a threat to the public if released</u>[b] • Aviation control towers, air traffic control centers, and emergency aircraft hangars • Buildings and other structures having critical national defense functions • Water storage facilities and pump structures required to maintain water pressure for fire suppression

(Portions of table not shown remain unchanged)

a. (no change to text)
b. <u>Where approved by the building official, the classification of buildings and other structures as Risk Category III or IV based on their quantities of toxic, highly toxic, or explosive materials is permitted to be reduced to Occupancy Category II, provided it can be demonstrated by a hazard assessment in accordance with Section 1.5.3 of ASCE 7 that a release of the toxic, highly toxic, or explosive materials is not sufficient to pose a threat to the public.</u>

Because the term "occupancy category" occurs in so many chapters of the code, the entire code change text is too extensive to be included here. Refer to Code Change S41-09/10 in the *2012 IBC Code Changes Resource Collection* for the complete text and history of the code change.

CHANGE SIGNIFICANCE: The term "occupancy category" is misleading because it implies something about the nature of the building occupants. The term "occupancy" relates primarily to the nonstructural fire and life safety provisions, not the risks associated with structural failure. In fact, some of the structures regulated by the IBC and IEBC are not even occupied but have an occupancy category assigned because their failure could pose a substantial risk to the public. Thus, the term "occupancy category" has been changed to "risk category" to better reflect the intent and provide consistency with the terminology used in ASCE 7-10. Although the terminology changed, the classifications continue to reflect the progression of the consequences of failure from the lowest (Risk Category I) to the highest (Risk Category IV). A detailed discussion of the risk categories is contained in Section C1.5 of the ASCE 7-10 commentary.

1605.2

Load Combinations Using Strength Design of Load and Resistance Factor Design

$$\Sigma\gamma_i Q_i \le R_n$$

LRFD method limit state

CHANGE TYPE: Modification

CHANGE SUMMARY: The strength design load combinations in the 2012 IBC have been coordinated with Section 2.3 of ASCE 7-10 and expanded to include loads due to fluids, F, and other lateral pressures, H, as well as ice loads.

2012 CODE: 1605.2.~~1~~ Load Combinations Using Strength Design of Load and Resistance Factor Design. Where strength design or load and resistance factor design is used, <u>buildings and other</u> structures, and portions thereof, shall <u>be designed to</u> resist the most critical effects resulting from the following combinations of factored loads:

$$1.4(D + F) \quad \text{(Equation 16-1)}$$

$$1.2(D + F + \cancel{T}) + 1.6(L + H) + 0.5(L_r \text{ or } S \text{ or } R) \quad \text{(Equation 16-2)}$$

$$1.2(D \underline{+ F}) + 1.6(L_r \text{ or } S \text{ or } R) \underline{+ 1.6H} + (f_1L \text{ or } \cancel{0.8}\ \underline{0.5}W) \quad \text{(Equation 16-3)}$$

$$1.2(D \underline{+ F}) + \cancel{1.6}\ \underline{1.0}W + f_1L \underline{+ 1.6H} + 0.5(L_r \text{ or } S \text{ or } R) \quad \text{(Equation 16-4)}$$

$$1.2(D \underline{+ F}) + 1.0E + f_1L \underline{+ 1.6H} + f_2S \quad \text{(Equation 16-5)}$$

$$0.9D + \cancel{1.6}\ \underline{1.0}W + 1.6H \quad \text{(Equation 16-6)}$$

$$0.9(D \underline{+ F}) + 1.0E + 1.6H \quad \text{(Equation 16-7)}$$

where:
- f_1 = 1 for ~~floors in~~ places of public assembly<u>,</u> ~~for~~ live loads in excess of 100 pounds per square foot (4.79 kN/m^2), and ~~for~~ parking garages,
- = and 0.5 for other live loads.
- f_2 = 0.7 for roof configurations (such as saw tooth) that do not shed snow off the structure, and
- = 0.2 for other roof configurations.

Exceptions:

1. Where other factored load combinations are specifically required by ~~the~~ <u>other</u> provisions of this code, such combinations shall take precedence.

2. <u>Where the effect of H resists the primary variable load effect, a load factor of 0.9 shall be included with H, where H is permanent, and H shall be set to zero for all other conditions.</u>

1605.2.1 ~~Flood~~ <u>Other</u> Loads. Where flood loads, F_a, are to be considered in the design, the load combinations of Section 2.3.3 of ASCE 7 shall be used. <u>Where self-straining loads, T, are considered in design, their structural effects in combination with other loads shall be determined in accordance with Section 2.3.5 of ASCE 7. Where an ice-sensitive structure is subjected to loads due to atmospheric icing, the load combinations of Section 2.3.4 of ASCE 7 shall be considered.</u>

CHANGE SIGNIFICANCE: The IBC load combinations have been coordinated with the strength design load combinations in Section 2.3 of ASCE 7-10 and loads due to fluids, F, and lateral earth pressures, ground water pressures, or the pressure of bulk materials, H, have been included. The load factor on the wind load, W, has been changed to 1.0 to account for the new ultimate design wind speed and strength level wind forces in ASCE 7-10. Note that F and H must be considered in ASCE 7-10, but they are indirectly combined with other loads as required by the text in Section 2.3.2. The self-straining load, T, was deleted from the load combinations because it is indirectly combined as described under Section 1605.2.2 for other loads. The load and resistance factor design (LRFD) load combinations in Section 1605.2.2 and the allowable stress design (ASD) load combinations in Section 1605.3.1.2 were modified to include ice loads for ice sensitive structures. Where atmospheric ice loads must be considered in the design of ice-sensitive structures, these sections cross reference ASCE 7 Section 2.3.4 for LRFD and Section 2.4.3 for ASD, respectively.

1605.3

Load Combinations Using Allowable Stress Design

$$\sum Q_i \leq R_n/\Omega$$

ASD method limit state

CHANGE TYPE: Modification

CHANGE SUMMARY: The allowable stress design load combinations in the 2012 IBC have been coordinated with Section 2.4 of ASCE 7-10 and expanded to include loads due to fluids, F, and other lateral pressures, H, as well as ice loads.

2012 CODE: **1605.3 Load Combinations Using Allowable Stress Design.**

1605.3.1 Basic Load Combinations. Where allowable stress design (working stress design), as permitted by this code, is used, structures and portions thereof shall resist the most critical effects resulting from the following combinations of loads:

$$D + F \tag{Equation 16-8}$$

$$D + H + F + L + T \tag{Equation 16-9}$$

$$D + H + F + (L_r \text{ or } S \text{ or } R) \tag{Equation 16-10}$$

$$D + H + F + 0.75(L + T) + 0.75(L_r \text{ or } S \text{ or } R) \tag{Equation 16-11}$$

$$D + H + F + (\underline{0.6}W \text{ or } 0.7E) \tag{Equation 16-12}$$

$$\begin{aligned}D + H + F + 0.75(\underline{0.6}W \text{ or } 0.7E) \\ + 0.75L + 0.75(L_r \text{ or } S \text{ or } R)\end{aligned} \tag{Equation 16-13}$$

$$\underline{D + H + F + 0.75(0.7E) + 0.75L + 0.75S} \tag{\underline{Equation 16-14}}$$

$$0.6D + \underline{0.6}W + H \tag{Equation 16-\cancel{14}\ \underline{15}}$$

$$0.6(\underline{D + F}) + 0.7E + H \tag{Equation 16-\cancel{15}\ \underline{16}}$$

Exceptions:

1.–2. (no changes to text)

3. <u>Where the effect of H resists the primary variable load effect, a load factor of 0.6 shall be included with H, where H is permanent, and H shall be set to zero for all other conditions.</u>

4. <u>In Equation 16-15, the wind load, W, is permitted to be reduced</u> 10 percent for design of the foundation other than anchorage of the structure to the foundation <u>in accordance with Exception 2 of Section 2.4.1 of ASCE 7.</u>

5. <u>In Equation 16-16, 0.6D is permitted to be increased to 0.9D for the design of special reinforced masonry shear walls complying with Chapter 21.</u>

1605.3.1.2 ~~Flood~~ Other Loads.   Where flood loads, F_a, are to be considered in design, the load combinations of Section 2.4.2 of ASCE 7 shall be used. <u>Where self-straining loads, T, are considered in design, their structural effects in combination with other loads shall be determined in accordance with Section 2.4.4 of ASCE 7. Where an ice-sensitive structure is subjected to loads due to atmospheric icing, the load combinations of Section 2.3.4 of ASCE 7 shall be considered.</u>

1605.3.2 Alternative Basic Load Combinations.   In lieu of the basic load combinations specified in Section 1605.3.1, structures and portions thereof shall be permitted to be designed for the most critical effects resulting from the following combinations. When using these alternative basic load combinations that include wind or seismic loads, allowable stresses are permitted to be increased or load combinations reduced where permitted by the material chapter of this code or the referenced standards. For load combinations that include the counteracting effects of dead and wind loads, only two-thirds of the minimum dead load likely to be in place during a design wind event shall be used. <u>When using allowable stresses which have been increased or load combinations which have been reduced as permitted by the material chapter of this code or the referenced standards,</u> where wind loads are calculated in accordance with <u>Chapters 26 through 31</u> ~~Chapter 6~~ of ASCE 7, the coefficient ω in the following equations shall be taken as 1.3. For other wind loads, ω shall be taken as 1. <u>When allowable stresses have not been increased or load combinations have not been reduced as permitted by the material chapter of this code or the referenced standards, ω shall be taken as 1.</u> When using these alternative load combinations to evaluate sliding, overturning, and soil bearing at the soil–structure interface, the reduction of foundation overturning from Section 12.13.4 in ASCE 7 shall not be used. When using these alternative basic load combinations for proportioning foundations for loadings, which include seismic loads, the vertical seismic *load effect*, *Ev*, in Equation 12.4-4 of ASCE 7, is permitted to be taken equal to zero.

$$D + L + (L_r \text{ or } S \text{ or } R) \qquad \textbf{(Equation 16-17)}$$

$$D + L + \cancel{(0.6\omega W)} \qquad \textbf{(Equation 16-18)}$$

$$D + L + \underline{0.6\omega W} + S/2 \qquad \textbf{(Equation 16-19)}$$

$$D + L + S + \underline{0.6\omega} \ W/2 \qquad \textbf{(Equation 16-20)}$$

$$D + L + S + E/1.4 \qquad \textbf{(Equation 16-21)}$$

$$0.9 \ D + E/1.4 \qquad \textbf{(Equation 16-22)}$$

Exceptions: (no changes to text)

1605.3.2.1 Other Loads.   Where F, H, or T are to be considered in design, each applicable load shall be added to the combinations specified in Section 1605.3.2. <u>Where self-straining loads, T, are considered in design, their structural effects in combination with other loads shall be determined in accordance with Section 2.4.4 of ASCE 7.</u>

1605.3 continues

1605.3 continued

CHANGE SIGNIFICANCE: The allowable stress design load (ASD) combinations in the IBC have been coordinated with the ASD load combinations in Section 2.4 of ASCE 7-10. The self-straining load, *T*, was deleted from the load combinations because it is indirectly combined as described under Section 1605.3.2.1 for other loads. To improve equivalency with the strength design load combinations and achieve consistency with ASCE 7-10, earthquake load effect, *E*, was removed from Equation 16-13, and the new load combination Equation 16-14 was added. This has the effect of retaining roof live load, L_r, and rain load, *R*, in combination with wind load, *W* (Equation 16-13), but removed these loads in combination with earthquake load, *E*, in Equation 16-14. This was done to achieve consistency between the ASD load combinations and the strength design or LRFD load combinations in Equations 16-4 and 16-5. The load factor on the wind load, *W*, has been changed to 0.6 in both the basic and alternative basic ASD load combinations to account for the new ultimate design wind speed and strength level wind forces in ASCE 7-10 ($W_{ASD} = 0.6W_{ult}$).

Based on deliberations with the Code Resource and Support Committee (CRSC) of the NEHRP, the ω factor in the alternative basic ASD load combinations has been modified to be either 1.3 or 1.0. When allowable stresses have been increased or load combinations have been reduced (as permitted by a material chapter in the code or a referenced standard), the coefficient ω is taken as 1.3. When allowable stresses have not been increased or load combinations have not been reduced, ω is to be taken as 1.0.

Note that although *F* and *H* are not in the ASD load combinations in ASCE 7-10, they are indirectly combined with other loads as described in the text in Section 2.4.1.

The ASD load combinations in Section 1605.3.1.2 and LRFD load combinations in Section 1605.2.2 were also modified to include ice loads for ice-sensitive structures. Where atmospheric ice loads must be considered in the design of ice-sensitive structures, these sections cross reference Section 2.4.3 for ASD and ASCE 7 Section 2.3.4 for LRFD, respectively.

CHANGE TYPE: Modification

CHANGE SUMMARY: The live loads established in IBC Section 1607 and Table 1607.1 have been modified and updated in order to coordinate with the live loads of Chapter 4 and Table 4-1 in ASCE 7-10.

Table 1607.1
Minimum Live Loads

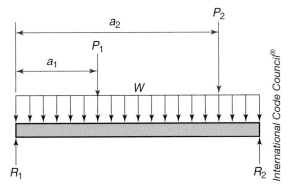

Simple beam load diagram

2012 CODE:

TABLE 1607.1 Minimum Uniformly Distributed Live Loads, L_o, and Minimum Concentrated Live Loads[g]

Occupancy or Use	Uniform (psf)	Concentrated (lb)
3. Armories and drill rooms	150[m]	—
4. Assembly areas ~~and theaters~~		
Fixed seats (fastened to floor)	60[m]	
Follow spot, projections, and control rooms	50	
Lobbies	100[m]	
Movable seats	100[m]	—
Stage~~s and~~ <u>floors</u>	~~125~~ <u>150</u>[m]	
Platforms <u>(assembly)</u>	~~125~~ <u>100</u>	
Other assembly areas	100[m]	
5. Balconies ~~(exterior)~~ and decks[h]	Same as occupancy served	
~~6. Bowling alleys~~	~~75~~	—
~~7.~~ <u>6.</u> Catwalks	40	300
~~9.~~ <u>8.</u> Corridors~~, except as otherwise indicated~~		
<u>First floor</u>	100	
<u>Other floors</u>	<u>Same as occupancy served except as indicated</u>	—
~~10. Dance halls and ballrooms~~	~~100~~	—
~~11~~ <u>9.</u> Dining rooms and restaurants	100[m]	—
~~13.~~ <u>11.</u> Elevator machine room grating (on area of ~~4 in²~~ <u>2 inches by 2 inches)</u>	—	300
~~14.~~ <u>12.</u> Finish light floor plate construction (on area of ~~1 in²~~ <u>1 inch by 1 inch)</u>	—	200
~~16.~~ <u>14.</u> Garages (passenger vehicles only) Trucks and buses	40[m]	Note a See Section 1607.7

Table 1607.1 continues

Table 1607.1 continued

Occupancy or Use	Uniform (psf)	Concentrated (lb)
~~17. Grandstands (see stadium and arena bleachers)~~	—	—
~~18. Gymnasiums, main floors and balconies~~	~~100~~	—
~~19.~~ 15. Handrails, guards and grab bars		See Section 1607.8
16. Helipads		See Section 1607.6
~~22~~ 19. Libraries		
Corridors above first floor	80	1000
Reading rooms	60	1000
Stack rooms	$150^{b, m}$	1000
~~23~~ 20. Manufacturing		
Heavy	250^{m}	3000
Light	125^{m}	2000
24. Recreational uses:		
Bowling alleys, poolrooms, and similar uses	75^{m}	
Dance halls and ballrooms	100^{m}	
Gymnasiums	100^{m}	
Reviewing stands, grandstands, and bleachers	$100^{c, m}$	
Stadiums and arenas with fixed seats (fastened to floor)	$60^{c, m}$	
~~27.~~ 25. Residential		
One- and two-family dwellings		
Uninhabitable attics without storage^{i}	10	
Uninhabitable attics with ~~limited~~ storage$^{i, j, k}$	20	
Habitable attics and sleeping areas^{k}	30	—
All other areas	40	
Hotels and multiple-family dwellings		
Private rooms and corridors serving them	40	
Public rooms^{m} and corridors serving them	100	
~~28. Reviewing stands, grandstands and bleachers~~		~~Note c~~
~~29.~~ 26. Roofs:		
All roof surfaces subject to maintenance workers		300
Awnings and canopies:		
Fabric construction supported by a ~~lightweight rigid~~ skeleton structure	5 Nonreducible	
All other construction	20	
Ordinary flat, pitched, and curved roofs (that are not occupiable)	20	
Where primary roof members, are exposed to a work floor, at single panel points of lower chord of roof trusses, or any point along primary structural members supporting roofs:		
Over manufacturing, storage warehouses, and repair garages		2000
All other ~~occupancies~~ primary roof members		300
~~Roofs used for other special purposes~~	~~Note l~~	~~Note l~~
~~Roofs used for promenade purposes~~	~~60~~	
~~Roofs used for roof gardens or~~ ~~assembly purposes~~	~~100~~	

Occupancy or Use	Uniform (psf)	Concentrated (lb)
Occupiable roofs:		
Roof gardens	100	
Assembly areas	100[m]	
All other similar areas	Note l	Note l
~~32.~~ 29. Sidewalks, vehicular driveways, and yards, subject to trucking	250[d, m]	8000[e]
~~33. Skating rinks~~	~~100~~	—
~~34. Stadiums and arenas~~		
~~Bleachers~~	~~100[c]~~	—
~~Fixed seats (fastened to floor)~~	~~60[c]~~	
~~35.~~ 30. Stairs and exits		~~Note f~~
One- and two-family dwellings	40	300[f]
All other	100	300[f]
~~36.~~ 31. Storage warehouses (shall be designed for heavier loads if required for anticipated storage)		
Heavy	250[m]	—
Light	125[m]	
~~37.~~ 32. Stores		
Retail		
First floor	100	1000
Upper floors	75	1000
Wholesale, all floors	125[m]	1000
~~38.~~ 33. Vehicle barriers ~~systems~~		See Section 1607.8.3
~~40.~~ 35. Yards and terraces, pedestrian	100 [m]	—

(Portions of table not shown are unchanged)

f. The minimum concentrated load on stair treads ~~(~~shall be applied on an area of ~~4 square~~ 2 inches by 2 inches~~)~~ ~~is 300 pounds~~. This load need not be assumed to act concurrently with the uniform load.

g. Where snow loads occur that are in excess of the design conditions, the structure shall be designed to support the loads due to the increased loads caused by drift buildup or a greater snow design determined by the building official (see Section 1608). ~~For special-purpose roofs, see Section 1607.11.2.2.~~

i. Uninhabitable attics without storage are those where the maximum clear height between the joists and rafters is less than 42 inches, or where there are not two or more adjacent trusses with web configurations capable of accommodating an assumed rectangle 42 inches in height by by 24 inches in width, or greater, within the plane of the trusses. ~~For attics without storage,~~ This live load need not be assumed to act concurrently with any other live load requirements.

j. ~~For attics with limited storage and constructed with trusses, this live load need only be applied to those portions of the bottom chord~~ Uninhabitable attics with storage are those where the maximum clear height between the joists and rafters is 42 inches or greater, or where there are two or more adjacent trusses with ~~the same~~ web configurations capable of ~~containing~~ accommodating an assumed rectangle 42 inches ~~high~~ in height by 24 inches ~~wide~~ in width, or greater, ~~located~~ within the plane of the trusses

~~The rectangle shall fit between the top of the bottom chord and the bottom of any other truss member, provided that each of the following criteria is met:~~ The live load need only be applied to those portions of the joists or truss bottom chords where both of the following conditions are met:

    i. The attic area is accessible ~~by a pull-dwon stairway or framed opening in accordance with Section 1209.2,~~ from an opening not less than 20 inches in width by 30 inches in length that is located where the clear height in the attic is a minimum of 30 inches; and

    ii. The slopes of the joists or truss ~~shall have a~~ bottom chords ~~pitch less than 2:12~~ are no greater than 2 units vertical to 12 units horizontal.

    ~~iii. Bottom chords of trusses shall be designed for the greater of actual imposed dead load or 10 psf, uniformly distributed over the entire span.~~ The remaining portions of the joists or bottom chords shall be designed for a uniformly distributed concurrent live load of not less than 10 lb/ft^2.

k. Attic spaces served by ~~a fixed stair~~ stairways other than pull-down type shall be designed to support the minimum live load specified for habitable attics and sleeping rooms.

l. ~~Roofs used for other special purposes~~ Areas of occupiable roofs, other than roof gardens and assembly areas, shall be designed for appropriate loads as approved by the building official. Unoccupied landscaped areas of roofs shall be designed in accordance with Section 1607.12.3.

m. Live load reduction is not permitted unless specific exceptions of Section 1607.10 apply.

(Footnotes not shown are unchanged)

Table 1607.1 continues

Table 1607.1 continued

CHANGE SIGNIFICANCE: Many live loads set forth in Chapter 4 of ASCE 7 were updated in the 2010 edition. To coordinate the changes in ASCE 7-10 with the 2012 IBC, corresponding modifications were made to Section 1607 and Table 1607.1. These changes are summarized as follows:

- Footnotes i, j, and k pertaining to residential attic live loads were updated to clarify the intent.

- The live load for stage floors was increased from 125 psf to 150 psf, and the live load for platforms in assembly areas was decreased from 125 psf to 100 psf.

- Various recreational type uses were consolidated under a new item called "recreational uses." These uses include bowling alleys, pool rooms, dance halls and ballrooms, gymnasiums, reviewing stands, grandstands and bleachers, and stadiums and arenas with fixed seats. No technical changes were made to the live loads. The factor, $f_1 = 1$ (See Section 1605.2.1) now applies to floors in places of public assembly areas and recreational uses for live loads in excess of 100 pounds per square foot. Skating rinks are deleted from Table 1607.1 because they are not listed in Table 4-1 of ASCE 7 and Table C4-1 of ASCE-7 specifies uniform live loads of 250 psf for ice skating rinks and 100 psf for roller skating rinks. Footnote m has been added to clarify that a live load reduction is not permitted unless specific exceptions of Section 1607.9 apply. The footnote has been added at each specific use or occupancy in Table 1607.1 where a live load reduction is restricted. With the addition of this footnote, Table 1607.1 clarifies limitations on live load reduction. References are added to Sections 1607.10.1 and 1607.10.2 to correlate with the footnote.

- The 300-pound concentrated load for stair treads has been relocated from footnote f to the table and the clarification is added that the 300-pound concentrated load need not act concurrently with the uniform load.

- New loading requirements for helipads have been added to Section 1607.6. (See a detailed discussion in the commentary to Section 1607.6.)

- The terminology associated with "occupiable roofs" has been clarified and coordinated with ASCE 7-10. Occupiable roof gardens and assembly areas have a live load of 100 psf. Occupiable roofs other than roof gardens and assembly areas must be designed for appropriate loads based on use or as required by the building official. Landscaped areas of roofs that are unoccupied must be designed for a live load of 20 psf plus the weight of the landscaping and saturated soil, which is considered a dead load.

CHANGE TYPE: Modification

CHANGE SUMMARY: The terminology and live load design requirements for helicopter landing areas (helipads) have been updated and coordinated with ASCE 7-10.

2012 CODE: 202 Definitions.

HELIPAD.   A structural surface that is used for the landing, taking off, taxiing, and parking of helicopters.

~~**1605.4**~~ **1607.6 Heliports and Helistops Helipads**   ~~Heliport and helistop landing areas~~ Helipads shall be designed for the following live loads~~, combined in accordance with Section 1605~~:

1. ~~Dead load, *D*, plus the gross weight of the helicopter, *Dh*, plus snow load, *S*.~~
   3.1. ~~Dead load, *D*, plus~~ A uniform live load, *L*, ~~of:~~ as specified below. This load shall not be reduced.
   1.1 ~~100~~ 40 psf (~~4.79~~ 1.92 kN/m2), where the design basis helicopter has a maximum take-off weight of 3000 pounds (13.35 kN) or less.
   1.2 60 psf (2.87 kN/m2), where the design basis helicopter has a maximum take-off weight greater than 3000 pounds (13.35 kN).
2. A single concentrated load, *L*, of 3000 pounds (13.35 kN) applied over an area of 4.5 inches by 4.5 inches (114 mm by 114 mm) and located so as to produce the maximum load effects on the structural elements under consideration. The concentrated load is not required to act concurrently with other uniform or concentrated live loads.
   ~~2~~ **3.** ~~Dead load, *D*, plus~~ Two single concentrated ~~impact~~ loads, *L*, ~~approximately~~ 8 feet (2438 mm) apart applied ~~anywhere~~ on the ~~touchdown~~ landing pad (representing ~~each of~~ the helicopter's two main landing gear, whether skid type or wheeled type), each having a magnitude of 0.75 times the ~~gross~~ maximum take-off weight of the helicopter, and located so as to produce the maximum load effects on the structural elements under consideration. ~~Both loads acting together total 1.5 times the gross weight of the helicopter.~~ The concentrated loads shall be applied over an area of 8 inches by 8 inches (203 mm by 203 mm) and are not required to act concurrently with other uniform or concentrated live loads.

~~**Exception:**~~   Landing areas designed for a design basis ~~helicopters~~ with a gross maximum take-off weight ~~not exceeding~~ of 3000 pounds (13.3_5_ kN) ~~in accordance with Items 1 and 2 shall be permitted to be designed using a 40 psf (1.92 kN/m2) uniform live load in Item 3, provided the landing area is~~ shall be identified with a 3000-pound (13.34-kN) weight limitation. ~~This 40 psf (1.92 kN/m2) uniform live load shall not be reduced.~~ The landing area weight limitation shall be indicated by the numeral "3" (kips) located in the bottom right corner

Helicopter helipad

International Code Council®

1607.6, 202 continues

1607.6, 202 continued

of the landing area as viewed from the primary approach path. The indication for the landing area weight limitation shall be a minimum 5 feet (1524 mm) in height.

CHANGE SIGNIFICANCE: The terms "heliport" and "helistop" have been changed to "helipad" to reflect the proper terminology used to describe helicopter landing areas. The previous provisions prescribed how to combine helicopter live loads but were unclear as to whether or not these combinations were intended to be separate or in addition to the load combinations required by Section 1605. The helipad loading requirements have been relocated from Section 1605 (load combinations) to Section 1607.6, which prescribes live loads. The actual loading criteria required to design helipads were updated and coordinated with the helicopter loads specified in ASCE 7-10 and references to the dead load, *D*, and snow load, *S*, which served no real purpose, were deleted.

1607.7
Heavy Vehicle Loads

Heavy vehicle *(Courtesy of Albuquerque Fire Department)*

CHANGE TYPE: Modification

CHANGE SUMMARY: Provisions relating to the design of structures that support heavy vehicle loads in excess of 10,000 pounds gross vehicle weight (GVW) have been updated.

2012 CODE: ~~**1607.6 Truck and Bus Garages.** Minimum live loads for garages having trucks or buses shall be as specified in Table 1607.6, but shall not be less than 50 psf (2.40 kN/m²), unless other loads are specifically justified and *approved* by the *building official*. Actual loads shall be used where they are greater than the loads specified in the table.~~

~~**1607.6.1 Truck and Bus Garage Live Load Application.** The concentrated load and uniform load shall be uniformly distributed over a 10-foot (3048 mm) width on a line normal to the centerline of the lane placed within a 12-foot-wide (3658 mm) lane. The loads shall be placed within their individual lanes so as to produce the maximum stress in each structural member. Single spans shall be designed for the uniform load in Table 1607.6 and one simultaneous concentrated load positioned to produce the maximum effect. Multiple spans shall be designed for the uniform load in Table 1607.6 on the spans and two simultaneous concentrated loads in two spans positioned to produce the maximum negative moment effect. Multiple span design loads, for other effects, shall be the same as for single spans.~~

~~Table 1607.6~~
~~UNIFORM AND CONCENTRATED LOADS~~

1607.7 Heavy Vehicle Loads. <u>Floors and other surfaces that are intended to support vehicle loads greater than a 10,000-pound (44.5-kN) gross vehicle weight rating shall comply with Sections 1607.7.1 through 1607.7.5.</u>

1607.7.1 Loads. <u>Where any structure does not restrict access for vehicles that exceed a 10,000-pound gross vehicle weight rating, those portions of the structure subject to such loads shall be designed using the vehicular live loads, including consideration of impact and fatigue, in accordance with the codes and specifications required by the jurisdiction having authority for the design and construction of the roadways and bridges in the same location of the structure.</u>

1607.7.2 Fire Truck and Emergency Vehicles. <u>Where a structure, or portions of a structure, are accessed and loaded by fire department access vehicles and other similar emergency vehicles, the structure shall be designed for the greater of the following loads:</u>

1. <u>The actual operational loads, including outrigger reactions and contact areas of the vehicles as stipulated and approved by the Building Official, or</u>
2. <u>The live loading specified in Section 1607.7.1.</u>

1607.7 continues

1607.7 continued

1607.7.3 Heavy Vehicle Garages. Garages designed to accommodate vehicles that exceed a 10,000-pound gross vehicle weight rating shall be designed using the live loading specified by Section 1607.7.1. For garages the design for impact and fatigue is not required.

Exception: The vehicular live loads and load placement are allowed to be determined using the actual vehicle weights for the vehicles allowed onto the garage floors, provided such loads and placement are based on rational engineering principles and are approved by the building official, but shall not be less than 50 psf (240 kN/m^2). This live load shall not be reduced.

1607.7.4 Forklifts and Movable Equipment. Where a structure is intended to have forklifts or other movable equipment present, the structure shall be designed for the total vehicle or equipment load and the individual wheel loads for the anticipated vehicles as specified by the owner of the facility. These loads shall be posted per Section 1607.7.5.

1607.7.4.1 Impact and Fatigue. Impact loads and fatigue loading shall be considered in the design of the supporting structure. For the purposes of design, the vehicle and wheel loads shall be increased by 30 percent to account for impact.

1607.7.5 Posting. The maximum weight of the vehicles allowed into or on a garage or other structure shall be posted by the owner in accordance with Section 106.1.

CHANGE SIGNIFICANCE: Structures intended to support heavy vehicles loads in excess of 10,000 pounds GVW must now be designed in accordance with the same specifications required by the jurisdiction for the design of roadways and bridges. The new requirements specifically apply to fire truck and emergency vehicles, heavy vehicle parking garages, forklifts, and movable equipment. The owner is required to post the maximum weight of the vehicles allowed in a garage or other structure in accordance with Section 106.1.

CHANGE TYPE: Modification

CHANGE SUMMARY: A definition of "susceptible bay" has been added to identify where ponding must be considered in the design of roof structures.

2012 CODE: 202 Definitions.

SUSCEPTIBLE BAY. A roof, or portion thereof, with:

(1) A slope less than ¼ inch per foot (0.0208 rad) or
(2) On which water is impounded upon it, in whole or in part, and the secondary drainage system is functional but the primary drainage system is blocked.

A roof surface with a slope of ¼ inch per foot (0.0208 rad) or greater toward points of free drainage is not a susceptible bay.

1608.3 Ponding Instability. Susceptible bays of roofs shall be evaluated for ponding instability in accordance with Section 7.11 of ASCE 7.

1611.2 Ponding Instability. Susceptible bays of roofs shall be ~~investigated by structural analysis to ensure that they possess adequate stiffness to preclude progressive deflection~~ evaluated for ponding instability in accordance with Section 8.4 of ASCE 7.

CHANGE SIGNIFICANCE: A definition of "susceptible bay" has been added to Section 202 to coordinate the ponding requirements in the IBC with those of ASCE 7-10. Only those portions of the roof where the slope is less than ¼ inch per foot or where water is impounded are considered susceptible bays and, as such, must be designed for ponding to avoid progressive deflection. Areas of the roof with a slope of ¼ inch per foot or greater toward points of free drainage are not considered susceptible and need not be designed for ponding.

1608.3, 1611.2, 202
Ponding Instability

International Code Council®

Rain water ponding on a roof

1609, 202

Determination of Wind Loads

$$V_{asd} = V_{ult}\sqrt{0.6}$$

Equation 16-33, conversion of wind speed from V_{ult} to V_{ASD}

CHANGE TYPE: Modification

CHANGE SUMMARY: The wind design requirements of Section 1609 have been updated and coordinated with the latest wind load provisions in ASCE/SEI 7 (ASCE 7-10) and the wind load maps in the IBC are now based on ultimate design wind speeds, V_{ult}, which produce a strength level wind load similar to seismic load effects.

2012 CODE: The following are excerpted portions of the subject code text. The entire code change is not shown here for brevity.

202 Definitions.

HURRICANE-PRONE REGIONS.   Areas vulnerable to hurricanes defined as:

1. The U. S. Atlantic Ocean and Gulf of Mexico coasts where the ~~basic~~ ultimate design wind speed, V_{ult}, for Risk Category II buildings is greater than 115 ~~90~~ mph (51.4 m/s).
2. Hawaii, Puerto Rico, Guam, Virgin Islands, and American Samoa.

WINDBORNE DEBRIS REGION.   Areas within ~~Portions of~~ hurricane-prone regions located: ~~that are~~

1. Within 1 mile (1.61 km) of the coastal mean high water line where the ~~basic~~ ultimate design wind speed, V_{ult}, is 130 ~~110~~ mph (58 m/s) or greater; or
2. In areas ~~portions of hurricane-prone regions~~ where the ~~basic~~ ultimate design wind speed, V_{ult}, is 140 ~~120~~ mph (63.6 m/s) or greater; or Hawaii.

For Risk Category II buildings and structures and Risk Category III buildings and structures, except health care facilities, the windborne debris region shall be based on Figure 1609A. For Risk Category IV buildings and structures and Risk Category III health care facilities, the windborne debris region shall be based on Figure 1609B.

WIND SPEED, V_{ult}.   Ultimate design wind speeds.

WIND SPEED, V_{ASD}.   Nominal design wind speeds.

1609.1.1 Determination of Wind Loads.   Wind loads on every building or structure shall be determined in accordance with Chapters ~~6~~ 26 to 30 of ASCE 7 or provisions of the alternate all-heights method in Section 1609.6. The type of opening protection required, the ~~basic~~ ultimate design wind speed, V_{ult}, and the exposure category for a site is permitted to be determined in accordance with Section 1609 or ASCE 7. Wind shall be assumed to come from any horizontal direction, and wind pressures shall be assumed to act normal to the surface considered.

Exceptions:

1. Subject to the limitations of Section 1609.1.1.1, the provisions of ICC 600 shall be permitted for applicable Group R-2 and R-3 buildings.

2. Subject to the limitations of Section 1609.1.1.1, residential structures using the provisions of the AF&PA WFCM.

3. Subject to the limitations of Section 1609.1.1.1, residential structures using the provisions of AISI S230.

4. Designs using NAAMM FP 1001.

5. Designs using TIA-222 for antenna-supporting structures and antennas, <u>provided the extent of Topographic Category 2, escarpments, in Section 2.6.6.2 of TIA-222 shall extend 16 times the height of the escarpment.</u>

6. Wind tunnel tests in accordance with ~~Section 6.6~~ <u>Chapter 31</u> of ASCE 7~~, subject to the limitations in Section 1609.1.1.2.~~

<u>The wind speeds in Figure 1609A, 1609B, and 1609C are ultimate design wind speeds, V_{ult}, and shall be converted in accordance with Section 1609.3.1 to nominal design wind speeds, V_{asd}, when the provisions of the standards referenced in Exceptions 1 through 5 are used.</u>

~~1609.1.1.2 Wind Tunnel Test Limitations.~~ ~~The lower limit on pressures for main wind-force-resisting systems and components and cladding shall be in accordance with Sections 1609.1.1.2.1 and 1609.1.1.2.2.~~

~~1609.1.1.2.1 Lower Limits on Main Wind-Force-Resisting System.~~ ~~Base overturning moments determined from wind tunnel testing shall be limited to not less than 80 percent of the design base overturning moments determined in accordance with Section 6.5 of ASCE 7, unless specific testing is performed that demonstrates it is the aerodynamic coefficient of the building, rather than shielding from other structures, that is responsible for the lower values. The 80-percent limit shall be permitted to be adjusted by the ratio of the frame load at critical wind directions as determined from wind tunnel testing without specific adjacent buildings, but including appropriate upwind roughness, to that determined in Section 6.5 of ASCE 7.~~

~~1609.1.1.2.2 Lower Limits on Components and Cladding.~~ ~~The design pressures for components and cladding on walls or roofs shall be selected as the greater of the wind tunnel test results or 80 percent of the pressure obtained for Zone 4 for walls and Zone 1 for roofs as determined in Section 6.5 of ASCE 7, unless specific testing is performed that demonstrates it is the aerodynamic coefficient of the building, rather than shielding from nearby structures, that is responsible for the lower values. Alternatively, limited tests at a few wind directions without specific adjacent buildings, but in the presence of an appropriate upwind roughness, shall be permitted to be used to demonstrate that the lower pressures are due to the shape of the building and not to shielding.~~

1609 continues

1609 continued

1609.1.2 Protection of Openings.   In windborne debris regions, glazing in buildings shall be impact resistant or protected with an impact-resistant covering meeting the requirements of an approved impact-resistant standard or ASTM E1996 and ASTM E1886 referenced herein as follows:

1. Glazed openings located within 30 feet (9144 mm) of grade shall meet the requirements of the Large Missile Test of ASTM E1996.

2. Glazed openings located more than 30 feet (9144 mm) above grade shall meet the provisions of the small missile test of ASTM E1996.

Exceptions:

1. Wood structural panels with a minimum thickness of $\frac{7}{16}$ inch (11.1 mm) and maximum panel span of 8 feet (2438 mm) shall be permitted for opening protection in one- and two-story buildings classified as Group R-3 or R-4 occupancy. Panels shall be precut so that they shall be attached to the framing surrounding the opening containing the product with the glazed opening. Panels shall be predrilled as required for the anchorage method and shall be secured with the attachment hardware provided. Attachments shall be designed to resist the components and cladding loads determined in accordance with the provisions of ASCE 7, with corrosion-resistant attachment hardware provided and anchors permanently installed on the building. Attachment in accordance with Table 1609.1.2 with corrosion-resistant attachment hardware provided and anchors permanently installed on the building is permitted for buildings with a mean roof height of 45 feet (13716 mm) or less where $\underline{V_{asd}}$ determined in accordance with Section 1609.3.1 ~~wind speeds do~~ does not exceed 140 mph (63 m/s).

2. Glazing in ~~Occupancy~~ Risk Category I buildings as defined in Section 1604.5, including greenhouses that are occupied for growing plants on a production or research basis, without public access shall be permitted to be unprotected.

3. Glazing in ~~Occupancy~~ Risk Category II, III, or IV buildings located over 60 feet (18288 mm) above the ground and over 30 feet (9144 mm) above aggregate surface roofs located within 1500 feet (458 m) of the building shall be permitted to be unprotected.

1609.3 Basic Wind Speed.   The ~~basic~~ underline{ultimate design} wind speed, $\underline{V_{ult}}$, in mph, for the determination of the wind loads shall be determined by ~~Figure 1609~~ Figures 1609A, 1609B, and 1609C. The ultimate design wind speed, V_{ult}, for use in the design of Risk Category II buildings and structures shall be obtained from Figure 1609A. The ultimate design wind speed, V_{ult}, for use in the design of Risk Category III and IV buildings and structures shall be obtained from Figure 1609B. The ultimate design wind speed, V_{ult}, for use in the design of Risk Category I buildings and structures shall be obtained from Figure 1609C. ~~Basic~~ The ultimate design wind speed, $\underline{V_{ult}}$, for the special wind regions indicated, near mountainous terrain and near gorges shall be in accordance with local jurisdiction requirements. ~~Basic~~ The ultimate design wind speeds, $\underline{V_{ult}}$, determined by the local jurisdiction shall be in accordance with Section 26.5.1 ~~6.5.4~~ of ASCE 7.

In nonhurricane-prone regions, when the ~~basic~~ <u>ultimate design</u> wind speed, V_{ult}, is estimated from regional climatic data, the ~~basic~~ <u>ultimate design</u> wind speed, V_{ult}, shall be ~~not less than the wind speed associated with an annual probability of 0.02 (50-year mean recurrence interval), and the estimate shall be adjusted for equivalence to a 3-second gust wind speed at 33 feet (10 m) above ground in Exposure Category C. The data analysis shall be performed~~ <u>determined</u> in accordance with Section <u>26.5.3</u> ~~6.5.4.2~~ of ASCE 7.

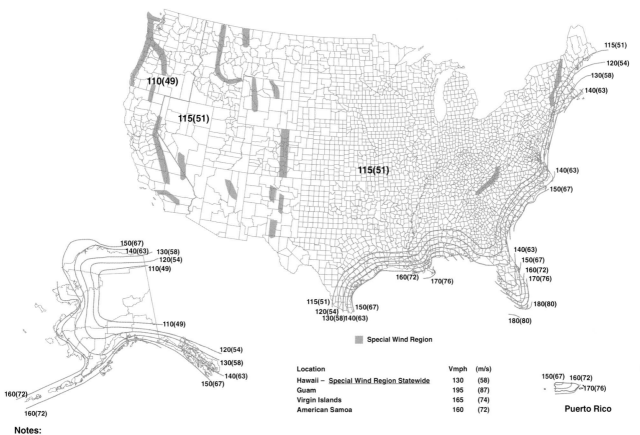

Special Wind Region

Location	Vmph	(m/s)
Hawaii – <u>Special Wind Region Statewide</u>	130	(58)
Guam	195	(87)
Virgin Islands	165	(74)
American Samoa	160	(72)

Puerto Rico

Notes:
1. Values are nominal design 3-second gust wind speeds in miles per hour (m/s) at 33 ft (10m) above ground for Exposure C category.
2. Linear interpolation between contours is permitted.
3. Islands and coastal areas outside the last contour shall use the last wind speed contour of the coastal area.
4. Mountainous terrain, gorges, ocean promontories, and special wind regions shall be examined for unusual wind conditions.
5. <u>Wind speeds correspond to approximately a 7% probability of exceedance in 50 years (Annual Exceedance Probability = 0.00143, MRI = 700 years).</u>

International Code Council®

<u>**Figure 1609A Ultimate Design Wind Speeds, V_{ult}, For Risk Category II Buildings and Other Structures**</u>

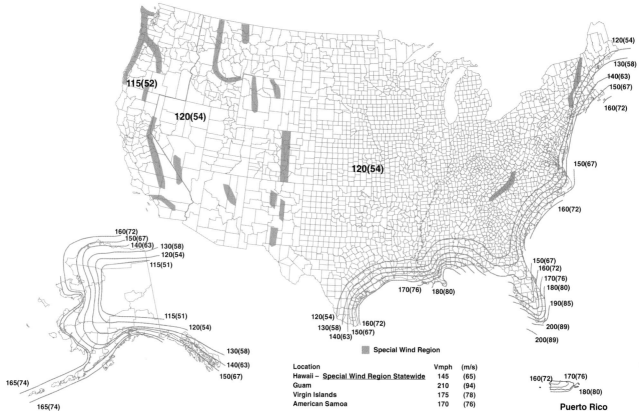

International Code Council®

Special Wind Region

Location	Vmph	(m/s)
Hawaii – Special Wind Region Statewide	145	(65)
Guam	210	(94)
Virgin Islands	175	(78)
American Samoa	170	(76)

Notes:

1. Values are nominal design 3-second gust wind speeds in miles per hour (m/s) at 33 ft (10m) above ground for Exposure C category.

2. Linear interpolation between contours is permitted.

3. Islands and coastal areas outside the last contour shall use the last wind speed contour of the coastal area.

4. Mountainous terrain, gorges, ocean promontories, and special wind regions shall be examined for unusual wind conditions.

5. Wind speeds correspond to approximately a 3% probability of exceedance in 50 years (Annual Exceedance Probability = 0.000588, MRI = 1700 years).

Figure 1609B Ultimate Design Wind Speeds, V_{ult}, For Risk Categories III And IV Buildings and Other Structures

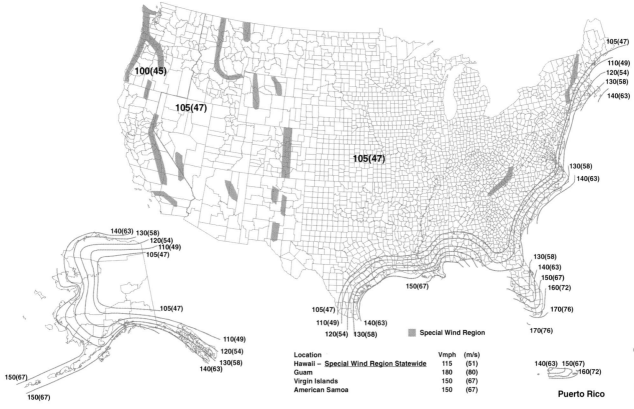

Notes:

1. Values are nominal design 3-second gust wind speeds in miles per hour (m/s) at 33 ft (10m) above ground for Exposure C category.
2. Linear interpolation between contours is permitted.
3. Islands and coastal areas outside the last contour shall use the last wind speed contour of the coastal area.
4. Mountainous terrain, gorges, ocean promontories, and special wind regions shall be examined for unusual wind conditions.
5. Wind speeds correspond to approximately a 15% probability of exceedance in 50 years (Annual Exceedance Probability = 0.00333, MRI = 300 years).

Figure 1609C Ultimate Design Wind Speeds, V_{ult}, For Risk Category I Buildings and Other Structures

1609 continued

1609.3.1 Wind speed conversion. When required, the ~~3-second gust basic~~ ultimate design wind speeds of Figure 1609A, B, and C shall be converted to nominal design wind speeds, V_{asd}, ~~fastest-mile wind speeds, Vfm~~, using Table 1609.3.1 or Equation 16-33.

$$\overline{V_{fm}} = \frac{V_{3S} - 10.5}{1.05}$$ (Equation 16-33)

~~where:~~

~~V_{3S} = 3-second gust basic wind speed from Figure 1609.~~
$V_{asd} = V_{ult}\sqrt{0.6}$

Where:
V_{asd} = nominal design wind speed applicable to methods specified in Exceptions 1 through 5 of Section 1609.1.1
V_{ult} = ultimate design wind speeds determined from Figures 1609A, 1609B, or 1609C

~~TABLE 1609.3.1~~ ~~Equivalent Basic Wind Speeds~~[a,b,c]

~~V_{3S}~~	~~85~~	~~90~~	~~100~~	~~105~~	~~110~~	~~120~~	~~125~~	~~130~~	~~140~~	~~145~~	~~150~~	~~160~~	~~170~~
~~V_{fm}~~	~~71~~	~~76~~	~~85~~	~~90~~	~~95~~	~~104~~	~~109~~	~~114~~	~~123~~	~~128~~	~~133~~	~~142~~	~~152~~

~~For SI: 1 mile per hour = 0.44 m/s.~~
~~a. Linear interpolation is permitted.~~
~~b. V3S is the 3-second gust wind speed (mph).~~
~~c. Vfm is the fastest mile wind speed (mph).~~

TABLE 1609.3.1 **Wind Speed Conversions[a,b,c]**

V_{ul}	100	110	120	130	140	150	160	170	180	190	200
V_{asd}	78	85	93	101	108	116	124	132	139	147	155

a. Linear interpolation is permitted
b. V_{asd} = nominal design wind speed applicable to methods specified in Exceptions 1 through 5 of Section 1609.1.1
c. V_{ult} = ultimate design wind speeds determined from Figures 1609A, 1609B, or 1609C

Because this code change affected substantial portions of Chapters 16 and 17, the entire code change text is too extensive to be included here. Refer to Code Change S84-09/10 in the *2012 IBC Code Changes Resource Collection* for the complete text and history of the code change.

CHANGE SIGNIFICANCE: The most significant aspect of the wind design change is that the wind speed maps in the 2012 IBC were updated to those adopted in ASCE 7-10. Over the past 10 years, new research has indicated that the hurricane wind speeds provided in ASCE 7-05 have been too conservative and should be adjusted downward. As more hurricane data became available, it was also recognized that substantial improvements could be made to the hurricane simulation model used to develop the wind speed maps. The new data resulted in an improved representation of the hurricane wind field, including the modeling of the sea–land transition and the hurricane boundary layer height; new models for hurricane weakening after landfall; and an improved statistical model for the Holland *B* parameter, which controls the wind pressure relationship. Although the new hurricane hazard model yields hurricane wind speeds that are lower than those given in the 2009 IBC and ASCE 7-05, the overall rate of intense storms (as defined by central pressure) produced by the new model increased compared to those produced by the hurricane simulation model used to develop previous wind speed maps.

In developing the new wind speed maps, it was decided to use multiple ultimate event or strength design based maps in conjunction with a wind load factor of 1.0 for strength design. For allowable stress design (ASD), the load factor has been reduced from 1.0 to 0.6, thus the load combinations in Section 1605 had to be modified accordingly. Several important factors related to more accurate wind load determination were considered that led to the decision to move to strength based ultimate event wind loads:

1. An ultimate event or strength design wind speed map makes the overall approach consistent with the well-established strength-based seismic design procedure in that both wind and seismic load effects are mapped as ultimate events and use a load factor of 1.0 for the strength design load combinations.

2. Utilizing different maps for the different risk categories eliminates previous issues associated with using importance factors that vary according to the risk (occupancy) category of the building. The different importance factors in ASCE 7-05 for hurricane prone versus non-hurricane prone regions for Risk (Occupancy) Category I structures prompted many questions by code users. This is no longer an issue in ASCE 7-10 because Risk Category I, Risk Category II, and Risk Category III and IV have separate wind speed maps, and the importance factor no longer appears in the velocity pressure equation. Note that the importance factor for wind in ASCE 7 Table 1.5-2 is now 1.00 for all risk categories.

3. The use of multiple maps based on risk category eliminates some confusion associated with the recurrence interval associated with the previous wind speed map in ASCE 7-05 because it was not a uniform 50-year return period map. This results in a situation where the level of safety achieved by the overall design was not consistent along the hurricane coast. The wind maps in ASCE/SEI 7-10 have a mean recurrence interval (MRI) of 300 years for Risk Category I, 700 years for Risk Category II, and 1700 years for Risk Categories III and IV.

As a result of the new strength-based wind speed, new terminology was introduced into the 2012 IBC. The former term "basic wind speed" has been changed to "ultimate design wind speed" and is designated V_{ult}. The wind speed that is equivalent to the former basic wind speed is now called the nominal design wind speed, V_{asd}, and the conversion between the two is given by Equation 16-33 as,

$$V_{asd} = V_{ult}\sqrt{0.6}$$

The conversion from V_{asd} to V_{ult} is a result of the wind load being proportional to the square of the velocity pressure and the ASD wind load being 0.6 times the strength level ultimate wind load. Thus,

$$W \cong V^2$$
$$W_{asd} = 0.6W_{ult}$$
$$V_{asd}^2 = 0.6V_{ult}^2$$
$$V_{asd} = \sqrt{0.6}V_{ult}$$

1609 continues

1609 continued

It should also be noted that the term "basic wind speed" in ASCE 7-10 corresponds to the "ultimate design wind speed" in the 2012 IBC.

Because many different code provisions in the code are based upon wind speed, it was necessary to modify the wind speed conversion section so that the many provisions triggered by wind speed were not changed. The terms "ultimate design wind speed" and "nominal design wind speed" were incorporated in numerous locations to help the code user distinguish between them. In cases where wind speed is used to trigger a requirement, the ultimate wind speed, V_{ult}, must be converted to an equivalent wind speed that corresponds to the former basic wind speed. Thus, a new table in the 2012 IBC converts V_{ult} to V_{ASD} so that the mapped wind speed thresholds in various parts of the code can still be used:

V_{ult}	100	110	120	130	140	150	160	170	180	190	200
V_{asd}	78	85	93	101	108	116	124	132	139	147	155

For example, in a case where the 2009 IBC imposed requirements where the basic wind speed exceeds 100 mph, the 2012 IBC imposes the requirements where V_{asd} exceeds 100 mph. A nominal design speed, V_{asd}, equal to 100 mph corresponds to an ultimate design wind speed, V_{ult}, equal to 129 mph. The following table (which is not in the IBC) may be more useful to the code user because it gives V_{ult} in terms of V_{asd} in increments of 10 mph:

V_{asd}	85	90	100	110	120	130	140	150
V_{ult}	110	115	126	139	152	164	177	190

For a comparison of ASCE 7-93 fastest mile wind speeds, ASCE 7-05 3-second gust ASD wind speeds, and ASCE 7-10 3-second gust wind speeds, refer to Table C26.5-6 of the ASCE 7-10 commentary. Note that the conversion in ASCE 7-10 is given by $V_{ult} = V_{asd}\sqrt{1.6}$, which produces slightly different values than IBC Equation 16-33.

Beyond the adoption of the new strength design wind speed maps, ASCE/SEI 7-10 also includes a new simplified method for use in the determination of wind loads for buildings up to 160 feet in height. In addition, the wind load calculation provisions that were contained in Chapter 6 of ASCE/SEI 7-05 have been reorganized into six separate chapters (26 through 31) for improved clarity and ease of use. This is similar to the reorganization in ASCE 7-05 where the seismic design provisions were divided into several chapters to facilitate use. This reorganization into multiple chapters required several coordination revisions to the code text.

A few other changes to the wind design provisions in Section 1609 are worth noting:

- To use any of the five standards referenced in the exception in Section 1609.1.1, the ultimate design wind speed must be determined based on the risk category of the building then converted to the nominal design wind speed.
- Wind tunnel test limitations in 2009 IBC Section 2309.1.2 were deleted from the IBC because they are incorporated into Chapter 31 of ASCE 7-10.

- The hurricane-prone region is redefined in terms of the ultimate design wind speed as shown on the Risk Category II wind speed map.

- The windborne debris region is now defined in terms of the ultimate design wind speed and determined from the appropriate risk category wind speed map. For example, for Risk Category II and III buildings and structures, except health care facilities, the windborne debris region is based on Figure 1609A. For Risk Category IV buildings and structures and Risk Category III health care facilities, the windborne debris region is based on Figure 1609B.

- The ultimate design wind speed, V_{ult}, for the special wind regions indicated, near mountainous terrain and near gorges is to be determined in accordance with local jurisdiction requirements and in accordance with Section 26.5.1 of ASCE 7. In nonhurricane-prone regions, when the ultimate design wind speed is estimated from regional climatic data, V_{ult}, is to be determined in accordance with Section 26.5.3 of ASCE 7.

It should be noted that the alternate all-heights wind design procedure is maintained in the 2012 IBC but was updated to conform to the new ultimate wind design procedure in ASCE 7-10.

1613.3.1, 202

Mapped Acceleration Parameters

CHANGE TYPE: Modification

CHANGE SUMMARY: The IBC seismic ground motion maps have been updated to reflect the 2008 maps developed by the United States Geological Survey (USGS) National Seismic Hazard Mapping Project and the technical changes adopted for the 2009 *NEHRP Recommended Seismic Provisions for New Buildings and Other Structures* (FEMA P750).

2012 CODE: 202 Definitions.

RISK-TARGETED MAXIMUM CONSIDERED EARTHQUAKE (MCE$_R$) GROUND MOTION RESPONSE ACCELERATIONS. The most severe earthquake effects considered by this code, <u>determined for the orientation that results in the largest maximum response to horizontal ground motions and, with adjustment for targeted risk.</u>

~~1613.5.1~~ **1613.3.1 Mapped Acceleration Parameters.** The parameters S_S and S_1 shall be determined from the 0.2- and 1-second spectral response accelerations shown on Figures 1613.5(1) through 1613.5(~~14~~6), respectively. Where S_1 is less than or equal to 0.04 and S_S is less than or equal to 0.15, the structure is permitted to be assigned to Seismic Design Category A. The parameters S_S and S_1 shall be, respectively, 1.5 and 0.6 for Guam and 1.0 and 0.4 for American Samoa.

Because this code change affected substantial portions of Chapters 16, the entire code change text is too extensive to be included here. Refer to Code Change S97-09/10 in the *2012 IBC Code Changes Resource Collection* for the complete text and history of the code change.

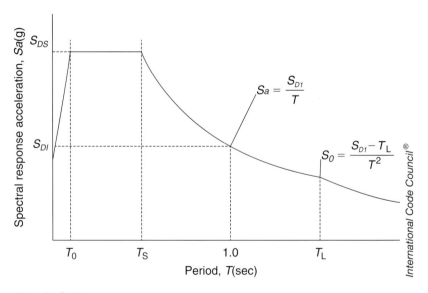

Seismic design response spectrum

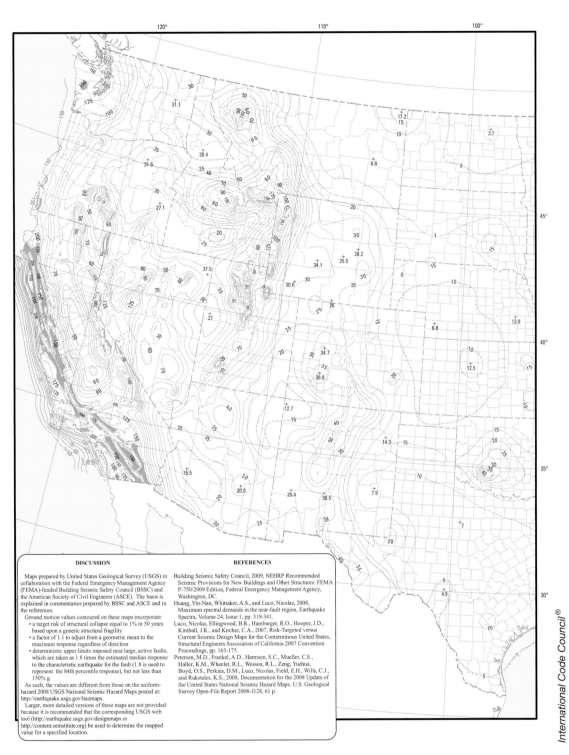

DISCUSSION

Maps prepared by United States Geological Survey (USGS) in collaboration with the Federal Emergency Management Agency (FEMA)-funded Building Seismic Safety Council (BSSC) and the American Society of Civil Engineers (ASCE). The basis is explained in commentaries prepared by BSSC and ASCE and in the references.

Ground motion values contoured on these maps incorporate:
• a target risk of structural collapse equal to 1% in 50 years based upon a generic structural fragility
• a factor of 1.1 to adjust from a geometric mean to the maximum response regardless of direction
• deterministic upper limits imposed near large, active faults, which are taken as 1.8 times the estimated median response to the characteristic earthquake for the fault (1.8 is used to represent the 84th percentile response), but not less than 150% g.

As such, the values are different from those on the uniform-hazard 2008 USGS National Seismic Hazard Maps posted at: http://earthquake.usgs.gov/hazmaps.

Larger, more detailed versions of these maps are not provided because it is recommended that the corresponding USGS web tool (http://earthquake.usgs.gov/designmaps or http://content.seinstitute.org) be used to determine the mapped value for a specified location.

REFERENCES

Building Seismic Safety Council, 2009, NEHRP Recommended Seismic Provisions for New Buildings and Other Structures: FEMA P-750/2009 Edition, Federal Emergency Management Agency, Washington, DC.

Huang, Yin-Nan, Whittaker, A.S., and Luco, Nicolas, 2008, Maximum spectral demands in the near-fault region, Earthquake Spectra, Volume 24, Issue 1, pp. 319-341.

Luco, Nicolas, Ellingwood, B.R., Hamburger, R.O., Hooper, J.D., Kimball, J.K., and Kircher, C.A., 2007, Risk-Targeted versus Current Seismic Design Maps for the Conterminous United States, Structural Engineers Association of California 2007 Convention Proceedings, pp. 163-175.

Petersen, M.D., Frankel, A.D., Harmsen, S.C., Mueller, C.S., Haller, K.M., Wheeler, R.L., Wesson, R.L., Zeng, Yuehua, Boyd, O.S., Perkins, D.M., Luco, Nicolas, Field, E.H., Wills, C.J., and Rukstales, K.S., 2008, Documentation for the 2008 Update of the United States National Seismic Hazard Maps: U.S. Geological Survey Open-File Report 2008-1128, 61 p.

International Code Council®

Figure ~~1613.5(1)~~ **1613.3.1(1)** Risk-Targeted Maximum Considered Earthquake (MCE$_R$) Ground Motion Response Accelerations **for the Conterminous United States of 0.2-Second Spectral Response Acceleration (5% of Critical Damping), Site Class B.**

1613.3.1, 202 continues

1613.3.1, 202 continued

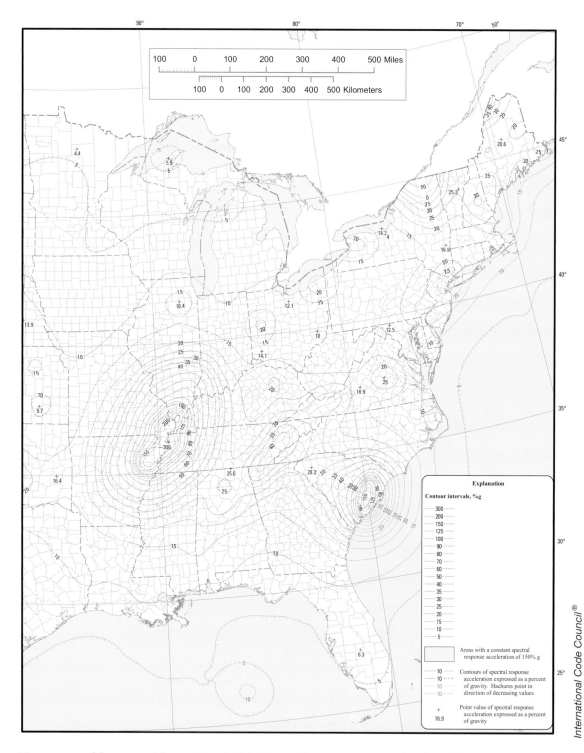

Figure ~~1613.5(1)~~ 1613.3.1(1) - continued Risk-Targeted Maximum Considered Earthquake (MCE_R) Ground Motion Response Accelerations for the Conterminous United States of 0.2-Second Spectral Response Acceleration (5% of Critical Damping), Site Class B.

International Code Council®

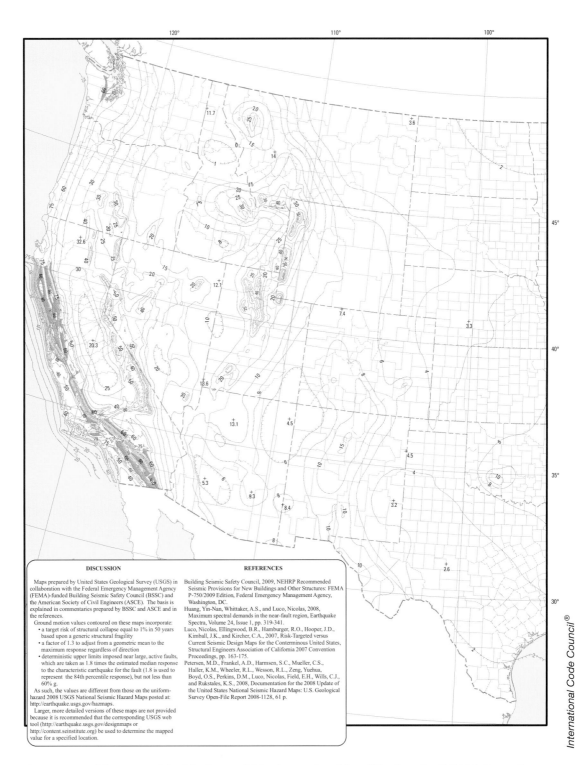

Figure ~~1613.5(2)~~ 1613.3.1(2) Risk-Targeted Maximum Considered Earthquake (MCE_R) Ground Motion Response Accelerations for the Conterminous United States of 1-Second Spectral Response Acceleration (5% of Critical Damping), Site Class B.

1613.3.1, 202 continues

1613.3.1, 202 continued

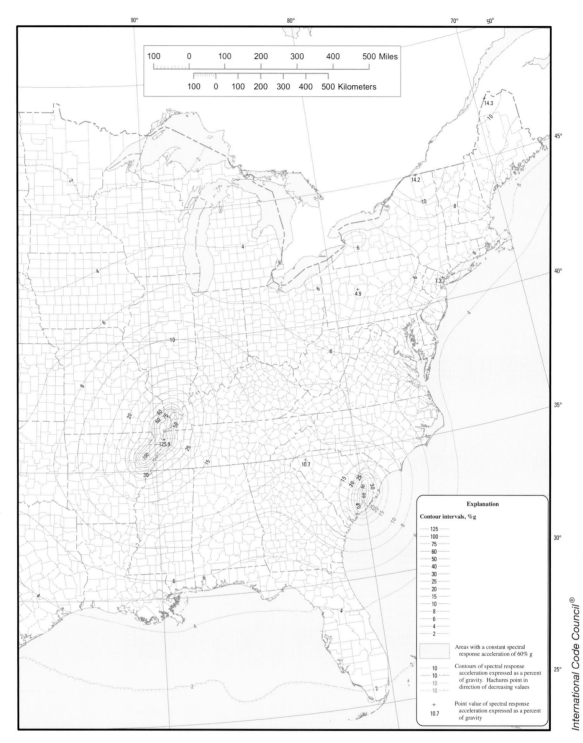

Figure ~~1613.5(2)~~ 1613.3.1(2) - continued <u>Risk-Targeted</u> Maximum Considered Earthquake (MCE_R) Ground Motion <u>Response Accelerations</u> for the Conterminous United States of 1-Second Spectral Response Acceleration (5% of Critical Damping), Site Class B.

International Code Council®

CHANGE SIGNIFICANCE: The USGS National Seismic Hazard Mapping Project and the technical changes adopted for the 2009 NEHRP (FEMA P750) are part of the ongoing federal effort to make the most current earthquake hazard information available to users of the IBC. The 2008 USGS seismic hazard maps incorporate new information on earthquake sources and ground motion prediction equations including the new Next-Generation Attenuation (NGA) relations. The ground motion maps in the 2012 IBC also incorporate technical changes adopted for the 2009 *NEHRP Provisions* that include: (1) use of risk-targeted ground motions, (2) the maximum direction ground motions, and (3) near-source 84th percentile ground motions.

The titles of the maps in the IBC were revised from the former "Maximum Considered Earthquake Ground Motion" to "Risk-Targeted Maximum Considered Earthquake (MCE_R) Ground Motion Response Accelerations" to reflect the changed titles in the 2009 NEHRP and the 2010 edition of *Minimum Design Loads for Buildings and Other Structures*, ASCE/SEI 7 (ASCE 7-10). The number of printed maps in the IBC was reduced from 14 to 6. Although the maps in the IBC are generally illustrative of the earthquake hazard, the contours in some regions cannot be read clearly enough to provide exact design values for specific sites. Precise seismic design values can be obtained from the USGS website (http://earthquake.usgs.gov/) using the longitude and latitude of the building site. Although the seismic design maps are based on corresponding USGS National Seismic Hazard Maps, their values typically differ from the hazard map values. Thus, engineers involved in seismic design of buildings and structures should generally use the maps, data, and tools presented here rather than other hazard map values presented elsewhere on the USGS website. The latitude and longitude of proposed building sites can be obtained from GPS mapping programs or websites such as topozone.com or trails.com.

Detailed descriptions of changes made for the 2009 *NEHRP Recommended Seismic Provisions* developed by the Building Seismic Safety Council and funded by the Federal Emergency Management Agency (FEMA) that served as the basis for the seismic design provisions in the 2012 IBC and ACSE/SEI 7-10 (by reference) are available at www.nibs.org or www.bssconline.org under the explanation of changes made for the 2009 edition of the *Provisions*.

1613.4

Alternatives to ASCE 7

CHANGE TYPE: Modification

CHANGE SUMMARY: Many of the alternatives to ASCE 7-05 that were in the 2009 IBC have been deleted because they have been incorporated into ASCE 7-10.

2012 CODE: ~~1613.6~~ __1613.4__ **Alternatives to ASCE 7.** The provisions of Section ~~1613.6~~ __1613.4__ shall be permitted as alternatives to the relevant provisions of ASCE 7.

~~**1613.6.1 Assumption of Flexible Diaphragm.**~~ ~~Add the following text at the end of Section 12.3.1.1 of ASCE 7.~~

~~Diaphragms constructed of wood structural panels or untopped steel decking shall also be permitted to be idealized as flexible, provided all of the following conditions are met:~~

1. ~~Toppings of concrete or similar materials are not placed over wood structural panel diaphragms except for nonstructural toppings no greater than 1-1/2 inches (38 mm) thick.~~

2. ~~Each line of vertical elements of the seismic-force-resisting system complies with the allowable story drift of Table 12.12-1.~~

3. ~~Vertical elements of the seismic-force-resisting system are light-framed walls sheathed with wood structural panels rated for shear resistance or steel sheets.~~

ASCE 7-10 *(Courtesy of American Society of Civil Engineers)*

4. Portions of wood structural panel diaphragms that cantilever beyond the vertical elements of the lateral-force-resisting system are designed in accordance with Section 4.2.5.2 of AF&PA SDPWS.

1613.6.3 Automatic Sprinkler Systems. Automatic sprinkler systems designed and installed in accordance with NFPA 13 shall be deemed to meet the requirements of Section 13.6.8 of ASCE 7.

1613.6.4 Autoclaved Aerated Concrete (AAC) Masonry Shear Wall Design Coefficients and System Limitations. Add the following text at the end of Section 12.2.1 of ASCE 7:

For ordinary reinforced AAC masonry shear walls used in the seismic force-resisting system of structures, the response modification factor, R, shall be permitted to be taken as 2, the deflection application factor, C_d, shall be permitted to be taken as 2, and the system overstrength factor, Ω_o, shall be permitted to be taken as 2-1/2.

Ordinary reinforced AAC masonry shear walls shall not be limited in height for buildings assigned to Seismic Design Category B, shall be limited in height to 35 feet (10 668 mm) for buildings assigned to Seismic Design Category C, and are not permitted for buildings assigned to Seismic Design Categories D, E and F.

For ordinary plain (unreinforced) AAC masonry shear walls used in the seismic-force-resisting system of structures, the response modification factor, R, shall be permitted to be taken as 1½, the deflection application factor, C_d, shall be permitted to be taken as 1-1/2, and the system overstrength factor, Ω_o, shall be permitted to be taken as 2-1/2. Ordinary plain (unreinforced) AAC masonry shear walls shall not be limited in height for buildings assigned to Seismic Design Category B and are not permitted for buildings assigned to Seismic Design Categories C, D, E and F.

1613.6.5 Seismic Controls for Elevators. Seismic switches in accordance with Section 8.4.10 of ASME A17.1 shall be deemed to comply with Section 13.6.10.3 of ASCE 7.

1613.6.6 Steel Plate shear wall height limits. Modify Section 12.2.5.4 of ASCE 7 to read as follows:

12.2.5.4 Increased Building Height limit for Steel-braced Frames, Special Steel Plate Shear Walls and Special Reinforced Concrete Shear walls. The height limits in Table 12.2-1 are permitted to be increased from 160 feet (48 768 mm) to 240 feet (75 152 mm) for structures assigned to *Seismic Design Category* D or E and from 100 feet (30 480 mm) to 160 feet (48 768 mm) for structures assigned to *Seismic Design Category* F that have steel-braced frames, special steel plate shear walls or special reinforced concrete cast-in-place shear walls and that meet both of the following requirements:

1. The structure shall not have an extreme torsional irregularity as defined in Table 12.2-1 (horizontal structural irregularity Type 1b).

2. The braced frames or shear walls in any one plane shall resist no more than 60 percent of the total seismic forces in each direction, neglecting accidental torsional effects.

1613.4 continues

1613.4 continued

1613.6.7 Minimum Distance for Building Separation. All buildings and structures shall be separated from adjoining structures. Separations shall allow for the maximum inelastic response displacement (δ_M). δ_M shall be determined at critical locations with consideration for both translational and torsional displacements of the structure using Equation 16-44.

$$\delta_M = \frac{C_d \cdot \delta_{max}}{I} \qquad \textbf{(Equation 16-44)}$$

Where

C_d = Deflection amplification factor in Table 12.2-1 of ASCE 7.
δ_{max} = Maximum displacement defined in Section 12.8.4.3 of ASCE 7.
I = Importance factor in accordance with Section 11.5.1 of ASCE 7

Adjacent buildings on the same property shall be separated by a distance not less than δ_{MT}, determined by Equation 16-45.

$$\delta_{MT} = \sqrt{(\delta_{M1})^2 + (\delta_{M2})^2} \qquad \textbf{(Equation 16-45)}$$

Where

δ_{M1}, δ_{M2} = The maximum inelastic response displacements of the adjacent buildings in accordance with Equation 16-44.
Where a structure adjoins a property line not common to a public way, the structure shall also be set back from the property line by not less than the maximum inelastic response displacement, δ_M, of that structure.

Exceptions:

1. Smaller separations or property line setbacks shall be permitted when justified by rational analyses.

2. Buildings and structures assigned to the Seismic Design Category A, B or C.

1613.6.8 HVAC Ductwork with I_p = 1.5. Seismic supports are not required for HVAC ductwork with I_p = 1.5 if either of the following conditions is met for the full length of each duct run:

1. HVAC ducts are suspended from hangers 12 inches (305 mm) or less in length with hangers detailed to avoid significant bending of the hangers and their attachments, or

2. HVAC ducts have a cross-sectional area of less than 6 square feet (0.557 m^2).

1613.7 ASCE 7, Section 11.7.5. Modify ASCE 7, Section 11.7.5 to read as follows:

11.7.5 Anchorage of Walls. Walls shall be anchored to the roof and all floors and members that provide lateral support for the wall or that are supported by the wall. The anchorage shall provide a direct connection between the walls and the roof or floor construction. The connections shall be capable of resisting the forces specified in Section 11.7.3 applied horizontally, substituted for E in the load combinations of Section 2.3 or 2.4.

CHANGE SIGNIFICANCE: Sections 1613.6.1, 1613.6.3 through 1613.6.8, and 1613.7 in the 2009 IBC provide alternative amendments to various requirements in ASCE 7-05. These alternatives are no longer necessary because similar provisions were incorporated into the 2010 edition of ASCE 7. The following table provides a cross reference to where the deleted sections in the 2009 IBC can be found in ASCE 7-10:

Section Deleted in 2009 IBC	Corresponding Section in ASCE 7-10
1613.6.1 Assumption of flexible diaphragm	12.3.1.1 Flexible Diaphragm Condition
1613.6.3 Automatic sprinkler systems	13.6.8.2 Fire Protection Sprinkler Piping Systems
1613.6.4 Autoclaved aerated concrete (AAC) masonry shear wall design coefficients and system limitations	Table 12.2-1 Design Coefficients and Factors for Seismic Force-Resisting Systems - Items 13 and 14
1613.6.5 Seismic controls for elevators	13.6.10 Elevator and Escalator Design Requirements
1613.6.6 Steel plate shear wall height limits	Section 12.2.5.4 Increased Structural Height Limit for Steel Eccentrically Braced Frames, Steel Concentrically Braced Frames, Steel Buckling-restrained Braced Frames, Steel Special Plate Shear Walls and Special Reinforced Concrete Shear Walls
1613.6.7 Minimum distance for building separation	12.12.3 Structural Separation
1613.6.8 HVAC Ductwork with $I_p = 1.5$	13.6.7 Ductwork
1613.7 ASCE 7, Section 11.7.5 Anchorage of walls	1.4.5 Anchorage of Structural Walls (SDC A)
	12.11 Structural Walls and Their Anchorage

1614, 202

Atmospheric Ice Loads

Load combinations involving ice loads

CHANGE TYPE: Addition

CHANGE SUMMARY: A new section, definition and notation for ice loads on ice sensitive structures have been added to the in order to provide consistency with ASCE 7-10.

2012 CODE: 202 Definitions.

ICE-SENSITIVE STRUCTURE   A structure for which the effect of an atmospheric ice load governs the design of a structure or portion thereof. This includes, but is not limited to, lattice structures, guyed masts, overhead lines, light suspension and cable-stayed bridges, aerial cable systems (e.g., for ski lifts or logging operations), amusement rides, open catwalks and platforms, flagpoles and signs.

1602 Definitions and Notations

D_i = Weight of ice in accordance with Chapter 10 of ASCE 7.

W_i = Wind-on-ice in accordance with Chapter 10 of ASCE 7.

SECTION 1614
ATMOSPHERIC ICE LOADS

1614.1 General.   Ice-sensitive structures shall be designed for atmospheric ice loads in accordance with Chapter 10 of ASCE 7.

CHANGE SIGNIFICANCE: Section 10.1 of ASCE 7-10 requires atmospheric ice loads to be considered in the design of ice-sensitive structures. The term "ice-sensitive structure" is defined in Section 10.2 of ASCE 7-10 and this definition has been added to the IBC. Having the definition in the IBC provides the technical basis for determining which structures are ice-sensitive and are required to be designed for ice loads in accordance with the applicable provisions in ASCE 7-10. The new Section 1614 references Chapter 10 of ASCE 7-10 for the determination ice loads on these structures. The LRFD load combinations in Section 1605.2.2 and ASD load combinations in Section 1605.3.1.2 have been modified to include ice loads where applicable. Where atmospheric ice loads must be considered in the design, these code sections cross reference ASCE 7 Section 2.3.4 for LRFD and Section 2.4.3 for ASD respectively.

CHANGE TYPE: Modification

CHANGE SUMMARY: The provisions requiring specific items to have special inspection and what information is required to be included in the statement of special inspections have been clarified and coordinated, with previous conflicts between the two being resolved.

2012 CODE:

~~SECTION 1705~~
~~STATEMENT OF SPECIAL INSPECTIONS~~

~~1705.1 General.~~ 1704.3 Statement of Special Inspections. Where special inspection or testing is required by Section ~~1704, 1707 or 1708~~ 1705, the registered design professional in responsible charge shall prepare a statement of special inspections in accordance with Section ~~1705~~ 1704.3.1 for submittal by the applicant ~~see~~ in accordance with Section ~~1704.1.1~~ 1704.2.3.

> **Exception:**   The statement of special inspections is permitted to be prepared by a qualified person approved by the building official for construction not designed by a registered design professional.

~~1705.2~~ 1704.3.1 Content of Statement of Special Inspections.   The statement of special inspections shall identify the following:

1. The materials, systems, components, and work required to have special inspection or testing by the building official or by the registered design professional responsible for each portion of the work.

2. The type and extent of each special inspection.

3. The type and extent of each test.

1704.3 continues

STATEMENT OF SPECIAL INSPECTIONS AGREEMENT

To permit applicants of projects requiring special inspection and/or testing per Secti___
Building Code (IBC):

Project Address: _____ Permit No.: _____

BEFORE A PERMIT CAN BE ISSUED, two (2) copies of th___
Inspection and the Special Inspection and Testing Schedule ___
by the owner, or registered design professional in responsib___
conference with the parties involved may be required to ___

APPROVAL OF SPECIAL INSPECTORS: Special i___
which they provide special inspection. Special inspectors ___
performing any duties. Special inspectors shall submit their ___
prequalification. Special inspectors shall display approved id___
performing the function of special inspector.

Special inspection and testing shall meet the minimum requir___
International Building Code. The following conditions are *also* ___

International Code Council®

Statement of special inspections agreement

1704.3 continued

4. Additional requirements for special inspection or testing for seismic or wind resistance as specified in Sections ~~1705.3~~, ~~1705.4~~, 1705.10, ~~1707~~ 1705.11, ~~or~~ and ~~1708~~1705.12.

5. For each type of special inspection, identification as to whether it will be continuous special inspection or periodic special inspection.

~~1705.3.6~~ <u>1704.3.2 Seismic Requirements in the Statement of Special Inspections.</u> <u>Where Section 1705.11 or 1705.12 specifies special inspection, testing, or qualification for seismic resistance, the statement of special inspections shall identify the designated seismic systems and seismic force-resisting systems that are subject to special inspections.</u>

~~1705.4.1~~ <u>1704.3.3 Wind Requirements in the Statement of Special Inspections.</u> <u>Where Section 1705.10 specifies special inspection for wind requirements, the statement of special inspections shall identify the main wind-force-resisting systems and wind-resisting components subject to special inspection.</u>

~~1705.3 Seismic Resistance.~~ ~~The statement of special inspections shall include seismic requirements for cases covered in Sections 1705.3.1 through 1705.3.5.~~

~~**Exceptions:** Seismic requirements are permitted to be excluded from the statement of special inspections for structures designed and constructed in accordance with the following:~~

1. ~~The structure consists of light-frame construction; the design spectral response acceleration at short periods, S_{DS}, as determined in Section 1613.5.4, does not exceed 0.5 g; and the height of the structure does not exceed 35 feet (10 668 mm) above grade plane; or~~

2. ~~The structure is constructed using a reinforced masonry structural system or reinforced concrete structural system; the design spectral response acceleration at short periods, S_{DS}, as determined in Section 1613.5.4, does not exceed 0.5 g, and the height of the structure does not exceed 25 feet (7620 mm) above grade plane; or~~

3. ~~Detached one- or two-family dwellings not exceeding two stories above grade plane, provided the structure does not have any of the following plan or vertical irregularities in accordance with Section 12.3.2 of ASCE 7:~~

   3.1 ~~Torsional irregularity.~~
   3.2 ~~Nonparallel systems.~~
   3.3 ~~Stiffness irregularity extreme soft story and soft story.~~
   3.4 ~~Discontinuity in capacity weak story.~~

~~1705.3.1 Seismic-Force-Resisting Systems.~~ ~~The seismic force-resisting systems in structures assigned to Seismic Design Category C, D, E or F, in accordance with Section 1613.~~

~~**Exceptions:** Requirements for the seismic-force-resisting system are permitted to be excluded from the statement of special inspections for steel systems in structures assigned to Seismic Design Category C that are not specifically detailed for seismic resistance, with a response modification coefficient, R, of 3 or less, excluding cantilever column systems.~~

1705.3.2 Designated Seismic Systems. ~~Designated seismic systems in structures assigned to Seismic Design Category D, E or F.~~

1705.3.3 Seismic Design Category C. ~~The following additional systems and components in structures assigned to Seismic Design Category C:~~

~~1. Heating, ventilating and air-conditioning (HVAC) ductwork containing hazardous materials and anchorage of such ductwork.~~

~~2. Piping systems and mechanical units containing flammable, combustible or highly toxic materials.~~

~~3. Anchorage of electrical equipment used for emergency or standby power systems.~~

1705.3.4 Seismic Design Category D. ~~The following additional systems and components in structures assigned to Seismic Design Category D:~~

~~1. Systems required for Seismic Design Category C.~~

~~2. Exterior wall panels and their anchorage.~~

~~3. Suspended ceiling systems and their anchorage.~~

~~4. Access floors and their anchorage.~~

~~5. Steel storage racks and their anchorage, where the importance factor is equal to 1.5 in accordance with Section 15.5.3 of ASCE 7.~~

1705.3.5 Seismic Design Category E or F. ~~The following additional systems and components in structures assigned to Seismic Design Category E or F:~~

~~1. Systems required for Seismic Design Categories C and D.~~

~~2. Electrical equipment.~~

1705.3.6 Seismic requirements in the Statement of Special Inspections. ~~When Sections 1705.3 through 1705.3.5 specify that seismic requirements be included, the statement of special inspections shall identify the following:~~

~~1. The designated seismic systems and seismic force-resisting systems that are subject to special inspections in accordance with Sections 1705.3 through 1705.3.5.~~

~~2. The additional special inspections and testing to be provided as required by Sections 1707 and 1708 and other applicable sections of this code, including the applicable standards referenced by this code.~~

1705.4 Wind Resistance. ~~The statement of special inspections shall include wind requirements for structures constructed in the following areas:~~

~~1. In wind Exposure Category B, where the 3-second-gust basic wind speed is 120 miles per hour (mph) (52.8m/s) or greater.~~

1704.3 continues

1704.3 continued

2. ~~In wind Exposure Category C or D, where the 3-secondgust basic wind speed is 110 mph (49 m/s) or greater.~~

~~1705.4.1 Wind Requirements in the Statement of Special Inspections.~~ ~~When Section 1705.4 specifies that wind requirements be included, the statement of special inspections shall identify the main wind-force-resisting systems and wind-resisting components subject to special inspections as specified in Section 1705.4.2.~~

~~1705.4.2 Detailed Requirements.~~ ~~The statement of special inspections shall include at least the following systems and components:~~

1. ~~Roof cladding and roof framing connections.~~
2. ~~Wall connections to roof and floor diaphragms and framing.~~
3. ~~Roof and floor diaphragm systems, including collectors, drag struts and boundary elements.~~
4. ~~Vertical wind-force-resisting systems, including braced frames, moment frames and shear walls.~~
5. ~~Wind-force-resisting system connections to the foundation.~~
6. ~~Fabrication and installation of systems or components required to meet the impact-resistance requirements of Section 1609.1.2.~~

~~Exception:~~ ~~Fabrication of manufactured systems or components that have a label indicating compliance with the wind-load and impact-resistance requirements of this code.~~

Because this code change deleted and revised substantial portions of Chapter 17, the entire code change text is too extensive to be included here. Refer to Code Changes S129-09/10, S131-09/10, S132-09/10, S133-09/10, and S134-09/10 in the *2012 IBC Code Changes Resource Collection* for the complete text and history of these code changes.

CHANGE SIGNIFICANCE: In the 2009 IBC, Section 1704 covered what specific items required special inspection and Section 1705 covered what is required to be included in the statement of special inspections. In the 2009 IBC, there are a variety of conflicts and inconsistencies between the two sections. The charging sentence of Section 1705 stated that where special inspection or testing was required by Section 1704, 1707, or 1708, the registered design professional in responsible charge must prepare a statement of special inspections in accordance with Section 1705. For example, suspended ceilings in Seismic Design Category D are required to be included in the statement of special inspections, yet Section 1704, 1707, and 1708 did not specifically require special inspection for suspended ceilings. Additionally, items that required special inspection or testing by Section 1704, 1707, or 1708 were not all covered in the requirements for the statement of special inspections in Section 1705. In other words, not all items that require special inspection under Sections 1704, 1707, or 1708 were listed in Section 1705 and not all items required to be in the statement of special inspection in Section 1705 require special inspection in Sections 1704, 1707, or 1708. To resolve these issues, many deletions, revisions, and a reorganization of the special inspection and statement of special inspections requirements were undertaken in an effort to clarify the intent and improve proper application and enforcement.

1705.2
Special Inspection of Steel Construction

CHANGE TYPE: Modification

CHANGE SUMMARY: Special inspection requirements for structural steel have been deleted from Chapter 17 because the new standard for structural steel buildings (ANSI/AISC 360-10) includes quality assurance provisions.

2012 CODE: ~~1704.3~~ **1705.2 Steel Construction.** The special inspections for steel elements of buildings and structures shall be as required <u>in this section</u> ~~by Section 1704.3 and Table 1704.3.~~

Exceptions:

1. ~~1.~~ Special inspection of the steel fabrication process shall not be required where the fabricator does not perform any welding, thermal cutting, or heating operation of any kind as part of the fabrication process. In such cases, the fabricator shall be required to submit a detailed procedure for material control that demonstrates the fabricator's ability to maintain suitable records and procedures such that, at any time during the fabrication process, the material specification~~,~~ <u>and</u> grade ~~and mill test reports~~ for the main stress-carrying elements are capable of being determined. <u>Mill test reports shall be identifiable to the main stress-carrying elements when required by the approved construction documents.</u>

2. ~~2. The special inspector need not be continuously present during welding of the following items, provided the materials, welding procedures and qualifications of welders are verified prior to the start of the work; periodic inspections are made of the work in progress and a visual inspection of all welds is made prior to completion or prior to shipment of shop welding.~~

    ~~2.1 Single-pass fillet welds not exceeding 5/16 inch (7.9 mm) in size.~~

    ~~2.2 Floor and roof deck welding.~~

1705.2 continues

Structural steel building (*Courtesy of Able Steel Fabricators*)

1705.2 continued

2.3 ~~Welded studs when used for structural diaphragm.~~
2.4 ~~Welded sheet steel for cold-formed steel members.~~
2.5 ~~Welding of stairs and railing systems.~~

1705.2.1 Structural Steel. Special inspection for structural steel shall be in accordance with the quality assurance inspection requirements of AISC 360.

1705.2.2 Steel Construction Other Than Structural Steel. Special inspection for steel construction other than structural steel shall be in accordance with Table 1705.2.2 and this section.

~~1704.3.1~~ **1705.2.2.1 Welding.** Welding inspection and welding inspector qualification shall be in accordance with this section.

~~1704.3.1.1 Structural Steel.~~ ~~Welding inspection and welding inspector qualification for structural steel shall be in accordance with AWS D1.1.~~

~~1704.3.1.2~~ **1705.2.2.1.1 Cold-Formed Steel.** Welding inspection and welding inspector qualification for cold-formed steel floor and roof decks shall be in accordance with AWS D1.3.

~~1704.3.1.3~~ **1705.2.2.1.2 Reinforcing Steel.** Welding inspection and welding inspector qualification for reinforcing steel shall be in accordance with AWS D1.4 and ACI 318.

~~1704.3.2 Details.~~ ~~The special inspector shall perform an inspection of the steel frame to verify compliance with the details shown on the *approved construction documents*, such as bracing, stiffening, member locations and proper application of joint details at each connection.~~

~~1704.3.3 High-Strength Bolts.~~ ~~Installation of high-strength bolts shall be inspected in accordance with AISC 360.~~

~~1704.3.3.1 General.~~ ~~While the work is in progress, the special inspector shall determine that the requirements for bolts, nuts, washers and paint; bolted parts and installation and tightening in such standards are met. For bolts requiring pretensioning, the special inspector shall observe the preinstallation testing and calibration procedures when such procedures are required by the installation method or by project plans or specifications; determine that all plies of connected materials have been drawn together and properly snugged and monitor the installation of bolts to verify that the selected procedure for installation is properly used to tighten bolts. For joints required to be tightened only to the snug-tight condition, the special inspector need only verify that the connected materials have been drawn together and properly snugged.~~

~~1704.3.3.2 Periodic Monitoring.~~ ~~Monitoring of bolt installation for pretensioning is permitted to be performed on a periodic basis when using the turn-of-nut method with matchmarking techniques, the direct tension indicator method or the alternate design fastener (twist-off bolt) method. Joints designated as snug tight need be inspected only on a periodic basis.~~

~~1704.3.3.3 Continuous Monitoring.~~ ~~Monitoring of bolt installation for pretensioning using the calibrated wrench method or the turn-of-nut method without matchmarking shall be performed on a continuous basis.~~

TABLE ~~1704.3~~ 1705.2.2   Required Verification and Inspection of Steel Construction Other Than Structural Steel

Verification and Inspection	Continuous	Periodic	Referenced Standard[a]	IBC Reference
~~1. Material verification of high-strength bolts, nuts and washers:~~				
~~a. Identification markings to conform to ASTM standards specified in the approved construction documents.~~	—	X	~~AISC 360, Section A3.3 and applicable ASTM material standards~~	—
~~b. Manufacturer's certificate of compliance required.~~	—	X	—	—
~~2. Inspection of high-strength bolting:~~				
~~a. Snug-tight joints.~~	—	X		
~~b. Pretensioned and slip-critical joints using turn-of-nut with matchmarking, twist-off bolt, or direct tension indicator methods of installation.~~	—	X	~~AISC 360, Section M2.5~~	~~1704.3.3~~
~~c. Pretensioned and slip-critical joints using turn-of-nut without matchmarking or calibrated wrench methods of installation.~~	X	—		
~~3~~ 1. Material verification of ~~structural steel and~~ cold-formed steel deck:				
~~a. For structural steel, identification markings to conform to AISC 360.~~	—	X	~~AISC 360, Section M5.5~~	
b. a. ~~For other steel, i~~Identification markings to conform to ASTM standards specified in the approved construction documents.	—	X	Applicable ASTM material standards	
~~c.~~ b. Manufacturers' certified test reports.	—	X		
~~4. Material verification of weld filler materials:~~				
~~a. Identification markings to conform to AWS specification in the approved construction documents.~~	—	X	~~AISC 360, Section A3.5 and Applicable AWS A5 documents~~	—
~~b. Manufacturer's certificate of compliance required.~~	—	X	—	—
~~5~~ 2. Inspection of welding:				
a. ~~Structural steel and c~~Cold-formed steel deck:				
~~1) Complete and partial joint penetration groove welds.~~	X	—	~~AWS D1.1~~	~~1704.3.1~~
~~2) Multipass fillet welds.~~	X	—		
~~3) Single-pass fillet welds >5/16"~~	X	—		
~~4) Plug and slot welds~~	X	—		
~~5) Single-pass fillet welds ≤5/16"~~	—	X		
~~6~~ 1) Floor and roof deck welds.	—	X	AWS D1.3	
b. Reinforcing steel:				
1) Verification of weldability of reinforcing steel other than ASTM A 706.	—	X	AWS D1.4 or ACI 318: Section 3.5.2	—

1705.2 continues

1705.2 continued

2) Reinforcing steel-resisting flexural and axial forces in intermediate and special moment frames, and boundary elements of special reinforced concrete shear walls and shear reinforcement.	X	—
3) Shear reinforcement.	X	—
4) Other reinforcing steel.	—	X

6. Inspection of steel frame joint details for compliance with approved construction documents:				
a. Details such as bracing and stiffening.	—	X	—	1704.3.2
b. Member locations.	—	X		
c. Application of joint details at each connection.	—	X		

a. Where applicable, see also Section ~~1707.1~~ 1705.11, Special Inspection for Seismic Resistance.

~~1704.3.4~~ 1705.2.2.2 Cold-Formed Steel Trusses Spanning 60 Feet or Greater. Where a cold-formed steel truss clear span is 60 feet (18288 mm) or greater, the special inspector shall verify that the temporary installation restraint/bracing and the permanent individual truss member restraint/bracing are installed in accordance with the approved truss submittal package.

CHANGE SIGNIFICANCE: Substantial portions of the special inspection requirements for structural steel were deleted from the code because the 2010 edition of ANSI/AISC 360, *Specification for Structural Steel Buildings,* incorporates a new Chapter N, which includes comprehensive quality control and quality assurance requirements for structural steel construction. AISC 360, Chapter N, covers quality control requirements pertaining to the structural steel fabricator and erector, as well as quality assurance requirements pertaining to the owner's inspecting and/or testing agencies. The requirements in ANSI/AISC 360-10 are similar to those that were incorporated into AISC 341-05, Appendix Q. AISC 360-10, Chapter N, provides the foundation for the quality control and quality assurance requirements for general structural steel construction, along with AISC 341-10, Chapter I, thereby extending specific requirements to high-seismic applications. The inspection requirements in AISC 360-10 of the Quality Assurance Inspector are esentially equivalent to those specified for the special inspector in IBC Chapter 17.

Section 1704.3 of the 2009 IBC addressed all forms of steel construction, but the majority of the requirements in the section and Table 1704.3 pertained to structural steel construction and have been deleted. However, some items apply to cold-formed steel construction and rebar welding, which are not covered by AISC 360. Requirements for special inspection of other forms of steel construction are in a separate section and in a reduced table titled, *Required Verification and Inspection of Steel Construction Other Than Structural Steel.* The exception in Section 1705.2 has been retained but modified to clarify the requirement. In practice, the "representative mill test reports" are supplied as described in the AISC Code of Standard Practice, so the added sentence in the exception on mill test reports allows traceability when required by the construction documents and defers to AISC 360 in other cases. For a correlation between the provisions that were deleted from 2009 IBC Section 1704.3 that are covered in AISC 360-10, Chapter N, refer to code change S121-09/10 in the *2012 IBC Code Changes Resource Collection.*

CHANGE TYPE: Modification

CHANGE SUMMARY: The type of special inspection required for anchors cast in concrete and post installed anchors in hardened concrete have been clarified.

Table 1705.3

Required Verification and Inspection of Concrete Construction

2012 CODE:

TABLE ~~1704.4~~ <u>1705.3</u>   Required Verification and Inspection of Concrete Construction

Verification and Inspection	Continuous	Periodic	Referenced Standard[a]	IBC Reference
3. Inspection of ~~bolts to be installed in concrete prior to and during placement of~~ <u>anchors cast in</u> concrete where allowable loads have been increased or where strength design is used.	~~X~~ <u>—</u>	<u>—</u> <u>X</u>	ACI 318: 8.1.3, 21.2.8	1908.5, 1909.1
4. Inspection of anchors post-installed in hardened concrete <u>members</u>[b]	—	X	ACI 318: 3.8.6, 8.1.3, 21.2.8	1909.1

(portions of table not shown are unchanged)

a. Where applicable, see also Section ~~1707.1~~ <u>1705.11</u>, Special Inspection for Seismic Resistance.

b. <u>Specific requirements for special inspection shall be included in the research report for the anchor issued by an approved source in accordance with ACI 355.2 or other qualification procedures. Where specific requirements are not provided, special inspection requirements shall be specified by the registered design professional and shall be approved by the building official prior to the commencement of the work.</u>

Table 1705.3 continues

Screw-bolt anchor *(Courtesy of Powers Fasteners)*

Table 1705.3 continued

CHANGE SIGNIFICANCE: Anchors cast into concrete are visible for inspection from the time of installation until the concrete is placed, similar to concrete reinforcement, which may have periodic special inspection. Because it is sufficient for special inspectors to be present intermittently during installation of the cast-in-place anchors, the code now allows cast-in-place anchors to have periodic special inspection. The new footnote b has been added to account for post-installed anchors approved through the alternate methods of construction provisions of Section 104.11, such as anchors installed in accordance with ICC Evaluation Service Reports. It is also intended to distinguish between the requirements for special inspection of anchors designed to comply with the IBC alone versus those qualified by approved research reports in accordance with ACI 355.2, *Qualification of Post-Installed Mechanical Anchors in Concrete.* Typically, items requiring special inspection that are approved under Section 104.11 are covered by Section 1705.1.1, Special Cases. Where special inspection requirements are not provided in a research report, the special inspection requirements must be specified by the registered design professional, who would indicate whether inspections are continuous or periodic, and be approved by the building official prior to commencement of the work.

CHANGE TYPE: Modification

CHANGE SUMMARY: Requirements pertaining to special inspection of masonry construction were deleted from Chapter 17 of the 2012 IBC because the 2011 edition of TMS 402/ACI 530/ASCE 5 and TMS 602/ACI 530.1/ASCE 6, includes requirements for quality assurance of masonry construction.

2012 CODE: ~~1704.5~~ 1705.4 Masonry Construction.   Masonry construction shall be inspected and verified in accordance with TMS 402/ ACI 530/ASCE 5 and TMS 602/ACI 530.1/ASCE 6 quality assurance program requirements. ~~the requirements of Sections 1704.5.1 through 1704.5.2, depending on the occupancy category of the building or structure.~~

> **Exceptions:**   Special inspections shall not be required for:
>
> 1. Empirically designed masonry, glass unit masonry, or masonry veneer designed by Section 2109, Section 2110, or Chapter 14, respectively, ~~or by Chapter 5, 7 or 6 of TMS 402/ ACI 530/ASCE 5, respectively,~~ where they are part of structures classified as ~~Occupancy~~ Risk Category I, II, or III in accordance with Section 1604.5.
> 2. Masonry foundation walls constructed in accordance with Table 1807.1.6.3(1), 1807.1.6.3(2), 1807.1.6.3(3), or 1807.1.6.3(4).
> 3. Masonry fireplaces, masonry heaters, or masonry chimneys installed or constructed in accordance with Section 2111, 2112, or 2113, respectively.

~~1704.5.1~~ 1705.4.1 Empirically Designed Masonry, Glass Unit Masonry, and Masonry Veneer in ~~Occupancy~~ Risk Category IV.   The minimum special inspection program for empirically designed masonry, glass unit masonry, or masonry veneer designed by Section 2109, Section 2110, or Chapter 14, respectively, ~~or by Chapter 5, 7 or 6 of TMS 402/ACI ASCE 5, respectively,~~ in structures classified as ~~Occupancy~~ Risk Category IV, in accordance with Section 1604.5, shall comply with TMS 402/ACI 530/ASCE 5 Level B Quality Assurance. ~~Table 1704.5.1.~~

~~1704.5.2 Engineered Masonry in Occupancy Category I, II or III.~~ ~~The minimum special inspection program for masonry designed by Section 2107 or 2108 or by chapters other than Chapter 5, 6 or 7 of TMS402/ ACI 530/ASCE 5 in structures classified as Occupancy Category I, II or III, in accordance with Section 1604.5, shall comply with Table 1704.5.1.~~

~~1704.5.3 Engineered Masonry in Occupancy Category IV.~~ ~~The minimum special inspection program for masonry designed by Section 2107 or 2108 or by chapters other than Chapter 5, 6 or 7 of TMS402/ACI 530/ ASCE 5 in structures classified as Occupancy Category IV, in accordance with Section 1604.5, shall comply with Table 1704.5.3.~~

1705.4 continues

1705.4
Special Inspection of Masonry Construction

Masonry special inspector *(Photo Courtesy of CTC Geotek)*

1705.4 continued

~~TABLE 1704.5.1~~ ~~Level 1 Required Verification and Inspection of Masonry Construction~~

(deleted table not shown for brevity)

~~TABLE 1704.5.3~~ ~~Level 2 Required Verification and Inspection of Masonry Construction~~

(deleted table not shown for brevity)

~~1704.11~~ 1705.4.2 Vertical Masonry Foundation Elements. Special inspection shall be performed in accordance with Section ~~1704.5~~ 1705.4 for vertical masonry foundation elements.

CHANGE SIGNIFICANCE: The basis for the design and construction of masonry structures in Chapter 21 of the 2012 IBC is the 2011 edition of TMS 402/ACI 530/ASCE 5 and TMS 602/ACI 530.1/ASCE 6 by reference. The special inspection provisions for masonry construction in Chapter 17 have been deleted and replaced with references to the standard for quality assurance of masonry construction. Section 1.19 of TMS 402/ACI 530/ASCE 5 and Article 1.6 of TMS 602/ACI 530/ASCE 6 include the requirements for tests, inspections, and verifications of masonry construction. All masonry designed in accordance with Chapter 5, 6, or 7 of TMS 402 is subject to a quality assurance program specified in Section 1.19 of TMS 402. The modifications made to Chapter 17 are as follows:

- For structures in Risk Category IV, reference is now made to Level B Quality Assurance requirements specified in TMS 402 for the list of tests, inspections, and verifications required for masonry designed in accordance with IBC Sections 2109 (empirical design), 2110 (glass unit masonry), and Chapter 14 (veneer).

- 2009 IBC Sections 1704.5.2 and 1704.5.3 have been deleted entirely because all masonry designed in accordance IBC Sections 2107 (allowable stress design) and 2108 (strength design) must comply with Chapter 1 of TMS 402/ACI 530/ASCE 5, which requires masonry construction to be tested, inspected, and verified.

- 2009 IBC Tables 1704.5.1 and 1704.5.3 have been deleted entirely because all tests, inspections, and verifications are identified in TMS 402/ACI 530/ASCE 5.

CHANGE TYPE: Addition

CHANGE SUMMARY: Where penetration firestop systems and fire-resistant joint systems are used in high-rise buildings and those buildings assigned to Risk Categories III and IV, it is now mandatory that they be inspected by an approved inspection agency as a part of the special inspection process.

2012 CODE: 1705.16 Fire-Resistant Penetrations and Joints. In high-rise buildings or in buildings assigned to Risk Category III or IV in accordance with Section 1604.5, special inspections for through penetrations, membrane penetration firestops, fire-resistant joint systems, and perimeter fire barrier systems that are tested and listed in accordance with Sections 714.3.1.2, 714.4.1.2, 715.3, and 715.4 shall be in accordance with Section 1705.16.1 or 1705.16.2.

1705.16.1 Penetration Firestops. Inspections of penetration firestop systems that are tested and listed in accordance with Sections 714.3.1.2 and 714.4.1.2 shall be conducted by an approved inspection agency in accordance with ASTM E2174.

1705.16.2 Fire-Resistant Joint Systems. Inspection of fire-resistant joint systems that are tested and listed in accordance with Sections 715.3 and 715.4 shall be conducted by an approved inspection agency in accordance with ASTM E 2393.

Chapter 35.

ASTM E 2174-09, *Standard Practice for On-Site Inspection of Installed Fire Stops*

ASTM E 2393-09, *Standard Practice for On-Site Inspection of Installed Fire Resistive Joint Systems and Perimeter Fire Barrier*

CHANGE SIGNIFICANCE: Through-penetration and membrane-penetration firestop systems, as well as fire-resistant joint systems and perimeter fire barrier systems, are critical to maintaining the fire-resistive integrity of fire-resistance-rated construction elements, including fire walls, fire barriers, fire partitions, smoke barriers, and horizontal assemblies. The proper selection and installation of such systems must be in compliance with the code and/or appropriate listing. With thousands of listed firestop systems available—each with variations that multiply possible systems for a building exponentially—the selection of the correct system is not a generic process. Where such systems are used in two types of buildings considered as "high risk," it is now mandatory that they be included as a part of the special inspection process. Such "high-risk" buildings have been identified as:

- Buildings assigned to Risk Category III or IV in accordance with Section 1604.5, and

- High-rise buildings.

Although the proper application of firestop and joint system requirements is very important in all types and sizes of buildings, the requirement for special inspection is limited to specific building types that represent a substantial hazard to human life in the event of a system failure or that are considered to be essential facilities. Inspection to ASTM E2174 for penetration firestop systems and ASTM E2393 for fire-resistant joint systems brings an increased level of review to this important discipline.

1705.16
Special Inspection of Fire-Resistant Penetration and Joint Systems

Penetration protection at fire-resistance-rated floor/ceiling assembly

International Code Council®

1803.5.12

Geotechnical Reports for Foundation Walls and Retaining Walls

Retaining wall *(Photo Courtesy of Alan D. Wilcox, P.E.)*

CHANGE TYPE: Modification

CHANGE SUMMARY: The requirement that geotechnical reports address earthquake loads on foundation walls and retaining walls in Seismic Design Categories D, E, and F has been modified so that it only applies to those walls supporting more than 6 feet of backfill.

2012 CODE: 1803.5.12 Seismic Design Categories D through F. For structures assigned to Seismic Design Category D, E, or F, the geotechnical investigation required by Section 1803.5.11 shall also include <u>all of the following as applicable</u>:

1. The determination of <u>dynamic seismic</u> lateral <u>earth</u> pressures on foundation walls and retaining walls <u>supporting more than 6 feet (1830 mm) of backfill height</u> due to <u>design</u> earthquake <u>ground</u> motions.

2.-4. (no significant changes to text)

CHANGE SIGNIFICANCE: Geotechnical reports have previously been required to address earthquake loads on foundation walls and retaining walls for buildings in Seismic Design Categories D, E, and F. In the application of the requirements, there was no exemption based on the height of the wall or the amount of soil supported by the wall. This was deemed to be overly restrictive for foundation walls supporting light-frame construction, small retaining walls, and swimming pools. Evidence from recent earthquakes and recent experimental research results, including work recently completed at the University of California–Berkeley, has demonstrated that retaining wall structures must move in order to develop the failure wedge postulated in the so-called Mononobe and Okabe method. However, the postulated condition can only occur when the wall has already failed due to other causes. The current body of field evidence does not provide any evidence for the existence of this mechanism of failure. It was determined that the requirement in the 2009 IBC and ASCE 7-05 imposed an unjustifiable burden on the permit applicant to investigate a site for small retaining structures such as foundation walls, retaining walls, and swimming pools that support no more than 6 feet of backfill.

CHANGE TYPE: Modification

CHANGE SUMMARY: The uplift capacity of pile groups is now permitted to include two-thirds of the shear resistance of the soil block.

2012 CODE: 1810.3.3.1.6 Uplift Capacity of Grouped Deep Foundation Elements. For grouped deep foundation elements subjected to uplift, the allowable working uplift load for the group shall be calculated by an approved method of analysis. Where the deep foundation elements in the group are placed at a center-to-center spacing of at least 2.5 times the least horizontal dimension of the largest single element, the allowable working uplift load for the group is permitted to be calculated as the lesser of:

1. The proposed individual uplift working load times the number of elements in the group.
2. Two-thirds of the effective weight of the group and the soil contained within a block defined by the perimeter of the group and the length of the element, plus two-thirds of the ultimate shear resistance along the soil block.

CHANGE SIGNIFICANCE: In the determination of uplift capacity of grouped deep foundation elements, previous editions of the code have allowed two-thirds of the effective weight of a pile group and the weight of the soil contained within the block defined by the perimeter of the group but did not include an allowance for the shear resistance of the soil block. This was determined to be unreasonably conservative because not only the weight of the soil within the pile group resists uplift, but also the shear resistance developed contributes to the resistance to uplift of the pile group. The code now allows use of two-thirds of the effective weight of the pile group, two-thirds of the weight of the soil contained within a block defined by the perimeter of the group and the length of the piles, plus two-thirds of the ultimate shear resistance along the soil block.

1810.3.3.1.6

Uplift Capacity of Grouped Deep Foundation Elements

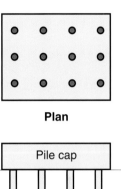

Plan

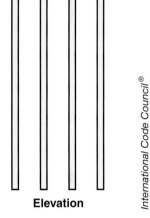

Elevation

Pile group

International Code Council®

Chapter 19
Concrete Construction

CHANGE TYPE: Modification

CHANGE SUMMARY: The provisions related to concrete construction were deleted from Chapter 19 because they are contained in the 2011 edition of ACI 318, *Building Code Requirements for Structural Concrete and Commentary.*

2012 CODE: ~~1901.3 Source and Applicability.~~ ~~The format and subject matter of Sections 1902 through 1907 of this chapter are patterned after, and in general conformity with, the provisions for structural concrete in ACI 318.~~

~~1901.4~~ **1901.3 Construction Documents.** The construction documents for structural concrete construction shall include:

1. The specified compressive strength of concrete at the stated ages or stages of construction for which each concrete element is designed.
2. The specified strength or grade of reinforcement.
3. The size and location of structural elements, reinforcement, and anchors.
4. Provision for dimensional changes resulting from creep, shrinkage, and temperature.
5. The magnitude and location of prestressing forces.

Reinforced concrete foundation

International Code Council®

6. Anchorage length of reinforcement and location and length of lap splices.

7. Type and location of mechanical and welded splices of reinforcement.

8. Details and location of contraction or isolation joints specified for plain concrete.

9. Minimum concrete compressive strength at time of posttensioning.

10. Stressing sequence for posttensioning tendons.

11. For structures assigned to Seismic Design Category D, E, or F, a statement if slab on grade is designed as a structural diaphragm ~~(see Section 21.12.3.4 of ACI 318)~~.

1903.3 Flat Wall Insulating Concrete Form (ICF) Systems. Insulating concrete form material used for forming flat concrete walls shall conform to ASTM E 2634.

~~**1904.1 Water-Cementitious Materials Ratio.** Where maximum water-cementitious materials ratios are specified in ACI 318, they shall be calculated in accordance with ACI 318, Section 4.1.~~

SECTION 1904
DURABILITY REQUIREMENTS

~~**1904.2**~~ **1904.1 Exposure Categories and Classes.** Concrete shall be assigned to exposure classes in accordance with the durability requirements of ACI 318~~, Section 4.2,~~ based on:

1. Exposure to freezing and thawing in a moist condition or deicer chemicals.

2. Exposure to sulfates in water or soil.

3. Exposure to water where the concrete is intended to have low permeability.

4. Exposure to chlorides from deicing chemicals, salt, saltwater, brackish water, seawater, or spray from these sources, where the concrete has steel reinforcement.

~~**1904.3**~~ **1904.2 Concrete Properties.** Concrete mixtures shall conform to the most restrictive maximum water-cementitious materials ratios, maximum cementitious admixtures, minimum air-entrainment, and minimum specified concrete compressive strength requirements of ACI 318~~, Section 4.3~~, based on the exposure classes assigned in Section ~~1904.2~~ 1904.1.

Exception: For occupancies and appurtenances thereto in Group R occupancies that are in buildings less than four stories above grade plane, normal-weight aggregate concrete is permitted to comply with the requirements of Table ~~1904.3~~ 1904.2 based on the weathering classification (freezing and thawing) determined from Figure ~~1904.3~~ 1904.2 in lieu of the durability requirements of ACI 318 ~~Table 4.3.1~~.

Chapter 19 continues

Chapter 19 continued

TABLE ~~1904.3~~ 1904.2 Minimum Specified Compressive Strength (f'$_c$)

<u>(no changes to table)</u>

a. Concrete in these locations that can be subjected to freezing and thawing during construction shall be of air-entrained concrete in accordance with Section ~~1904.4.1~~ <u>1904.2</u>

b. Concrete shall be air entrained in accordance with ~~Section 1904.4.1~~ <u>ACI 318</u>.

c. Structural plain concrete basement walls are exempt from the requirements for exposure conditions of Section ~~1904.3~~ <u>1904.2</u> ~~(see Section 1909.6.1)~~.

d. For garage floor slabs where a steel trowel finish is used, the total air content required by ~~Section 1904.4.1~~ <u>ACI 318</u> is permitted to be reduced to not less than 3 percent, provided the minimum specified compressive strength of the concrete is increase to 4000 psi.

FIGURE ~~1904.3~~ 1904.2
WEATHERING PROBABILITY MAP FOR CONCRETE [a, b, c]

(no changes to map and footnotes)

~~1904.4 Freezing and Thawing Exposures.~~ ~~Concrete that will be exposed to freezing and thawing, in the presence of moisture, with or without deicing chemicals being present, shall comply with Sections 1904.4.1 and 1904.4.2.~~

~~1904.4.1 Air Entrainment.~~ ~~Concrete exposed to freezing and thawing while moist shall be air entrained in accordance with ACI 318, Section 4.4.1.~~

~~1904.4.2 Deicing Chemicals.~~ ~~For concrete exposed to freezing and thawing in the presence of moisture and deicing chemicals, the maximum weight of fly ash, other pozzolans, silica fume or slag that is included in the concrete shall not exceed the percentages of the total weight of cementitious materials permitted by ACI 318, Section 4.4.2.~~

~~1904.5 Alternative Cementitious Materials for Sulfate Exposure.~~ ~~Alternative combinations of cementitious materials for use in sulfate-resistant concrete to those listed in ACI 318, Table 4.3.1 shall be permitted in accordance with ACI 318, Section 4.5.1.~~

~~SECTION 1905~~
~~CONCRETE QUALITY, MIXING AND PLACING~~

(deleted code text not shown for brevity)

~~SECTION 1906~~
~~FORMWORK, EMBEDDED PIPES AND CONSTRUCTION JOINTS~~

(deleted code text not shown for brevity)

~~SECTION 1907~~
~~DETAILS OF REINFORCEMENT~~

(deleted code text not shown for brevity)

Because this code change deleted or revised substantial portions of Chapter 19, the entire code change text is too extensive to be included here. Refer to Code Change S160-09/10 in the *2012 IBC Code Changes Resource Collection* for the complete text and history of the code change.

CHANGE SIGNIFICANCE: Sections 1901 through 1907 of the 2009 IBC, which contained concrete construction requirements, did not provide any technical content but merely referenced corresponding sections in ACI 318. For example, in the 2009 IBC, Section 1906.1, Formwork, reads as follows, "The design, fabrication and erection of forms shall comply with ACI 318, Section 6.1." Therefore, IBC Sections 1905 (Concrete Quality, Mixing, and Placing), 1906 (Formwork, Imbedded Pipes, and Construction Joints), and 1907 (Details of Reinforcement) have been deleted entirely because they do not provide any information other than referencing the corresponding section in the ACI 318 standard. As stated in Section 1901.2, structural concrete is required to be designed and constructed in accordance with ACI 318 as amended in Section 1905.

1905.1.3

Seismic Detailing of Wall Piers

Precast concrete wall panel (*Courtesy of Tilt-Up Concrete Association*)

CHANGE TYPE: Addition

CHANGE SUMMARY: ACI 318 Section 21.4 provides seismic requirements for intermediate precast structural walls. Section 1905.1.3 amends ACI 318 Section 21.4 by adding seismic detailing requirements for wall piers in Seismic Design Categories D, E, and F.

2012 CODE: 1905.1.3 ACI 318, Section 21.4. Modify ACI 318, Section 21.4, by renumbering Section 21.4.3 to become 21.4.4 and adding new Sections 21.4.3, 21.4.5, ~~and~~ 21.4.6, <u>and 21.4.7</u> to read as follows:

21.4.3 *Connections that are designed to yield shall be capable of maintaining 80 percent of their design strength at the deformation induced by the design displacement or shall use Type 2 mechanical splices.*

21.4.4 Elements of the connection that are not designed to yield shall develop at least 1.5 S_y.

21.4.5 <u>*Wall piers in Seismic Design Category D, E, or F shall comply with Section 1905.1.4.*</u>

~~***21.4.5***~~ ***21.4.6*** *Wall piers not designed as part of a moment frame <u>in buildings assigned to SDC C</u> shall have transverse reinforcement designed to resist the shear forces determined from 21.3.3. Spacing of transverse reinforcement shall not exceed 8 inches (203 mm). Transverse reinforcement shall be extended beyond the pier clear height for at least 12 inches (305 mm).*

Exceptions:

1. *Wall piers that satisfy 21.13.*
2. *Wall piers along a wall line within a story where other shear wall segments provide lateral support to the wall piers and such segments have a total stiffness of at least six times the sum of the stiffnesses of all the wall piers.*

~~***21.4.6***~~ ***21.4.7*** *Wall segments with a horizontal length-to-thickness ratio less than 2.5 shall be designed as columns.*

CHANGE SIGNIFICANCE: ASCE 7 permits intermediate precast structural wall systems in Seismic Design Categories D, E, or F. Section 1908.1.3 of the 2009 IBC had no specific seismic detailing requirements for wall piers in Seismic Design Categories D, E, or F. ACI 318 Commentary R 21.1.1 emphasizes that is essential that structures assigned to higher seismic design categories possess a higher degree of toughness and ductility and encourages practitioners to use special structural wall systems in regions of high seismic risk. Commercial buildings constructed using precast panel wall systems often have large window and door openings and narrow wall piers. Wall panels varying in height up to three stories high with openings resembling wall frames that have not been recognized under any of the defined seismic-force resisting systems other than by considering them structural wall systems. By requiring wall piers in Seismic Design Category D, E, or F to comply with Section 1905.1.4 for special structural walls, the seismic design and detailing of wall piers will ensure better performance when subjected to the design earthquake. The transverse reinforcing requirements for wall piers in Section 21.9.8.2 will enhance ductile response. The modification clarifies the intent by separating wall piers in structures assigned to Seismic Design Category C from those assigned to Seismic Design Category D, E, or F.

1905.1.8

Plain Concrete Footings in Dwelling Construction

CHANGE TYPE: Modification

CHANGE SUMMARY: Plain concrete footings may now only support detached one- and two-family dwellings with light frame stud walls where such structures are located in Seismic Design Categories A, B, and C.

2012 CODE: 1905.1.8 ACI 318, Section 22.10. *Delete ACI 318, Section 22.10, and replace with the following:*

22.10 *Plain concrete in structures assigned to Seismic Design Category C, D, E, or F.*

22.10.1 *Structures assigned to Seismic Design Category C, D, E, or F shall not have elements of structural plain concrete, except as follows:*

a. *Structural plain concrete basement, foundation, or other walls below the base are permitted in detached one- and two-family dwellings three stories or less in height constructed with stud-bearing walls. In dwellings assigned to Seismic Design Category D or E, the height of the wall shall not exceed 8 feet (2438 mm), the thickness shall not be less than 7½ inches (190 mm), and the wall shall retain no more than 4 feet (1219 mm) of unbalanced fill. Walls shall have reinforcement in accordance with 22.6.6.5.*

b. *Isolated footings of plain concrete supporting pedestals or columns are permitted, provided the projection of the footing beyond the face of the supported member does not exceed the footing thickness.*

    Exception: *In detached one- and two-family dwellings three stories or less in height, the projection of the footing beyond the face of the supported member is permitted to exceed the footing thickness.*

c. *Plain concrete footings supporting walls are permitted, provided the footings have at least two continuous longitudinal reinforcing bars. Bars shall not be smaller than No. 4 and shall have a total area of not less than 0.002 times the gross cross-sectional area of*

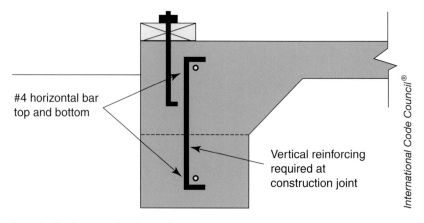

#4 horizontal bar top and bottom

Vertical reinforcing required at construction joint

International Code Council®

Longitudinal reinforcing in footing

the footing. For footings that exceed 8 inches (203 mm) in thickness, a minimum of one bar shall be provided at the top and bottom of the footing. Continuity of reinforcement shall be provided at corners and intersections.

Exceptions:

1. *In <u>Seismic Design Categories A, B, and C</u>, detached one- and two-family dwellings three stories or less in height ~~and~~ constructed with stud-bearing walls, <u>are permitted to have plain concrete footings without longitudinal reinforcement</u> ~~supporting walls are permitted~~.*

2. *For foundation systems consisting of a plain concrete footing and a plain concrete stemwall, a minimum of one bar shall be provided at the top of the stemwall and at the bottom of the footing.*

3. *Where a slab on ground is cast monolithically with the footing, one No. 5 bar is permitted to be located at either the top of the slab or bottom of the footing.*

CHANGE SIGNIFICANCE: An exception in the 2009 IBC allowed detached one- and two-family dwellings three stories or less in height with stud walls to have plain concrete footings without longitudinal reinforcement. The exception has been modified in the 2012 IBC to only apply to Seismic Design Categories A, B, and C because there is a real potential that footings in Seismic Design Categories D, E, and F could experience high flexural demands during the design earthquake. The exception was also in conflict with other provisions that specifically require longitudinal reinforcing in footings. For example, the prescriptive provisions for alternate braced wall panels covered in Sections 2308.9.3.1 and 2308.9.3.2 require a continuous footing with one No. 4 longitudinal bar top and bottom. Having one No. 4 bar at the top of the footing and one No. 4 bar at the bottom of the footing ensures both positive and negative flexural capacity if the footing is subjected to reversing earthquake loads.

1905.1.9

Shear Wall to Concrete Foundation Connection

Anchor bolts with square plate washers

CHANGE TYPE: Modification

CHANGE SUMMARY: Portions of ACI 318 Appendix D have been amended in recognition of the fact that failure of the wood sill plate or the cold formed steel track control the capacity of the connection of the shear wall to the concrete foundation.

2012 CODE: ~~1908.1.9~~ 1905.1.9 **ACI 318, Section D.3.3.** Delete ACI 318 Sections D.3.3.4 through D.3.3.7 and replace with the following:

D.3.3.4 - The anchor design strength associated with concrete failure modes shall be taken as $0.75\phi N_n$ and $0.75\phi V_n$, where ϕ is given in D4.3 or D4.4 and N_n and V_n are determined in accordance with D5.2, D5.3, D5.4, D6.2, and D6.3, assuming the concrete is cracked unless it can be demonstrated that the concrete remains uncracked

D.3.3.5 *Anchors shall be designed to be governed by the steel strength of a ductile steel element as determined in accordance with D.5.1 and D.6.1, unless either D.3.3.6 or D.3.3.7 is satisfied.*

Exceptions:

1. *Anchors designed to resist wall out-of-plane forces with design strengths equal to or greater than the force determined in accordance with ASCE 7 Equation 12.11-1 or 12.14-10 need not satisfy Section D.3.3.5.*

2. *D.3.3.5 need not apply and the design shear strength in accordance with D.6.2.1(c) need not be computed for anchor bolts attaching wood sill plates of bearing or nonbearing walls of light-frame wood structures to foundations or foundation stem walls provided all of the following are satisfied:*
    2.1. *The allowable in-plane shear strength of the anchor is determined in accordance with AF&PA NDS Table 11E for lateral design values parallel to grain.*
    2.2. *The maximum anchor nominal diameter is ⅝ inches (16 mm).*
    2.3. *Anchor bolts are embedded into concrete a minimum of 7 inches (178 mm).*
    2.4. *Anchor bolts are located a minimum of 1¾ inches (45 mm) from the edge of the concrete parallel to the length of the wood sill plate.*
    2.5. *Anchor bolts are located a minimum of 15 anchor diameters from the edge of the concrete perpendicular to the length of the wood sill plate.*
    2.6. *The sill plate is 2-inch or 3-inch nominal thickness.*

3. *Section D.3.3.5 need not apply and the design shear strength in accordance with Section D.6.2.1(c) need not be computed for anchor bolts attaching cold-formed steel track of bearing or nonbearing walls of light-frame construction to foundations or foundation stem walls provided all of the following are satisfied:*
    3.1. *The maximum anchor nominal diameter is ⅝ inches (16 mm).*

> **3.2.** _Anchors are embedded into concrete a minimum of 7 inches (178 mm)._
>
> **3.3.** _Anchors are located a minimum of 1¾ inches (45 mm) from the edge of the concrete parallel to the length of the track._
>
> **3.4.** _Anchors are located a minimum of 15 anchor diameters from the edge of the concrete perpendicular to the length of the track._
>
> **3.5.** _The track is 33- to 68-mil designation thickness._
> _Allowable in-plane shear strength of exempt anchors, parallel to the edge of concrete shall be permitted to be determined in accordance with AISI S100 Section E3.3.1._
>
> **4.** _In light-frame construction, design of anchors in concrete shall be permitted to satisfy D.3.3.8._

D.3.3.6 _Instead of D.3.3.5, the attachment that the anchor is connecting to the structure shall be designed so that the attachment will undergo ductile yielding at a force level corresponding to anchor forces no greater than the design strength of anchors specified in D.3.3.4._

Exceptions:

> **1.** _Anchors in concrete designed to support nonstructural components in accordance with ASCE 7 Section 13.4.2 need not satisfy Section D.3.3.6._
>
> **2.** _Anchors designed to resist wall out-of-plane forces with design strengths equal to or greater than the force determined in accordance with ASCE 7 Equation 12.11-1 or 12.14-10 need not satisfy Section D.3.3.6._

D.3.3.7 _As an alternative to D.3.3.5 and D.3.3.6, it shall be permitted to take the design strength of the anchors as 0.4 times the design strength determined in accordance with D.3.3.4._

D.3.3.8 _In light-frame construction, bearing or nonbearing walls, shear strength of concrete anchors less than or equal to 1 inch [25 mm] in diameter of sill plate or track to foundation or foundation stem wall need not satisfy D.3.3.7 when the design strength of the anchors is determined in accordance with D.6.2.1(c)._

CHANGE SIGNIFICANCE: To appreciate the significance of the code change, some historical discussion is warranted. Until this new modification, the design procedure for determining the capacity of sill plate anchor bolts in wood frame shear walls had not changed since the allowable values were first tabulated and introduced in the 1979 _Uniform Building Code_ (UBC). In the past, the sill plate anchorage design method was an allowable stress design procedure that compared the capacity of the anchor bolts in the wood sill member (loaded in single shear parallel to grain) and the capacity of the anchor bolts embedded in concrete and loaded parallel to the edge of the foundation. For typical wood frame shear walls using 2X sill plates, the capacity of the wood sill plate member governed the design of the shear wall sill anchor bolts.

1905.1.9 continues

1905.1.9 continued

Although there were some changes to the NDS and the anchorage table in the 1994 UBC, this procedure continued to be used to design wood frame shear wall anchor bolts up to and including the 1997 UBC. In the IBC, anchor bolt design is addressed in Sections 1911 (Allowable Stress Design) and 1912 (Strength Design). Section 1911 specifically requires that where anchor bolts resist seismic forces, anchor bolt capacities must be designed by the strength design procedure in Section 1912, which references ACI 318 Appendix D for anchorage to concrete. Appendix D is a complicated process and was not specifically intended to apply to relatively small-diameter anchor bolts (dowels) connecting wood sill plates loaded parallel to the edge of the concrete. Additionally, it has been reported that the results of nonductile anchor design capacities based on Appendix D are approximately one-third of the capacity historically found for typical wood sill plates anchors. Although ductile connections do have increased design capacities in ACI 318 Appendix D, the bending yield behavior of dowels in wood-to-concrete connections is not specifically recognized. A recent study sponsored by the Structural Engineers Association of Northern California (SEAONC) Special Projects Initiative showed that ductile yielding in accordance with the NDS Mode IIIs or Mode IV is consistently achieved prior to concrete failure. Based on these findings, Section 1905.1.9 in the 2012 IBC amends portions of Appendix D in recognition of the fact that failure of the wood sill plate or cold formed steel track controls the capacity of the connection of the shear wall to the concrete foundation. The following summarizes the new provisions:

- ACI 318 D.3.3.5 need not apply and the design shear strength in accordance with D.6.2.1(c) need not be computed for anchor bolts attaching wood sill plates or cold-formed steel track of bearing or nonbearing walls of light-frame wood structures to foundations or foundation stem walls provided all of the following conditions are satisfied:
  - The maximum anchor nominal diameter is ⅝ inches.
  - Anchor bolts are embedded into concrete a minimum of 7 inches.
  - Anchor bolts are located a minimum of 1¾ inches from the edge of the concrete parallel to the length of the wood sill plate or steel track.
  - Anchor bolts are located a minimum of 15 anchor diameters from the edge of the concrete perpendicular to the length of the wood sill plate or steel track.
  - Wood sill plate are 2-inch or 3-inch nominal thickness and steel tracks are 33- to 68-mil designation thickness.
  - The allowable in-plane shear strength of the anchor parallel to the edge of concrete is determined in accordance with AF&PA NDS Table 11E for wood plates and AISI S100 Section E3.3.1 for steel tracks.

For a detailed discussion of the basis for the code changes related to wood shear wall sill anchorage and Appendix D, refer to code change S168-09/10 in the *2012 IBC Code Changes Resource Collection*. The report, *Testing of Anchor Bolts Connecting Wood Sill Plates to Concrete with Minimum Edge Distances* (March 29, 2009), is available for download from the SEAONC website at www.seaonc.org. An excellent article on the subject of anchorage of wood frame shear walls to concrete can be found in the 2009 edition of the *SEAONC Blue Book: Seismic Design Recommendations of the Seismology Committee,* which is available from the International Code Council at www.iccsafe.org.

2101.2

Design Methods for Masonry Structures

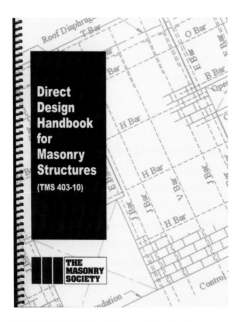

TMS 403-10 *(The Masonry Society)*

CHANGE TYPE: Addition

CHANGE SUMMARY: The new TMS 403-10 masonry design standard, now referenced in the IBC, provides a direct design method for simple, single-story, concrete masonry bearing-wall structures.

2012 CODE: 2101.2 Design Methods. Masonry shall comply with the provisions of one of the following design methods in this chapter as well as the requirements of Sections 2101 through 2104. Masonry designed by the allowable stress design provisions of Section 2101.2.1, the strength design provisions of Section 2101.2.2, ~~or~~ the prestressed masonry provisions of Section 2101.2.3, or the direct design requirements of Section 2101.2.7 shall comply with Section 2105.

2101.2.7 Direct Design. Masonry designed by the direct design method shall comply with the provisions of TMS 403.

Chapter 35. TMS 403—10 Direct Design Handbook for Masonry Structures

CHANGE SIGNIFICANCE: Chapter 21 now includes a simplified design method for single-story, concrete masonry buildings based on the new referenced standard TMS 403, *Direct Design Handbook for Masonry Structures.* The methodology used to develop the standard is based upon the strength design provisions of the 2005 and 2008 editions of TMS 402/ACI 530/ASCE 5 and the factored load combinations for dead, roof live, wind, seismic, snow, and rain loads in accordance with ASCE 7. The new design standard was developed by the masonry industry in response to concerns from the design community that structural loads and design requirements have become too complicated, particularly for relatively small, simple structures. The direct design procedure is a table-based structural design method that permits the user, following a specific series of steps, to design and specify relatively simple, single-story, concrete masonry bearing-wall structures. The method is simple to implement compared to conventional design approaches, but it limits the design to only those configurations addressed by the standard. It introduces slightly more conservatism compared to conventional design procedures as a result of the conditions and assumptions inherent to the design method. Some of the key design limitations in the standard are:

- Snow—ground snow load is limited to 60 psf.
- Wind—3-second gust basic wind speed is limited to 150 mph; wind exposure category is limited to B or C; and site topography is limited to $K_{zt} = 1.0$.
- Seismic—mapped spectral accelerations S_S and S_1 are limited to $3.0g$ and $1.25g$, respectively, and Site Classes are limited to A, B, C, or D.
- Walls—walls are limited to single-story, single-wythe, 8-inch concrete masonry with a maximum height of 30 feet.
- Roof—roof diaphragms are required to be flexible, have rectangular dimensions with an aspect ratio not exceeding 4:1, and a maximum plan dimension of 200 feet.
- Reinforcement—all reinforcing bars are limited to No. 5 and Grade 60.

The *Direct Design Handbook for Masonry Structures* is intended to capture many of the simple load-bearing masonry structures commonly designed today.

2206
Composite Structural Steel and Concrete Structures

CHANGE TYPE: Addition

CHANGE SUMMARY: The requirement that composite structures in Seismic Design Categories D, E, and F provide substantiating evidence demonstrating that they will perform as intended by Part II of AISC 341 has been deleted because these structures are now addressed in the 2010 edition of AISC 341, and a new section for composite structures of structural steel and concrete has been added.

2012 CODE: ~~**2205.3.1 Seismic Design Categories D, E and F.**~~ ~~Composite structures are permitted in~~ *~~Seismic Design Categories~~* ~~D,~~ ~~E and F, subject to the limitations in Section 12.2.1 of ASCE 7, where~~ ~~substantiating evidence is provided to demonstrate that the proposed~~ ~~system will perform as intended by AISC 341, Part II. The substantiating~~ ~~evidence shall be subject to~~ *~~building official~~* ~~approval. Where composite~~ ~~elements or connections are required to sustain inelastic deformations,~~ ~~the substantiating evidence shall be based on cyclic testing.~~

SECTION 2206
COMPOSITE STRUCTURAL STEEL AND CONCRETE STRUCTURES

2206.1 General.   Systems of structural steel acting compositely with reinforced concrete shall be designed in accordance with AISC 360 and ACI 318, excluding ACI 318 Chapter 22. Where required, the seismic design of composite steel and concrete systems shall be in accordance with the additional provisions of Section 2206.2.

2206 continues

Simon headquarters office building *(Courtesy of Haris Engineering)*

2206 continued

~~2205.3~~ 2206.2 Seismic Requirements for Composite <u>Structural Steel and Concrete</u> Construction. <u>Where a response modification coefficient, R, in accordance with ASCE 7, Table 12.2-1, is used for the design of systems of structural steel acting compositely with reinforced concrete, the structures shall be designed and detailed in accordance with the requirements of AISC 341.</u> ~~The design, construction and quality of composite steel and concrete components that resist seismic forces shall conform to the requirements of the AISC 360 and ACI 318. An R factor as set forth in Section 12.2.1 of ASCE 7 for the appropriate composite steel and concrete system is permitted where the structure is designed and detailed in accordance with the provisions of AISC 341, Part II. In Seismic Design Category B or above, the design of such systems shall conform to the requirements of AISC 341, Part II.~~

CHANGE SIGNIFICANCE: The requirement that composite structures in Seismic Design Categories D, E, and F provide substantiating evidence demonstrating that the proposed system will perform as intended by Part II of AISC 341 is no longer necessary because the 2010 edition of AISC 341 contains detailed provisions for testing composite special moment frames, composite partially restrained moment frames, and composite eccentrically braced frames. A new IBC section is added specifically for composite structures of structural steel and concrete, patterned after Section 2305 on structural steel. However, unlike structural steel structures regulated by Section 2305, composite structures must be designed and detailed in accordance with AISC 341 regardless of seismic design category. Therefore, the new section makes no specific reference to Seismic Design Categories D, E, and F.

CHANGE TYPE: Addition

CHANGE SUMMARY: A reference to the new AISI S110 standard for seismic design of cold-formed steel special moment frames has been added to Chapter 22. The new standard includes design provisions for a new cold-formed steel seismic force resisting system called Cold-Formed Steel—Special Bolted Moment Frames (CFS-SBMF).

2012 CODE: **2210.2 Seismic Requirements for Cold-Formed Steel Structures.** Where a response modification coefficient, R, in accordance with ASCE 7, Table 12.2-1, is used for the design of cold-formed steel structures, the structures shall be designed and detailed in accordance with the requirements of AISI S100, ASCE 8, and, for cold-formed steel special bolted moment frames, AISI S110.

Chapter 35. AISI S110-07, *Standard for Seismic Design of Cold-Formed Steel Structural Systems—Special Bolted Moment Frames.*

CHANGE SIGNIFICANCE: The 2012 IBC references the new AISI S110, *Standard for Seismic Design of Cold-Formed Steel Structural Systems—Special Bolted Moment Frames.* The standard was developed by AISI as a result of research conducted at the University of California at San Diego by Professors Chia-Ming Uang and Atsushi Sato. CFS-SBMF systems experience substantial inelastic deformation during a design seismic event, with most of the inelastic deformation occurring at the bolted connections due to slip and bearing. The CFS-SBMF system was vetted through the Building Seismic Safety Council (BSSC) process for

2210.2 continues

Cold-formed steel special bolted moment frame *(Photo Courtesy of FCP Inc. Structures)*

2210.2

Seismic Requirements for Cold-Formed Steel Structures

2210.2 continued   inclusion in the 2009 NEHRP Provisions and subsequently incorporated into the ASCE 7-10 standard. Cyclic testing has shown that the system has large ductility capacity and significant hardening. A capacity design procedure is provided in the AISI S110 Commentary so that the designer can explicitly calculate the seismic load effect with overstrength, E_{mh}, at the design story drift. Alternatively, a conservative system overstrength factor of 3.0 is also provided to be compatible with the conventional approach to compute E_{mh} in accordance with ASCE 7. To develop the expected mechanism, requirements based on capacity design principles are provided in the standard for the design of the beams, columns, and their connections. Table 12.2-1 of ASCE 7-10 includes seismic design parameters for CFS-SBMF system of $R = 3.5$, $\Omega_o = 3.0$ and $C_d = 3.5$. It should be noted that ASCE 7 limits structures using CFS-SBMF systems to one story in height and 35 feet. AISI S110 also includes specific requirements for quality assurance and quality control procedures.

2305

General Design Requirements for Lateral-Force-Resisting Systems

CHANGE TYPE: Modification

CHANGE SUMMARY: The provisions in Section 2305 for the lateral design of wood structures have been coordinated with those set forth in the 2008 edition of the AF&PA standard, *Special Design Provisions for Wind and Seismic* (SDPWS-08). Design and deflection values for stapled wood-frame diaphragms and shear walls remain in the code.

2012 CODE: 2305.1 General. Structures using wood ~~frame~~ shear walls ~~and~~ or wood frame diaphragms to resist wind, seismic, ~~and~~ or other lateral loads shall be designed and constructed in accordance with AF&PA SDPWS and the applicable provisions of Sections 2305, 2306, and 2307.

2305.2 Diaphragm Deflection. The deflection of wood frame diaphragms shall be determined in accordance with AF&PA SDPWS. The deflection (Δ) of a blocked wood structural panel diaphragm uniformly fastened throughout with staples is permitted to be calculated in accordance with ~~the following~~ Equation 23-1. If not uniformly fastened, the constant 0.188 (For SI: 1/1627) in the third term shall be modified ~~accordingly~~ by an approved method.

$$A = \frac{5vL^3}{8EAb} + \frac{vL}{4Gt} + 0.188Le_n + \frac{\sum(\Delta_c X)}{2b}$$ **(Equation 23-1)**

$$\text{For SI:} \quad A = \frac{0.052vL^3}{EAb} + \frac{vL}{4Gt} + \frac{Le_n}{1627} + \frac{\sum(\Delta_c X)}{2b}$$

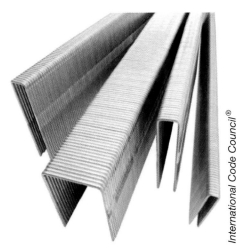

16-gage staples

where:

A = Area of chord cross section, in square inches (mm²).
B = Diaphragm width, in feet (mm).
E = Elastic modulus of chords, in pounds per square inch (N/mm²).
e_n = Staple deformation, in inches (mm) [see Table 2305.2(1)].
Gt = Panel rigidity through the thickness, in pounds per inch (N/mm) of panel width or depth [see Table 2305.2(2)].
L = Diaphragm length, in feet (mm).
V = Maximum shear due to design loads in the direction under consideration, in pounds per linear foot (plf) (N/mm).
Δ = The calculated deflection, in inches (mm).
$\sum(\Delta_c X)$ = Sum of individual chord-splice slip values on both sides of the diaphragm, each multiplied by its distance to the nearest support.

2305.3 Shear Wall Deflection. The deflection of wood-frame shear walls shall be determined in accordance with AF&PA SDPWS. The deflection (Δ) of a blocked wood structural panel shear wall uniformly fastened

2305 continues

2305 continued

throughout with staples is permitted to be calculated <u>in accordance with</u> <u>Equation 23-2:</u>

$$\Delta = \frac{8vh^3}{EAb} + \frac{vh}{Gt} + 0.75he_n + d_a\frac{h}{b}$$ **(Equation 23-2)**

$$For\ SI:\ \Delta = \frac{vh^3}{3EAb} + \frac{vh}{Gt} + \frac{he_n}{407.6} + d_a\frac{h}{b}$$

where:

A = Area of boundary element cross section in square inches (mm²) (vertical member at shear wall boundary).

b = Wall width, in feet (mm).

d_a = Vertical elongation of overturning anchorage (including fastener slip, device elongation, anchor rod elongation, etc.) at the design shear load (v).

E = Elastic modulus of boundary element (vertical member at shear wall boundary), in pounds per square inch (N/mm²).

e_n = Staple deformation, in inches (mm) [see Table 2305.2(1)].

Gt = Panel rigidity through the thickness, in pounds per inch (N/mm) of panel width or depth [see Table 2305.2(2)].

h = Wall height, in feet (mm).

v = Maximum shear due to design loads at the top of the wall, in pounds per linear foot (N/mm).

Δ = The calculated deflection, in inches (mm).

CHANGE SIGNIFICANCE: Section 2305 references the 2008 edition of the AF&PA standard, *Special Design Provisions for Wind and Seismic* (*SDPWS*) for lateral design of wood structures. Design values for nailed diaphragms and shear walls have been deleted from the tables in Section 2306 because the values are in the SDPWS standard. However, design values for stapled shear walls and diaphragms still remain in the code. Although the deflection of nailed wood-frame diaphragms and shear walls is determined in accordance with AF&PA SDPWS, the deflection of stapled diaphragms and shear walls is not covered in the standard. Section 2305 provides the formulae and parameters required to calculate the deflection of blocked wood structural panel diaphragms and shear walls fastened with staples.

2306
Allowable Stress Design

CHANGE TYPE: Modification

CHANGE SUMMARY: The provisions in Section 2306 addressing the allowable stress design of wood structures have been coordinated with those in the 2008 edition of the AF&PA standard, *Special Design Provisions for Wind and Seismic* (*SDPWS-08*).

2012 CODE: 2306.1 Allowable Stress Design. The ~~structural analysis~~ design and construction of wood elements in structures using allowable stress design shall be in accordance with the following applicable standards:
    (no change to list of allowable stress design standards)

2306.2 Wood-Frame Diaphragms. Wood-frame diaphragms shall be designed and constructed in accordance with AF&PA SDPWS. Where panels are fastened to framing members with staples, requirements and limitations of AF&PA SDPWS shall be met and the allowable shear values set forth in Table 2306.2(1) or 2306.2(2) shall be permitted. The allowable shear values in Tables 2306.2(1) and 2306.2(2) are permitted to be increased 40 percent for wind design.

2306.2.1 Wood Structural Panel Diaphragms. ~~Wood structural panel diaphragms shall be designed and constructed in accordance with AF&PA SDPWS. Wood structural panel diaphragms are permitted to resist horizontal forces, using the allowable shear capacities set forth in Table 2306.2.1(1) or 2306.2.1(2). The allowable shear capacities in Tables 2306.2.1(1) and 2306.2.1(2) are permitted to be increased 40 percent for wind design.~~

2306.2.2 Single Diagonally Sheathed Lumber Diaphragms. ~~Single diagonally sheathed lumber diaphragms shall be designed and constructed in accordance with AF&PA SDPWS.~~

2306.2.3 Double Diagonally Sheathed Lumber Diaphragms. ~~Double diagonally sheathed lumber diaphragms shall be designed and constructed in accordance with AF&PA SDPWS.~~

~~2306.2.4~~ 2306.2.1 Gypsum Board Diaphragm Ceilings. Gypsum board diaphragm ceilings shall be in accordance with Section 2508.5.

TABLE ~~2306.2.1(1)~~ 2306.2(1)   ALLOWABLE SHEAR VALUES (POUNDS PER FOOT) FOR WOOD STRUCTURAL PANEL DIAPHRAGMS UTILIZING STAPLES WITH FRAMING OF DOUGLAS-FIR-LARCH, OR SOUTHERN PINE[a] FOR WIND OR SEISMIC LOADING [h] [f]

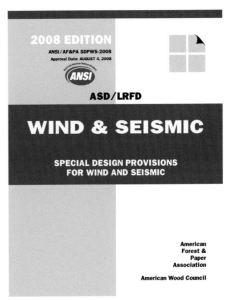

2008 edition of hte AF&PA standard
(American Wood Council, Leesburg, VA)

2306 continues

2306 continued

TABLE ~~2306.2.1(2)~~ 2306.2(2)   ALLOWABLE SHEAR VALUES (POUNDS PER FOOT) FOR WOOD STRUCTURAL PANEL BLOCKED DIAPHRAGMS UTILIZING MULTIPLE ROWS OF ~~FASTENERS~~ STAPLES (HIGH LOAD DIAPHRAGMS) WITH FRAMING OF DOUGLAS FIR, LARCH, OR SOUTHERN PINE[a] FOR WIND OR SEISMIC LOADING[b, g, h]

~~2306.3 Wood Structural Panel Shear Walls.~~   ~~Wood structural panel shear walls shall be designed and constructed in accordance with AF&PA SDPWS. Wood structural panel shear walls are permitted to resist horizontal forces, using the allowable capacities set forth in Table 2306.3. Allowable capacities in Table 2306.3 are permitted to be increased 40 percent for wind design.~~

2306.3 Wood-Frame Shear Walls.   Wood-frame shear walls shall be designed and constructed in accordance with AF&PA SDPWS. Where panels are fastened to framing members with staples, requirements and limitations of AF&PA SDPWS shall be met and the allowable shear values set forth in Table 2306.3(1), 2306.3(2) or 2306.3(3) shall be permitted. The allowable shear values in Tables 2306.3(1) and 2306.3(2) are permitted to be increased 40 percent for wind design. Panels complying with ANSI/APA PRP-210 shall be permitted to use design values for Plywood Siding in the AF&PA SDPWS.

TABLE ~~2306.3~~ 2306.3(1)   ALLOWABLE SHEAR VALUES (POUNDS PER FOOT) FOR WOOD STRUCTURAL PANEL SHEAR WALLS UTILIZING STAPLES WITH FRAMING OF DOUGLAS FIR-LARCH OR SOUTHERN PINE[a] FOR WIND OR SEISMIC LOADING [b,h,i,j,l] [f,g]

~~2306.4 Lumber Sheathed Shear Walls.~~   ~~Single and double diagonally sheathed lumber shear walls shall be designed and constructed in accordance with AF&PA SDPWS. Single and double diagonally sheathed lumber walls shall not be used to resist seismic forces in structures assigned to *Seismic Design Category* E or F.~~

~~2306.5 Particleboard Shear Walls.~~   ~~Particleboard shear walls shall be designed and constructed in accordance with AF&PA SDPWS. Particleboard shear walls shall be permitted to resist horizontal forces using the allowable shear capacities set forth in Table 2306.5. Allowable capacities in Table 2306.5 are permitted to be increased 40 percent for wind design. Particleboard shall not be used to resist seismic forces in structures assigned to *Seismic Design Category* D, E or F.~~

~~2306.6 Fiberboard Shear Walls.~~   ~~Fiberboard shear walls shall be designed and constructed in accordance with AF&PASDPWS. Fiberboard shear walls are permitted to resist horizontal forces, using the allowable shear capacities set forth in Table 2306.6. Allowable capacities in Table 2306.6 are permitted to be increased 40 percent for wind design. Fiberboard shall not be used to resist seismic forces in structures assigned to *Seismic Design Category* D, E or F.~~

~~TABLE 2306.5~~ ~~ALLOWABLE SHEAR FOR PARTICLEBOARD SHEAR WALL SHEATHING~~[b]

PANEL GRADE	MINIMUM NOMINAL PANEL THICKNESS (inch)	MINIMUM NAIL PENETRATION IN FRAMING (inches)	Nail size (common or galvanized box)	PANELS APPLIED DIRECT TO FRAMING Allowable shear (pounds per foot) nail spacing at panel edges (inches)[a]			
				6	4	3	2
~~M-S "Exterior Glue" and M-2 "Exterior" Glue"~~	~~3/8~~	~~1 1/2~~	~~6d~~	~~120~~	~~180~~	~~230~~	~~300~~
	~~3/8~~	~~1 1/2~~	~~8d~~	~~130~~	~~190~~	~~240~~	~~315~~
	~~1/2~~			~~140~~	~~210~~	~~270~~	~~350~~
	~~1/2~~	~~1 5/8~~	~~10d~~	~~185~~	~~275~~	~~360~~	~~460~~
	~~5/8~~			~~200~~	~~305~~	~~395~~	~~520~~

TABLE ~~2306.6~~ 2306.3(2) **ALLOWABLE SHEAR VALUES (plf) FOR WIND OR SEISMIC LOADING ON SHEAR WALLS OF FIBERBOARD SHEATHING BOARD CONSTRUCTION UTILIZING STAPLES FOR TYPE V CONSTRUCTION ONLY**[a,b,c,d,e]

~~2306.7 Shear Walls Sheathed With Other Materials.~~ ~~Shear walls sheathed with portland cement plaster, gypsum lath, gypsum sheathing or gypsum board shall be designed and constructed in accordance with AF&PA SDPWS. Shear walls sheathed with these materials p are permitted to resist horizontal forces using the allowable shear capacities set forth in Table 2306.7. Shear walls sheathed with portland cement plaster, gypsum lath, gypsum sheathing or gypsum board shall not be used to resist seismic forces in structures assigned to *Seismic Design Category* E or F.~~

TABLE ~~2306.7~~ 2306.3(3) **ALLOWABLE SHEAR VALUES FOR WIND OR SEISMIC FORCES FOR SHEAR WALLS OF LATH AND PLASTER OR GYPSUM BOARD WOOD FRAMED WALL ASSEMBLIES UTILIZING STAPLES**

Because this code change affected many tables in Section 2306, the entire code change text is too extensive to be included here. Refer to Code Change S208-09/10 in the *2012 IBC Code Changes Resource Collection* for the complete text and history of the code change.

CHANGE SIGNIFICANCE: Section 2306 references the 2008 edition of the AF&PA standard, *Special Design Provisions for Wind and Seismic (SDPWS)* for lateral design of wood structures. The general term "wood frame" has been added as a clarification of the intent so the code now refers to wood frame diaphragms and shear walls. Design values for nailed diaphragms and shear walls were deleted from the 2009 IBC tables because the values are in the SDPWS-08 standard. Design values for stapled shear walls and diaphragms remain in the tables in the code. Table footnotes have been revised to account for removal of allowable design values for nailed diaphragms and shear walls. Sections 2306.2 and 2306.3 have been revised to clarify that design and construction as well as limitations in the SDPWS are applicable to stapled diaphragms and shear walls. The

2306 continues

2306 continued

sections referring to particleboard, fiberboard, and lumber-sheathed shear walls have been deleted because they are covered under the general term of "wood frame shear walls" and their design provisions are included in the SDPWS standard.

A new national consensus standard, APA PRP-210, has been added to address wood structural panel siding products that were formerly covered under several national standards such as APA PRP-108. Siding products manufactured to the ANSI/APA PRP-210 standard have been developed specifically for wall-covering/weatherproofing applications, carry an exterior exposure durability classification, and have equivalent shear performance on a thickness-by-thickness basis when nailed in accordance with Table 2306.3. The code permits panels complying with ANSI/APA PRP-210 to be designed using the values for plywood siding in the AF&PA SDPWS.

To clarify the intent, the figure that accompanies the diaphragms table has been modified in the 2012 IBC. The figure in previous editions of the code has been difficult to interpret because of improper placement of the annotation lines. The new figure has a legend to better differentiate between blocking and framing members, and the annotation lines are more accurately placed in the figure. The design engineer is concerned with a specific diaphragm sheathing layout pattern with two loading cases, one for each orthogonal direction. Instead of six separate diaphragm configurations, the new figure shows the three diaphragm layout patterns and two load cases for each configuration. Although no technical changes were made to the figure, the new figure better illustrates the intent of the diaphragm design table.

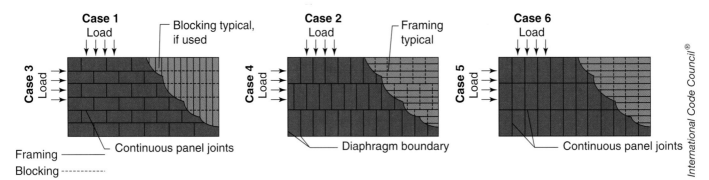

Layout patterns and loading for wood structural panel diaphragms

2307

Load and Resistance Factor Design

CHANGE TYPE: Modification

CHANGE SUMMARY: The provisions in Section 2307 dealing with the load and resistance factor design of wood structures are now coordinated with the 2008 edition of the AF&PA standard, *Special Design Provisions for Wind and Seismic* (SDPWS-08).

2012 CODE: 2307.1 Load and Resistance Factor Design.   The ~~structural analysis~~ <u>design</u> and construction of wood elements and structures using load and resistance factor design shall be in accordance with AF&PA NDS and AF&PA SDPWS.

~~**2307.1.1 Wood Structural Panel Shear Walls.**~~   ~~In structures assigned to Seismic Design Category D, E or F, where shear design values exceed 490 pounds per lineal foot (7154 N/m), all framing members receiving edge nailing from abutting panels shall not be less than a single 3-inch (76 mm) nominal member or two 2-inch (51 mm) nominal members fastened together in accordance with AF&PA NDS to transfer the design shear value between framing members. Wood structural panel joint and sill plate nailing shall be staggered at all panel edges. See Sections 4.3.6.1 and 4.3.6.4.2 of AF&PA SDPWS for sill plate size and anchorage requirements.~~

CHANGE SIGNIFICANCE: The requirements for 3X members at abutting panel joints in 2009 IBC Section 2307.1.1 are no longer necessary because similar provisions are contained in Section 4.3.7.1 of SDPWS-08. The SDPWS requires 3X framing at adjoining panel edges and staggered nailing where nail spacing is 2 inches or less on center at adjoining panel edges,

2307 continues

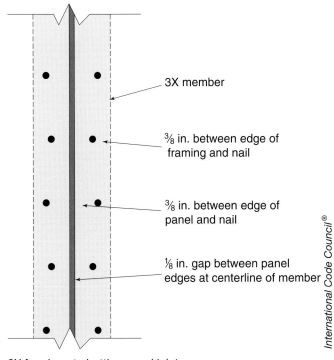

3X framing at abutting panel joints

- 3X member
- ⅜ in. between edge of framing and nail
- ⅜ in. between edge of panel and nail
- ⅛ in. gap between panel edges at centerline of member

International Code Council®

2307 continued   or where 10d common nails having penetration into framing members and blocking of more than 1½ inches are nailed at 3 inches or less on center at adjoining panel edges, or where the required nominal unit shear capacity on either side of the shear wall exceeds 700 plf for buildings in Seismic Design Category D, E, or F. An exception permits two 2X framing members provided they are fastened together in accordance with the NDS to transfer the shear between the members. When the fasteners connecting the two framing members are spaced less than 4 inches on center, the fasteners are required to be staggered.

CHANGE TYPE: Clarification

CHANGE SUMMARY: Table 2308.12.4 has been revised to provide a minimum percentage rather than a minimum length of wall bracing for conventionally constructed buildings in Seismic Design Categories D and E. New Section 2308.12.4.1 has been added to clarify that the substitution of alternate braced wall panels in buildings located in Seismic Design Categories D and E is permitted. Footnote "a" was also modified so that the 2:1 height-to-width ratio limitation does not apply to substituted alternate braced wall panels.

2012 CODE: 2308.12.4 Braced Wall Line Sheathing. Braced wall lines shall be braced by one of the types of sheathing prescribed by Table 2308.12.4, as shown in Figure 2308.9.3. The sum of lengths of braced wall panels at each braced wall line shall conform to <u>the required percentage of wall length required to be braced per braced wall line in</u> Table 2308.12.4. Braced wall panels shall be distributed along the length of the braced wall line and start at not more than 8 feet (2438 mm) from each end of the braced wall line. Panel sheathing joints shall occur over studs or blocking. Sheathing shall be fastened to studs, to top and bottom plates, and at panel edges occurring over blocking. Wall framing to which sheathing used for bracing is applied shall be nominal 2-inch-wide [actual 1½ inch (38 mm)] or larger members.

Cripple walls having a stud height exceeding 14 inches (356 mm) shall be considered a story for the purpose of this section and shall be braced as required for braced wall lines in accordance with <u>the required percentage of wall length required to be braced per braced wall line in</u> Table 2308.12.4. Where interior braced wall lines occur without a continuous foundation below, the length of parallel exterior cripple wall

2308.12 continues

2308.12
Braced Wall Line Sheathing

International Code Council®

Braced wall sheathing

2308.12 continued

bracing shall be one-and-one-half times the lengths required by Table 2308.12.4. Where the cripple wall sheathing type used is Type S-W and this additional length of bracing cannot be provided, the capacity of Type S-W sheathing shall be increased by reducing the spacing of fasteners along the perimeter of each piece of sheathing to 4 inches (102 mm) o.c.

2308.12.4.1 Alternative Bracing. An alternate braced wall panel constructed in accordance with Section 2308.9.3.1 or 2308.9.3.2 is permitted to be substituted for a braced wall panel in Section 2308.9.3 Items 2 through 8. For methods 2, 3, 4, 6, 7, and 8, each 48-inch (1219-mm) section or portion thereof required by Table 2308.12.4 is permitted to be replaced by one alternate braced wall panel constructed in accordance with Section 2308.9.3.1 or 2308.9.3.2. For method 5, each 96-inch (2438-mm) section (applied to one face) or 48-inch (1219-mm) section (applied to both faces) or portion thereof required by Table 2308.12.4 is permitted to be replaced by one alternate braced wall panel constructed in accordance with Section 2308.9.3.1 or 2308.9.3.2.

TABLE 2308.12.4   **Wall Bracing in Seismic Design Categories D and E (Minimum ~~Percentage Length~~ of Wall Bracing per Each ~~25 Linear Feet~~ of Braced Wall Line[a])**

Condition	SHEATHING TYPE[b]	$S_{DS} < 0.50$	$0.50 \leq S_{DS} < 0.75$	$0.75 \leq S_{DS} \leq 1.00$	$S_{DS} > 1.00$
One Story	G-P[c]	~~10 feet 8 inches~~	~~14 feet 8 inches~~	~~18 feet 8 inches~~	~~25 feet 0 inches~~
		43%	59%	75%	100%
	S-W	~~5 feet 4 inches~~	~~8 feet 0 inches~~	~~9 feet 4 inches~~	~~12 feet 0 inches~~
		21%	32%	37%	48%

a. Minimum length of panel bracing of one face of the wall for S-W sheathing or both faces of the wall for G-P sheathing; h/w ratio shall not exceed 2:1. For S-W panel bracing of the same material on two faces of the wall, the minimum length is permitted to be one-half the tabulated value but the h/w ratio shall not exceed 2:1 and design for uplift is required. The 2:1 h/w ratio limitation does not apply to alternate braced wall panels constructed in accordance with Section 2308.9.3.1 or 2308.9.3.2.

b.-c. *no changes to text*

CHANGE SIGNIFICANCE: Three modifications were made to Section 2308.12.4 that clarify the intent of the provisions regarding prescriptive wall bracing of buildings in Seismic Design Categories D and E. Conventional wood frame buildings in Seismic Design Category C require braced wall panels spaced every 25 feet, and such panels must comprise at least 25 percent of the braced wall line according to Table 2308.9.3(1). Conventional wood frame buildings in Seismic Design Category D and E are regulated by Table 2308.12.4 in determining the amount of required wall bracing. The table previously provided the minimum length of wall bracing per 25 feet of wall but did not give a minimum percentage. Because there was no percentage given, it was not clear how to properly apply the requirements of the table. For example, for a building sited where $S_{DS} > 1.00$g, the table required 12 feet of wall bracing for each 25 feet of wall. It was clear that a 25-foot-long wall required 12 feet of bracing and a 50-foot-long wall required 24 feet of bracing, but it was not explicitly clear what amount of bracing would be required for a 40-foot-long wall. To resolve this, Section 2308.12.4 and Table 2308.12.4 were revised to specify the minimum percentage of wall bracing instead of a minimum length per 25 feet. For example, for a building sited where $S_{DS} > 1.00g$, the table now requires 48 percent of wall bracing. Thus, a 40-foot-long braced wall line requires $0.48 \times 40 = 19$ feet of wall bracing.

Second, for buildings in Seismic Design Category A, B, and C, Sections 2308.9.3.1 and 2308.9.3.2 are quite clear that alternate braced wall panels can be substituted for the wall bracing required by Section 2308.9.3. However, for buildings in Seismic Design Category D and E, the code was not clear that alternate braced wall panels could be substituted for the wall bracing required by Table 2308.12.4. Although it was the intent of the code, it was difficult to conclude this by simply reading the language in the code. To resolve this issue, a new Section 2308.12.4.1 has been added that clearly states that alternate braced wall panels constructed in accordance with Section 2308.9.3.1 or 2308.9.3.2 are permitted to be substituted for braced wall panels required by Table 2308.12.4.

Third, footnote "a" of Table 2308.12.4 states that the height-to-width ratio for braced wall panels cannot exceed 2:1. For a typical 8-foot-high wall, this means the minimum length of the braced wall panel must be 4 feet. The alternate braced wall panel described in Section 2308.9.3.1 is 2 feet 8 inches in length, and the alternate braced wall panels adjacent to an opening in Section 2308.9.3.2 can be only 16 inches in length for a one-story building. It was not clear whether footnote "a" in effect means that the alternate braced wall panels of Section 2308.9.3.1 and 2308.9.3.2 cannot be used in Seismic Design Category D and E because of the restriction on height-to-width ratio. This table is derived from Section 12.4 and Table 12.4-2 of the 2003 NEHRP Provisions. Table 2308.12.4 originated with the NEHRP provisions, which do not require overturning restraint and do not include alternate braced wall panels. According to the NEHRP Commentary, it appears that the primary concern that led to the aspect ratio requirement is aimed at minimizing overturning demand due to the lack of overturning restraint. The alternate braced wall panels address overturning directly by requiring overturning restraint devices; thus they should not be subject to the 2:1 h/w limit. To resolve this, footnote "a" was modified so that the 2:1 height-to-width ratio limitation does not apply to alternate braced wall panels.

2406.1, 2406.4

Safety Glazing—Hazardous Locations

CHANGE TYPE: Modification

CHANGE SUMMARY: The hazardous locations identified in the safety glazing provisions have been reorganized and clarified in order to provide better consistency between the IBC and IRC.

2012 CODE: 2406.1 Human Impact Loads. Individual glazed areas, including glass mirrors, in hazardous locations as defined in Section 2406.4 shall comply with Sections 2406.1.1 through 2406.1.4.

> **Exception:** <u>Mirrors and other glass panels mounted or hung on a surface that provides a continuous backing support.</u>

2406.4 Hazardous Locations. The ~~following~~ <u>locations specified in Sections 2306.4.1 through 2406.4.7</u> shall be considered specific hazardous locations requiring safety glazing materials<u>.</u>~~:~~

> 1. ~~Glazing in swinging doors except jalousies (see Section 2406.4.1).~~
> 2. ~~Glazing in fixed and sliding panels of sliding door assemblies and panels in sliding and bifold closet door assemblies.~~
> 3. ~~Glazing in storm doors.~~
> 4. ~~Glazing in unframed swinging doors.~~
> 5. ~~Glazing in doors and enclosures for hot tubs, whirlpools, saunas, steam rooms, bathtubs and showers. Glazing in any portion of a building wall enclosing these compartments where the bottom exposed edge of the glazing is less than 60 inches (1524 mm) above a standing surface.~~

2406.4.1 Glazing in Doors. <u>Glazing in all fixed and operable panels of swinging, sliding, and bifold doors shall be considered a hazardous location.</u>

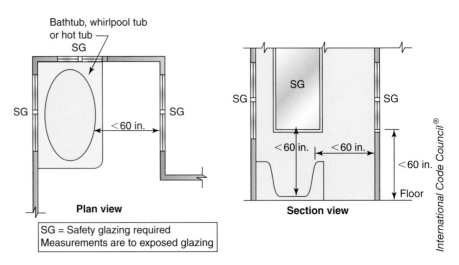

Hazardous locations near wet surfaces

Glazing near wet surfaces is considered as being in a hazardous location.

Exceptions:

1. Glazed openings of a size through which a 3-inch (76-mm)-diameter sphere is unable to pass.

2. Decorative glazing.

3. Glazing materials used as curved glazed panels in revolving doors.

4. Commercial refrigerated cabinet glazed doors.

6. **2406.4.2 Glazing Adjacent to Doors.** Glazing in an individual fixed or operable panel adjacent to a door where the nearest ~~exposed~~ vertical edge of the glazing is within a 24-inch (610-mm) arc of either vertical edge of the door in a closed position and where the bottom exposed edge of the glazing is less than 60 inches (1524 mm) above the walking surface shall be considered a hazardous location.

Exceptions:

1. Decorative glazing.

2. ~~Panels where~~ Where there is an intervening wall or other permanent barrier between the door and glazing.

3. Where access through the door is to a closet or storage area 3 feet (914 mm) or less in depth. Glazing in this application shall comply with Section ~~2406.4, Item 7~~ 2406.4.3.

4. Glazing in walls ~~perpendicular to the plane of the door in a closed position, other than the wall towards which the door~~

2406.1, 2406.4 continues

2406.1, 2406.4 continued

~~swings when opened,~~ <u>on the latch side of and perpendicular to the plane of the door in a closed position</u> in one- and two-family dwellings or within dwelling units in Group R-2.

~~7.~~ <u>**2406.4.3 Glazing in Windows.**</u>   Glazing in an individual fixed or operable panel~~, other than in those locations described in preceding Items 5 and 6, which~~ <u>that</u> meets all of the following conditions <u>shall be considered a hazardous location</u>:

~~7.~~1. ~~Exposed~~ <u>The exposed</u> area of an individual pane <u>is</u> greater than 9 square feet (0.84 m^2).

~~7.~~2. ~~Exposed~~ <u>The</u> bottom edge <u>of the glazing is</u> less than 18 inches (457 mm) above the floor.

~~7.~~3. ~~Exposed~~ <u>The</u> top edge <u>of the glazing is</u> greater than 36 inches (914 mm) above the floor.

~~7.~~4. One or more walking surface(s) <u>are</u> within 36 inches (914 mm), <u>measured</u> horizontally ~~of the plane~~ <u>and in a straight line,</u> of the glazing.

Exception<u>s</u>:   ~~Safety glazing for Item 7 is not required for the following installations:~~

~~1.~~   ~~A protective bar 11/2 inches (38 mm) or more in height, capable of withstanding a horizontal load of 50 pounds plf (730 N/m) without contacting the glass, is installed on the accessible sides of the glazing 34 inches to 38 inches (864 mm to 965 mm) above the floor.~~

1.   Decorative glazing.

<u>**2.**</u>   <u>When a horizontal rail is installed on the accessible side(s) of the glazing 34 to 38 inches above the walking surface. The rail shall be capable of withstanding a horizontal load of 50 pounds per linear foot (730 N/m) without contacting the glass and be a minimum of 1½ inches (38 mm) in cross-sectional height.</u>

~~2~~<u>3</u>.   ~~The outboard~~ <u>Outboard</u> panes in insulating glass units or multiple glazing where the bottom exposed edge of the glass is 25 feet (7620 mm) or more above any grade, roof, walking surface, or other horizontal or sloped (within 45 degrees of horizontal) (0.78 rad) surface adjacent to the glass exterior.

~~8.~~ <u>**2406.4.4 Glazing in Guards and Railings.**</u>   Glazing in guards and railings, including structural baluster panels and nonstructural in-fill panels, regardless of area or height above a walking surface, <u>shall be considered a hazardous location.</u>

~~9.   Glazing in walls and fences enclosing indoor and outdoor swimming pools, hot tubs and spas where all of the following conditions are present:~~

~~9.1.   The bottom edge of the glazing on the pool or spa side is less than 60 inches (1524 mm) above a walking surface on the pool or spa side of the glazing; and~~

~~9.2.   The glazing is within 60 inches (1524 mm) horizontally of the water's edge of a swimming pool or spa.~~

2406.4.5 Glazing and Wet Surfaces. Glazing in walls, enclosures, or fences containing or facing hot tubs, spas, whirlpools, saunas, steam rooms, bathtubs, showers, and indoor or outdoor swimming pools, where the bottom exposed edge of the glazing is less than 60 inches (1524 mm) measured vertically above any standing or walking surface, shall be considered a hazardous location. This shall apply to single glazing and all panes in multiple glazing.

> **Exception:** Glazing that is more than 60 inches (1524 mm), measured horizontally and in a straight line, from the water's edge of a bathtub, hot tub, spa, whirlpool, or swimming pool.

10. **2406.4.6 Glazing Adjacent Stairs and Ramps.** Glazing ~~adjacent to~~ where the bottom exposed edge of the glazing is less than 60 inches (1524 mm) above the plane of the adjacent walking surface of stairways, landings between flights of stairs, and ramps shall be considered a hazardous location ~~within 36 inches (914 mm) horizontally of a walking surface; when the exposed surface of the glass is less than 60 inches (1524 mm) above the plane of the adjacent walking surface.~~

Exceptions:

1. The side of a stairway, landing, or ramp that has a guard complying with the provisions of Sections 1013 and 1607.8, and the plane of the glass is greater than 18 inches (457 mm) from the railing.

2. Glazing 36 inches (914 mm) or more measured horizontally from the walking surface.

11. **2406.4.7 Glazing Adjacent The Bottom Stair Landing.** ~~Glazing adjacent to stairways within 60 inches (1524 mm) horizontally of the bottom tread of a stairway in any direction when the exposed surface of the glass is less than 60 inches (1524 mm) above the nose of the tread.~~ Glazing adjacent the landing at the bottom of a stairway where the glazing is less than 36 inches (914 mm) above the landing and within 60 inches (1524 mm) horizontally of the bottom tread shall be considered a hazardous location.

> **Exception:** ~~Safety glazing for Item 10 or 11 is not required for the following installations where:~~
>
> 1. ~~The side of a stairway, landing or ramp which has~~ Glazing that is protected by a guard ~~or handrail, including balusters or in-fill panels,~~ complying with ~~the provisions of~~ Sections 1013 and 1607.8~~; and 2. The~~ the plane of the glass is greater than 18 inches (457 mm) from the ~~railing~~ guard.

2406.4.1 Exceptions. ~~The following products, materials and uses shall not be considered specific hazardous locations:~~

1. ~~Openings in doors through which a 3-inch (76 mm) sphere is unable to pass.~~

2. ~~Decorative glass in Section 2406.4, Item 1, 6 or 7.~~

2406.1, 2406.4 continues

2406.1, 2406.4 continued

3. ~~Glazing materials used as curved glazed panels in revolving doors.~~

4. ~~Commercial refrigerated cabinet glazed doors.~~

5. ~~Glass-block panels complying with Section 2101.2.5.~~

6. ~~Louvered windows and jalousies complying with the requirements of Section 2403.5.~~

7. ~~Mirrors and other glass panels mounted or hung on a surface that provides a continuous backing support.~~

CHANGE SIGNIFICANCE: An effective reorganization of the hazardous locations for safety glazing purposes has been accomplished, resulting in the elimination of conflicts, creation of consistency, and ease of use. By taking the 11 hazardous locations and seven exceptions that previously existed in Section 2604.4 and reformatting them into seven individual provisions with the appropriate exceptions located directly within the applicable provision, the understanding of the intent should be much easier. Code users should be aware that although this was predominately a reorganization effort, some technical changes do result from the relocation or combination of provisions. As an example, see the discussion related to Section 2406.4.5.

The point-by-point explanation that follows should assist in understanding the reorganization of the various requirements.

The exception to Section 2406.1 was relocated from Item 7 in Section 2406.4.1 with no change in application because these items were previously exempted.

The "glazing in doors" requirements of new Section 2406.4.1 now include Items 1 through 4 from previous Section 2406.4. In a technical change, jalousie windows were previously exempted from the safety glazing requirement. Because jalousies are no longer listed among the exceptions, they are now required to be safety glazing unless exempted by the limited size or decorative glazing provisions of Exception 1 or 2. The four exceptions that are listed in this section were previously listed as the first four exceptions in Section 2406.4.1.

Section 2406.4.2 dealing with "glazing adjacent to doors" and several of the exceptions were previously found in Item 6 of Section 2406.4. New Exception 1 was previously Exception 2 in Section 2406.4.1. Exception 4 was revised in order to clarify the provisions and to coordinate with similar text in the IRC.

The glazed window requirements of Section 2406.4.3 now combine the provisions of previous Section 2406.4, Item 7, and Exception 2 from Section 2406.4.1. The provisions regarding protecting the window from impact by the use of a horizontal rail have been revised in order to coordinate with the language of the IRC.

Section 2406.4.5 addressing glazing adjacent to wet surfaces is essentially a combination of the previous provisions related to glazing adjacent to hot tubs, bathtubs, and showers (Item 5 in Section 2406.4) as well as pools and spas (Item 9 in Section 2406.4). A single section relating to hazardous glazing adjacent to water will include the criteria that previously applied to walls and fences around a pool as the means to determine if the glazing is in a hazardous location. This revision will affect the application of the requirements to the bathtubs, showers, and other items that were previously included in Item 5 in Section 2406.4. The issue of glazing adjacent to a freestanding bathtub is addressed in the same manner as a hot tub, spa, or whirlpool. To illustrate the most significant impact of the change, consider a bathroom with a shower that is enclosed on three sides by solid walls and on the fourth side by a set of glass doors. Previously, these four sides created the "enclosure" for the shower and the only location regulated for safety glazing purposes was the glass doors. Under the revised provisions, if a person would step out of the enclosed shower and a window in the wall of the bathroom is located within the established 60-inch height and 60-inch horizontal distance, that window is regulated. Previously, because the bathroom window was considered outside of the shower "enclosure," it would have been regulated by the general window requirements (Section 2406.4, Item 7) and not by the shower enclosure provisions of Section 2406.4, Item 5.

The provisions of Sections 2406.4.6 and 2406.4.7 will replace what had previously been Section 2406.4, Items 10 and 11, addressing two different locations related to glazing near stairways. The primary distinction is that Section 2406.4.7 will only regulate the glazing that is adjacent to the bottom landing on a stair. Therefore, when a stairway terminates at a floor level, Section 2406.4.7 would be applicable within 60 inches of the bottom tread, but if the landing were located between two adjacent flights of stairs, then Section 2406.4.6 would be the applicable provision. Code users should note that the provisions dealing with glazing at the bottom of the stair will apply when the bottom edge of the glazing is less than 36 inches above the landing. Previously, any glazing that was less than 60 inches above the nosing of the last tread was regulated. The reduction down to the 36-inch height was made based on an exception within the IRC exempting safety glazing where a solid wall or panel that places the glazing at or above the handrail or guard height is capable of withstanding the guard loading requirements.

2406.2

Safety Glazing—Impact Test

CHANGE TYPE: Modification

CHANGE SUMMARY: The default impact test criteria have been revised to impose the more restrictive test methodology. The higher impact requirements will apply unless the tables in Section 2406.2 allow for a lower impact test to be used.

2012 CODE: **2406.2 Impact Test.** Where required by other sections of this code, glazing shall be tested in accordance with CPSC 16 CFR 1201. Glazing shall comply with the test criteria for Category ~~I or~~ II, ~~as,~~ unless otherwise indicated in Table 2406.2(1).

> **Exception:** Glazing not in doors or enclosures for hot tubs, whirlpools, saunas, steam rooms, bathtubs, and showers shall be permitted to be tested in accordance with ANSI Z97.1. Glazing shall comply with the test criteria for Class A, ~~or B as~~ unless otherwise indicated in Table 2406.2(2).

CHANGE SIGNIFICANCE: Previously, a reference to Tables 2406.2(1) and 2406.2(2) was simply provided in order to establish the appropriate test criteria for safety glazing based on the size and location of the glazing. However, the tables did not address all of the hazardous locations listed in Section 2406.4. Without any specific test criteria assigned to many of the safety glazing locations, the requirements were confusing and difficult to enforce. As an example, the locations that were previously listed in Items 8, 9, 10, and 11 of Section 2406.4 of the 2009 IBC were not addressed within in either of the two glazing classification tables and therefore no appropriate test criteria were identified.

By modifying the base paragraph and exception as indicated, the more severe impact test becomes the default requirement (Class II, or Class A where applicable). Then, by referencing the tables, the lower

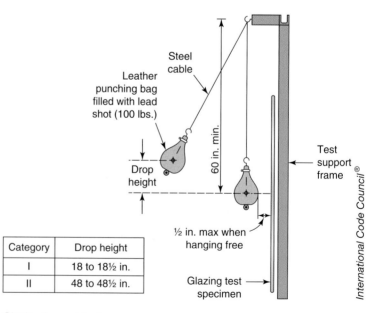

Category	Drop height
I	18 to 18½ in.
II	48 to 48½ in.

Glazing impact test

impact testing procedures can now be used when the tables indicate those criteria are appropriate, ensuring that the test criteria are addressed for every hazardous location.

The use of the proper test criteria is important to ensure that the safety glazing can truly withstand the anticipated human impact loads it may face.

2510.6

Water-Resistive Barriers for Stucco Applications

CHANGE TYPE: Modification

CHANGE SUMMARY: In order to reduce the likelihood of moisture getting into the building, detailed requirements have been provided for the installation of the two layers of weather-resistive barriers that are required behind stucco-covered exterior walls.

2012 CODE: 2510.6 Water-Resistive Barriers. Water-resistive barriers shall be installed as required in Section 1404.2 and, where applied over wood-based sheathing, shall include a water-resistive vapor-permeable barrier with a performance at least equivalent to two layers of Grade D paper. <u>The individual layers shall be installed independently such that each layer provides a separate continuous plane and any flashing (installed in accordance with Section 1405.4) intended to drain to the water-resistive barrier is directed between the layers.</u>

> **Exception:** Where the water-resistive barrier that is applied over wood-based sheathing has a water resistance equal to or greater than that of 60-minute Grade D paper and is separated from the stucco by an intervening, substantially non-water-absorbing layer or drainage space.

CHANGE SIGNIFICANCE: When installing stucco, a weather-resistive barrier has been required to have a performance level "at least equivalent to two layers" of Grade D paper, however there has been no specific information to indicate how the layers are to be installed. When installing two layers, there are two separate methods of installation, which provide different levels of performance, installing the layers together or installing them separately. These two options are often recognized as a "two-ply system" or a "two-layer system," respectively. Where each layer of the water-resistive barrier is installed individually (the two-layer system), a better level of moisture protection is provided. When installed individually, the interior layer is configured to form a continuous drainage plane and is integrated with the flashing. The independent outboard layer serves to separate and protect the inner layer from the stucco and allows a space

Two-Layer System

- Each layer of water-resistive barrier is individually installed in a ship lapped fashion

- Interior layer forms continuous drainage plane and integrated with flashing

Two-Ply System

- Both layers of water-resistive barrier installed and lapped together

- Exterior layer integrated with flashing

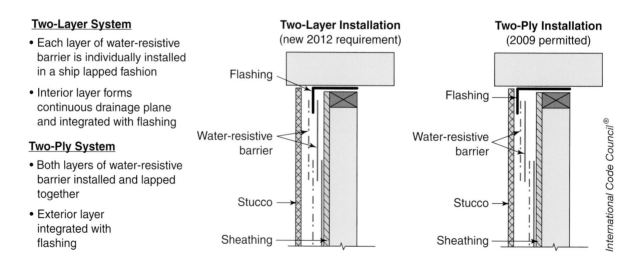

Installation of water-resistive barrier

between the two layers to improve drainage. If the layers are installed together (the two-ply system), then they function as a single layer, and the only benefit of using two layers is the additional water resistance.

The primary benefit of using two layers of water-resistive barrier (WRB) can only be realized if the method and manner of the installation establishes a continuous drainage plane, separated from the stucco. In a two-layer system, each layer provides a separate and distinct function. The primary function of the inboard layer is to resist water penetration into the building cavity. This interior layer should be integrated with window and door flashings, the weep screed at the bottom of the wall, and any through-wall flashings or expansion joints. The inner layer becomes the drainage plane for any incidental water that gets through the outer layer or at one of the joints or openings or where the outer layer is damaged. The primary function of the outboard layer (layer that comes in contact with the stucco) is to separate the stucco from the water-resistive barrier. This layer has historically been called a sacrificial layer, intervening layer, or bond break layer.

Where each layer is installed independently as the code now requires, it becomes possible to install each layer to meet its intended function. A continuous drainage plane can be established on the inboard layer. However, this is not the case with a two-ply system that functions as a single layer with additional water resistance as the only benefit. If additional water holdout provides no other benefit, then the installation of a superior single-layer WRB would be sufficient.

The new language should ensure that the two required layers are installed "independently" to provide the best level of protection possible based on the two installation options.

2603.4.1.14

Foam Plastic Insulation Installed in Floor Assemblies

CHANGE TYPE: Addition

CHANGE SUMMARY: The use of ½-inch wood structural panels installed on the walking surface side of a floor assembly is now permitted as an alternative to the thermal barrier typically required where foam plastic insulation is installed within a floor assembly.

2012 CODE: <u>2603.4.1.14 Floors.</u> <u>The thermal barrier specified in Section 2603.4 is not required to be installed on the walking surface of a structural floor system that contains foam plastic insulation when the foam plastic is covered by a minimum nominal ½-inch (12.7-mm)-thick wood structural panel or approved equivalent. The thermal barrier specified in Section 2603.4 is required on the underside of the structural floor system that contains foam plastic insulation when the underside of the structural floor system is exposed to the interior of the building.</u>

<u>**Exception:**   Foam plastic used as part of an interior floor finish.</u>

CHANGE SIGNIFICANCE: A viable means to protect foam plastic insulation when it is installed within a floor system is now provided. The thermal barrier typically required where foam plastic insulation is installed beneath a walking surface must not only be an adequate barrier to protect the foam plastic but must also be durable enough to withstand the load and wear-and-tear that is needed for the floor. The new allowance is consistent with the minimum protection required for attics and crawl spaces and should eliminate confusion about what level of protection is needed for floor assemblies containing foam insulation, such as those where structural insulated panels (SIPs) are used.

With society's focus on energy efficiency and conservation, many new types of products are being used that incorporate foam plastic insulation for energy reasons. One example is the use of structural insulated panels where the foam plastic is laminated between two structural wood facings.

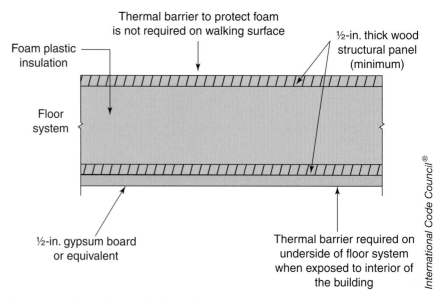

Requirements for floors with foam plastic insulation

Insulated panel system being installed

These types of panels can be used as a wall, floor, or roof. Regardless of where or why the material is installed, the code requires that foam plastic insulation materials be adequately separated from the interior of the building for both occupant protection and to prevent ignition sources from reaching the insulation.

Foam plastic is generally required to be protected by a thermal barrier, which typically consists of ½-inch gypsum wallboard. In the case of flooring, gypsum wallboard or other common thermal barrier materials cannot be used on the walking surfaces due to their friability. This concern is now addressed because ½-inch-thick plywood, or equivalent, when installed on the walking surface is considered to provide sufficient protection to the foam plastic insulation. While a ½-inch wood structural panel (i.e., plywood or oriented strand board) is not by itself considered as a complying "thermal barrier" as required by Section 2603, it will fulfill the dual need for structural strength and thermal protection of the foam insulation. In the case of a floor, the panels will provide sufficient protection because, in the event of an interior fire, the floor faces a reduced exposure and is typically the last building element to be significantly exposed by the fire.

It is important to note that the use of the ½-inch wood structural panels is only accepted on the walking surface side of the floor. If the floor is used in multi-story construction, then the underside of the floor system (ceiling of the room below) must be covered by the typically required thermal barrier. The required thermal barrier protection on the bottom side of the assembly cannot be reduced because it does not face the problems of a walking surface and it will face a more severe exposure to an interior fire.

As an additional note, the exception has been added to address items such as carpet padding, etc., that do not need to be covered by a thermal barrier.

2603.7, 2603.8

Interior Finish in Plenums

CHANGE TYPE: Modification

CHANGE SUMMARY: Three different options for separating foam plastic insulation used in plenum spaces are now provided. These options vary in relationship to the maximum permitted flame-spread and smoke-developed rating index.

2012 CODE: 2603.7 <u>Interior Finish in Plenums.</u> ~~Foam plastic insulation shall not be used as interior wall or ceiling finish in plenums except as permitted in Section 2604 or when protected by a thermal barrier in accordance with Section 2603.4.~~ <u>Foam plastic insulation used as interior wall or ceiling finish in plenums shall comply with one or more of the following:</u>

1. <u>The foam plastic insulation shall be separated from the plenum by a thermal barrier complying with Section 2603.4 and shall exhibit a flame-spread index of 75 or less and a smoke-developed index of 450 or less when tested in accordance with ASTM E84 or UL 723 at the thickness and density intended for use.</u>

2. <u>The foam plastic insulation shall exhibit a flame-spread index of 25 or less and a smoke-developed index of 50 or less when tested in accordance with ASTM E84 or UL 723 at the thickness and</u>

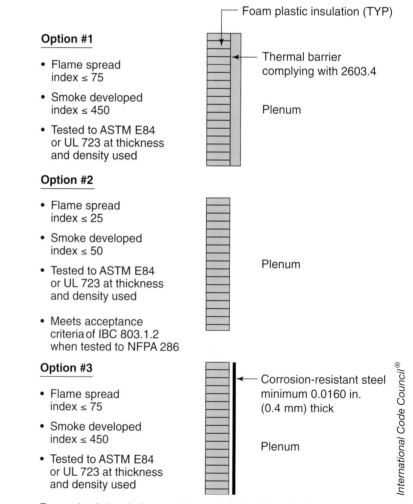

Option #1

- Flame spread index ≤ 75
- Smoke developed index ≤ 450
- Tested to ASTM E84 or UL 723 at thickness and density used

Option #2

- Flame spread index ≤ 25
- Smoke developed index ≤ 50
- Tested to ASTM E84 or UL 723 at thickness and density used
- Meets acceptance criteria of IBC 803.1.2 when tested to NFPA 286

Option #3

- Flame spread index ≤ 75
- Smoke developed index ≤ 450
- Tested to ASTM E84 or UL 723 at thickness and density used

Foam plastic insulation (TYP)

Thermal barrier complying with 2603.4

Plenum

Plenum

Corrosion-resistant steel minimum 0.0160 in. (0.4 mm) thick

Plenum

International Code Council®

Foam plastic insulation used as an interior finish in plenums density intended for use and shall meet the acceptance criteria of Section 803.1.2 when tested in accordance with NFPA 286.

   3. The foam plastic insulation shall be covered by corrosion-resistant steel having a base metal thickness of not less than 0.0160 inch (0.4 mm) and shall exhibit a flame-spread index of 75 or less and a smoke-developed index of 450 or less when tested in accordance with ASTM E84 or UL 723 at the thickness and density intended for use.

2603.8 Interior Trim in Plenums. Foam plastic insulation used as interior trim in plenums shall comply with the requirements of Section 2603.7.

CHANGE SIGNIFICANCE: The installation of foam plastic insulation within concealed plenum spaces creates a unique challenge due to the general requirements for protection of the insulation and how the protection is accomplished. Historically, the IBC and IMC provisions for materials and insulation exposed in plenums has required that they be either noncombustible or that they have a flame-spread index rating of 25 or less and a smoke-developed index of no more than 50 when using the ASTM E 84 test standard.

The requirements for the use of foam plastic insulation as interior wall and ceiling finish or as interior trim within plenums have been revised by clarifying the different protection methods and establishing the various alternatives that are available to provide the protection.

Foam plastic insulation that is exposed within the plenum as either an interior finish material or as interior trim must have a flame-spread index of 25 or less and a maximum smoke-developed index of 50. In addition, it must meet the requirements of the full-scale room-corner fire test (NFPA 286) with provisions for flame spread, heat release, smoke release, and inability for flashover. Previously, the use of foam plastic was deemed acceptable if in compliance with any of the alternate special approval standards listed in Section 2603.9 of the 2009 IBC (Section 2603.10 in 2012 edition). However, when using the 2009 IBC, only the NFPA 286 standard provided a means to determine the pass/fail criteria for smoke development. The reference in Section 2603.7 of the 2009 IBC to Section 2604 and, therefore, Section 2604.2 would have also allowed the use of foam plastic interior trim, which created an unreasonable risk.

In lieu of leaving the foam plastic insulation exposed and imposing fairly restrictive flame and smoke ratings, two alternate means for protection are now available. If either of these alternatives are used, the material is permitted to have a higher flame-spread and smoke-developed index rating. Item 1 will allow the use of the thermal barrier as a means of protection. This will allow the foam plastic to have a maximum flame-spread rating of 75 and a smoke-developed rating of 450. Item 3 will allow the same increased rating values if the insulation is covered by a layer of corrosion-resistant steel.

As mentioned previously, the use of trim made of foam plastic insulation should be regulated the same as where the insulation is used as a wall or ceiling finish material within the plenum. Section 2603.8 will refer to the three protection options that are allowed by Section 2603.7 and will therefore eliminate the possibility that the trim may be allowed by any type of reference to Section 2604 and 2604.2.

2603.10, 2603.10.1

Special Approval of Foam Plastics

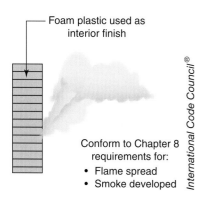

Special approval testing

CHANGE TYPE: Modification

CHANGE SUMMARY: The specific approval requirements now ensure that the smoke development of all assemblies that contain foam plastic is evaluated, regardless of the test standard used.

2012 CODE: ~~2603.9~~ 2603.10 Special Approval.   Foam plastic shall not be required to comply with the requirements of Sections 2603.4 through ~~2603.7~~ 2603.8, where specifically approved based on large-scale tests such as, but not limited to, NFPA 286 (with the acceptance criteria of Section 803.2), FM 4880, UL 1040, or UL 1715. Such testing shall be related to the actual end-use configuration and be performed on the finished manufactured foam plastic assembly in the maximum thickness intended for use. Foam plastics that are used as interior finish on the basis of special tests shall also conform to the flame-spread and smoke-developed requirements of Chapter 8. Assemblies tested shall include seams, joints, and other typical details used in the installation of the assembly and shall be tested in the manner intended for use.

2603.10.1 Exterior Walls.   Testing based on Section 2603.10 shall not be used to eliminate any component of the construction of an exterior wall assembly when that component was included in the construction that has met the requirements of Section 2603.5.5.

CHANGE SIGNIFICANCE: The imposition of the smoke-development requirements when using any of the special approval test standards for assemblies contain foam plastic now ensures that assemblies tested using these special approval options will provide a comparable level of performance and safety as would be required by the general provisions of Chapters 8 and 26. When using the test methods allowed by Section 2603.10, only NFPA 286 by virtue of the reference to Section 803.2 previously imposed any limitation on the smoke development. The other tests—FM 4880, UL 1040, and UL 1715—do not contain any smoke-development criterion and lack any reference to them. Therefore, compliance with the smoke-development requirements of Chapters 7, 8, and 26 was not specified.

In general, all thermal and sound-insulating materials addressed by the code, including noncombustible insulation materials and foam plastics, must meet certain minimum performance levels for the material to be used in the building, including those criteria related to both flame spread (fire growth) and smoke production. When using the special approval tests for foam plastics, testing must relate to the actual end-use configuration based on the finished assembly in the maximum thickness intended for use. But, as mentioned earlier, in the 2009 IBC Section 2603.9 did not clearly state whether it included or excluded smoke development requirements.

Chapter 8 is now referenced and along with the flame-spread and smoke-developed ratings using the ASTM E84 and UL 723 test methods. As a means to illustrate the application of this code change, consider the following example. Under the 2009 code, Section 2603.9 was allowed as an alternate to the plenum provisions of Sections 2603.7 and 2604 and would permit foam plastics in a plenum without the installation of a thermal barrier and without any limitation on the smoke-development rating, even if the thermal barrier was eliminated.

It has now been clarified that the "special approval" testing of Section 2603.10.1 cannot be used to eliminate any component of an assembly that was included in the test required by Section 2603.5.5. This ensures that the assemblies used for both of these tests are identical. Section 2603.5.5 requires that exterior walls with foam plastic insulation are tested to the NFPA 285 standard. This standard uses a more severe fire exposure and is run for a longer period of time than most of the tests listed in Section 2603.10. To illustrate this provision, consider the following situation. A single layer of gypsum wallboard is used to cover the interior surface of a wall assembly containing foam plastic insulation when it is tested to the NFPA 285 standard. Because one of the purposes of Section 2603.10 is to allow for the thermal barrier to be removed, it is proposed to test the wall to one of the four listed standards, but in this test, the gypsum would be removed to determine if the thermal barrier could be eliminated. Testing different assemblies to different requirements will not reflect the actual end-use configuration or show that the same assembly can meet both of the code's requirements (Sections 2603.5.5 and 2603.10). This would be an "apples to oranges" comparison and would be of little value in evaluating the acceptability of the wall assembly.

2610.3

Slope Requirements of a Dome Skylight

CHANGE TYPE: Modification

CHANGE SUMMARY: The dimension that is used for determining the minimum slope requirements of a dome skylight has been revised.

2012 CODE: 2610.3 Slope. Flat or corrugated light-transmitting plastic skylights shall slope at least four units vertical in 12 units horizontal (4:12). Dome-shaped skylights shall rise above the mounting flange a minimum distance equal to 10 percent of the maximum ~~span~~ width of the dome but not less than 3 inches (76 mm).

> **Exception:** Skylights that pass the Class B Burning Brand Test specified in ASTM E108 or UL 790.

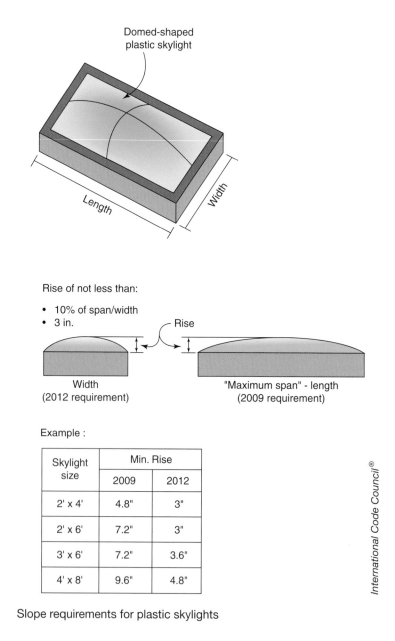

Domed-shaped plastic skylight

Length

Width

Rise of not less than:

- 10% of span/width
- 3 in.

Rise

Width
(2012 requirement)

"Maximum span" - length
(2009 requirement)

Example :

Skylight size	Min. Rise	
	2009	2012
2' x 4'	4.8"	3"
2' x 6'	7.2"	3"
3' x 6'	7.2"	3.6"
4' x 8'	9.6"	4.8"

Slope requirements for plastic skylights

International Code Council®

International Code Council®

Dome-shaped plastic skylight

CHANGE SIGNIFICANCE: Because plastic skylights represent a potential means for a fire to spread to a structure, the skylights must be provided with a minimum slope so burning brands or embers will roll off of them and not ignite the plastic material. The amount of slope required for a domed skylight has previously been determined based upon the maximum span of the skylight but with a minimum rise of 3 inches.

The size and aspect ratios (length/width) of early-generation plastic dome skylights have typically been less than current skylights. The length-to-width ratio was often 1:1 or 1.5:1, and seldom exceeded 2:1. When combined with the smaller skylight sizes that were typical, basing the dome's required rise on the "maximum span" with a minimum of 3 inches was not considered to be excessive or difficult to accomplish. With more recent changes in the materials and manufacturing process, it is now possible for the skylight size to be larger and for the length of dome shaped skylights to be much greater than the width; with aspect ratios (length/width) of 4 or more not uncommon. For these larger sizes and ratios, basing the required rise/slope on 10 percent of the maximum span was deemed excessive. The revision will simply allow the rise to be based upon the maximum width (the smaller of the two dimensions) while still retaining the minimum rise of 3 inches. Larger size dome-shaped skylights can now be accomodated and the slope continues to be adequate in order to shed burning brands or embers.

2612, 202

Fiber-Reinforced Polymer

CHANGE TYPE: Modification

CHANGE SUMMARY: Fiber-reinforced polymer installed on an exterior wall must now be a Class A material and is limited to 10 percent of the exterior wall area for any individual element or group of nonseparated elements.

2012 CODE: 202 Definitions.

FIBER REINFORCED POLYMER. A polymeric composite material consisting of reinforcement fibers, <u>such as glass,</u> impregnated with a fiber-binding polymer which is then molded and hardened. <u>Fiber-reinforced polymers are permitted to contain cores laminated between fiber-reinforced polymer facings.</u>

~~**FIBERGLASS REINFORCED POLYMER.** Polymeric composite material consisting of glass reinforcement fibers impregnated with a fiber-binding polymer which is then molded and hardened.~~

SECTION 2612
FIBER-REINFORCED POLYMER ~~AND FIBERGLASS REINFORCED POLYMER~~

2612.1 General. The provisions of this section shall govern the requirements and uses of fiber-reinforced polymer ~~or fiberglass reinforced polymer~~ in and on buildings and structures.

2612.2 Labeling and Identification. Packages and containers of fiber-reinforced polymer ~~or fiberglass reinforced polymer~~ and their components delivered to the job site shall bear the label of an approved agency showing the manufacturer's name, product listing, product identification, and information sufficient to determine that the end use will comply with the code requirements.

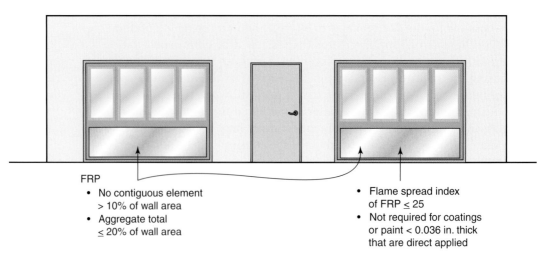

FRP
- No contiguous element > 10% of wall area
- Aggregate total ≤ 20% of wall area

- Flame spread index of FRP ≤ 25
- Not required for coatings or paint < 0.036 in. thick that are direct applied

Requirements for fiber-reinforced polymer (FRP)

International Code Council®

2612.3 Interior Finish~~es~~. Fiber-reinforced polymer ~~or fiberglass re-~~ ~~inforced polymer~~ used as interior finish<u>es, decorative materials, or trim</u> shall comply with Chapter 8.

<u>2612.3.1 Foam Plastic Cores.</u> <u>Fiber-reinforced polymer used as interior finish and which contain foam plastic cores shall comply with Chapter 8 and Chapter 26.</u>

~~2612.4 Decorative Materials and Trim.~~ ~~Fiber reinforced polymer or fiberglass reinforced polymer used as decorative materials or trim shall comply with Section 806.~~

2612.4 Light-Transmitting Materials. Fiber-reinforced polymer ~~or~~ ~~fiberglass reinforced polymer~~ used as light-transmitting materials shall comply with Sections 2606 through 2611 as required for the specific application.

2612.5 Exterior Use. Fiber-reinforced polymer ~~or fiberglass reinforced~~ ~~polymer~~ shall be permitted to be installed on the exterior walls of buildings of any type of construction when such polymers meet the requirements of Section 2603.5 ~~and is fireblocked~~. <u>Fireblocking shall be installed in accordance with Section 718.</u> ~~The fiber reinforced polymer or the fiberglass reinforced polymer shall be designed for uniform live loads as required in Table 1607.1 as well as for snow loads, wind loads and earthquake loads as specified in Sections 1608, 1609 and 1613, respectively.~~

Exceptions:

1. <u>Compliance with Section 2603.5 is not required</u> when all of the following conditions are met:

   1.1. <u>The fiber-reinforced polymer shall not exceed an aggregate total of 20 percent of the area of the specific wall to which it is attached, and no single architectural element shall exceed 10 percent of the area of the specific wall to which it is attached, and no contiguous set of architectural elements shall exceed 10 percent of the area of the specific wall to which they are attached.</u> ~~When the area of the fiber reinforced polymer or the fiberglass reinforced polymer does not exceed 20 percent of the respective wall area, the fiber reinforced polymer or the fiberglass reinforced polymer shall have a flame-spread index of 25 or less or when the area of the fiber reinforced polymer or the fiberglass reinforced polymer does not exceed 10 percent of the respective wall area, the fiber reinforced polymer or the fiberglass reinforced polymer shall have a flame-spread index of 75 or less. The flame-spread index requirement shall not be required for coatings or paints having a thickness of less than 0.036 inch (0.9 mm) that are applied directly to the surface of the fiber reinforced polymer or the fiberglass reinforced polymer.~~

2612, 202 continues

2612, 202 continued

1.2. The fiber-reinforced polymer shall have a flame-spread index of 25 or less. The flame-spread index requirement shall not be required for coatings or paints having a thickness of less than 0.036 inch (0.9 mm) that are applied directly to the surface of the fiber-reinforced polymer.

~~1.2.~~ **1.3.** Fireblocking complying with Section 718.2.6 shall be installed.

~~1.3.~~ **1.4.** The fiber-reinforced polymer ~~or the fiberglass reinforced polymer~~ shall be installed directly to a noncombustible substrate or be separated from the exterior wall by one of the following materials: corrosion-resistant steel having a minimum base metal thickness of 0.016 inch (0.41 mm) at any point, aluminum having a minimum thickness of 0.019 inch (0.5 mm), or other approved noncombustible material.

~~1.4. The fiber-reinforced polymer or the fiberglass reinforced polymer shall be designed for uniform live loads as required in Table 1607.1 as well as for snow loads, wind loads, and earthquake loads as specified in Sections 1608, 1609, and 1613, respectively.~~

2. Compliance with Section 2603.5 is not required when the fiber-reinforced polymer is ~~When~~ installed on buildings that are 40 feet (12,190 mm) or less above grade when all of the following conditions are met:

2.1. The fiber-reinforced polymer ~~or the fiberglass reinforced polymer~~ shall meet the requirements of Section 1406.2. ~~and shall comply with all of the following conditions:~~

~~2.1.~~ **2.2.** Where the fire separation distance is 5 feet (1524 mm) or less, the area of the fiber-reinforced polymer ~~or the fiberglass reinforced polymer~~ shall not exceed 10 percent of the wall area. Where the fire separation distance is greater than 5 feet (1524 mm), there shall be no limit on the area of the exterior wall coverage using fiber-reinforced polymer ~~or the fiberglass reinforced polymer~~.

~~2.2.~~ **2.3.** The fiber-reinforced polymer ~~or the fiberglass reinforced polymer~~ shall have a flame-spread index of 200 or less. The flame-spread index requirements do not apply to ~~shall not be required for~~ coatings or paints having a thickness of less than 0.036 inch (0.9 mm) that are applied directly to the surface of the fiber-reinforced polymer ~~or the fiberglass reinforced polymer~~.

~~2.3.~~ **2.4.** Fireblocking complying with Section 718.2.6 shall be installed.

~~2.4. The fiber reinforced polymer or the fiberglass reinforced polymer shall be designed for uniform live loads as required in Table 1607.1 as well as for snow loads, wind loads and earthquake loads as specified in Sections 1608, 1609 and 1613, respectively.~~

CHANGE SIGNIFICANCE: There have been a number of revisions related to the regulation of fiber reinforced polymers used in building construction. First, the definition and the phrasing within the code text to "fiberglass-reinforced polymer" has been deleted. This was somewhat of an editorial change because "fiberglass-reinforced polymer" is simply a type of "fiber-reinforced polymer" (FRP). In addition, the two terms always appeared together, and the provisions of Section 2612 contained identical regulations for polymers reinforced with glass as well as for other types of fibers. Another editorial change can be seen in the deletion of Section 2612.4 and the revision to Section 2612.3. Because Chapter 8 regulates interior finishes and includes provisions for decorative materials and trim, it was considered redundant to have a separate section addressing the different ways the material could be used but then directing the code users back to the same location of Chapter 8.

The revision to the definition regarding the use of cores between the FRP facing is in recognition that these materials are sometimes used in the construction of sandwich panels. These panels would have the fiber-reinforced polymer facing on both sides of a core material. The core material provides additional support to the FRP material. The core can be a variety of materials with balsa wood, plywood, foam plastics, or other products being fairly typical. The new Section 2612.3.1 will serve as a reminder that when using a foam plastic core, the material is expected to comply with the provisions of Chapter 8 as well as those of Chapter 26.

The most significant technical change is found within Items 1.1 and 1.2 of Section 2612.5, Exception 1. Previously, the maximum amount of FRP permitted to be used on exterior walls varied depending on the flame-spread-index of the material. If a Class A material was used, a maximum of 20 percent of the wall area could be FRP, while a Class B material was limited to only 10 percent. As revised, all fiber-reinforced polymer systems must have a Class A flame-spread index of 25 or less. In addition, while an aggregate 20 percent of the wall may be covered, the material must now be separated so that no single element or area is more than 10 percent of the wall area. This 10 percent maximum requirement will limit the potential for large amounts of fuel loading on any given portion of the exterior wall.

The minimum amount of separation required between the 10 percent areas is not specified, only that individual elements must be within that limit and that separate elements that are "contiguous" must also be less than 10 percent. The designer should consult with the building official to determine what amount of separation is required between individual elements or groups of elements so they are not considered "contiguous."

PART 7

Building Services, Special Devices, and Special Conditions

Chapters 27 through 34

Although building services such as electrical systems (Chapter 27), mechanical systems (Chapter 28), and plumbing systems (Chapter 29) are regulated primarily through separate and distinct codes, a limited number of provisions are set forth in the International Building Code. Chapter 30 regulates elevators and similar conveying systems to a limited degree, as most requirements are found in American Society of Mechanical Engineers standards. The special construction provisions of Chapter 31 include those types of elements or structures that are not conveniently addressed in other portions of the code. By special construction, the code is referring to membrane structures, pedestrian walkways, tunnels, awnings, canopies, marquees, and similar building features that are unregulated elsewhere. Chapter 32 governs the encroachment of structures into the public right-of way, while Chapter 33 addresses safety during construction and the protection of adjacent public and private prop-

erties. Provisions regulating the alteration, repair, addition, and change of occupancy of existing structures are established in Chapter 34. ■

2902.2
Single-User Toilet Facilities

2902.3
Toilet Facilities in Parking Garages

2902.3.5
Locking of Toilet Room Doors

2902.5
Required Drinking Fountains

3007
Fire Service Access Elevators

3008
Occupant Evacuation Elevators

3108
Telecommunication and Broadcast Towers

3302.3, 3303.7, 3313
Fire Safety during Construction

3401.3
Compliance for Existing Buildings

3411
Type B Units in Existing Buildings

CHANGE TYPE: Modification

CHANGE SUMMARY: Where separate sex toilet facilities are required and only one water closet is required in each facility, two family or assisted-use toilet rooms may now be provided as an acceptable alternative.

2012 CODE: 2902.2 Separate Facilities.   Where plumbing fixtures are required, separate facilities shall be provided for each sex.

Exceptions:

1.   Separate facilities shall not be required for dwelling units and sleeping units.

2.   Separate facilities shall not be required in structures or tenant spaces with a total occupant load, including both employees and customers, of 15 or less.

3.   Separate facilities shall not be required in mercantile occupancies in which the maximum occupant load is 5̶0̶ 100 or less.

2902.2.1 Family or Assisted-Use Toilet Facilities Serving as Separate Facilities.   Where a building or tenant space requires a separate toilet facility for each sex and each toilet facility is required to have only one water closet, two family/assisted-use toilet facilities shall be permitted to serve as the required separate facilities. Family or assisted-use toilet facilities shall not be required to be identified for exclusive use by either sex as required by Section 2902.4.

CHANGE SIGNIFICANCE: In most buildings, separate sex toilet rooms are mandated by Chapter 29. Required separate toilet rooms are now permitted to be designated as "family or assisted-use" toilet facilities as an alternative to requiring the facilities to be designated for each sex. Allowing each of the toilet facilities to be used by either sex will provide greater flexibility and increase overall availability. The alternative for family or assisted-use toilet rooms is only applicable where each of the separate sex facilities is only required to have a single water closet.

Historically, if the toilet room for one of the sexes is occupied or being cleaned, then a person needing that facility is typically forced to wait even though there is an empty toilet room available. Two separate facilities must still be provided, but by eliminating the separate-sex designations, both of the toilet rooms will be available to anyone and increase the overall availability in small establishments.

The selection of the term "family or assisted-use" toilet facilities is important because it will influence how the toilet rooms are designed and constructed. Sections 1109.2.1.1 through 1109.2.1.7 provide several details that affect the construction of these toilet rooms. While all of the requirements of these sections are applicable, the main provisions to review are Sections 1109.2.1.1, 1109.2.1.2, 1109.2.1.6, and 1109.2.1.7.

Exception 3 of Section 2902.2 has also been revised regarding when separate sex toilet facilities are required in mercantile occupancies. Separate facilities for each sex are no longer required until the occupant

2902.2
Single-User Toilet Facilities

Single-user restrooms may be used by either women or men.

2902.2 continues

2902.2 continued

Separate restrooms that are restricted based on the sex of the user.

load of the mercantile occupancy exceeds 100 people. Previously, mercantile uses were to be provided with separate facilities for occupant loads greater than 50. Based upon this increased occupant load threshold, a single toilet room is permitted for retail spaces up to about 3,000 square feet based on the occupant load factors specified in Chapter 10. Previously when the occupant load of 50 was used for determining when separate-sex restrooms were needed, retail spaces as small as 1,500 square feet were required to be provided with two toilet rooms. Because Table 2902.1 mandates one separate-sex water closet for each 500 people in a retail space, providing a single toilet room for up to 100 occupants was deemed reasonable.

The increase in the occupant load threshold also distinguishes the toilet room requirements from the egress provisions. As an example, consider a small 2,400-square-foot floral or jewelry shop. Based on Chapter 10, the calculated occupant load is 80, exceeding the 50 occupants that were previously allowed for a single toilet room. Because providing a second toilet room uses a larger percentage of the floor area, there were often requests of the building official that a smaller occupant load as allowed in Section 1004.1.2 be applied simply to reduce the toilet room requirements. With the revision to Exception 3, it is now possible to require a second exit based on the occupant load exceeding 50 and yet not require a second toilet room because the space is unlikely to need it. Therefore, this change significantly differentiates mercantile occupancy plumbing requirements (which are based primarily on convenience) from the egress requirements that are based on life safety needs.

CHANGE TYPE: Modification

CHANGE SUMMARY: Public toilet facilities are no longer required in open and enclosed parking garages and employee toilet facilities are not required in those garages that do not have parking attendants.

2012 CODE: 2902.3 Employee and Public Toilet Facilities. Customers, patrons, and visitors shall be provided with public toilet facilities in structures and tenant spaces intended for public utilization. The number of plumbing fixtures located within the required toilet facilities shall be provided in accordance with Section 2902.1 for all users. Employees shall be provided with toilet facilities in all occupancies. Employee toilet facilities shall either be separate or combined employee and public toilet facilities.

> **Exception:** Public toilet facilities shall not be required in open or enclosed parking garages. Toilet facilities shall not be required in parking garages where there are no parking attendants.

CHANGE SIGNIFICANCE: Public toilet facilities have historically been required in open and enclosed parking garages for the use of "customers, patrons, and visitors" because such garages were regulated in a manner consistent with any other Group S-2 occupancy. A new exception, based on the recognition that parking garages are somewhat unique in their occupancy, permits the omission of public toilet facilities in both open and enclosed parking garages. The allowance will also help resolve difficulties that could arise if an unheated garage was required to have plumbing fixtures and restroom facilities installed within it.

The permissible omission of all toilet facilities is limited to those parking garages that do not have parking attendants. Where the garage is attended, the attendants must have access to employee toilet facilities as required for all other occupancies.

2902.3
Toilet Facilities in Parking Garages

International Code Council®

Public toilet rooms are not required in parking garages.

2902.3.5

Locking of Toilet Room Doors

Restroom door without interior lock

CHANGE TYPE: Addition

CHANGE SUMMARY: In other than family or assisted-use toilet rooms, the door from a toilet room can no longer be lockable from the inside unless the toilet room is a single-user facility.

2012 CODE: <u>**2902.3.5 Door Locking.** Where a toilet room is provided for the use of multiple occupants, the egress door for the room shall not be lockable from the inside of the room. This section does not apply to family or assisted-use toilet rooms.</u>

CHANGE SIGNIFICANCE: Where a toilet room is equipped with multiple fixtures and is intended for the use of several people at the same time, a locking device is now prohibited on the egress door that would allow people within the restroom to lock the door and prevent others from entering. The limitation is intended to ensure that access to a toilet room that is intended to satisfy the sanitation needs of multiple persons cannot be restricted by the users within the toilet room. In addition, a toilet room that could be locked from the inside by the users could increase the potential for occupants to be attacked or could encourage various illegal activities.

It is important to recognize that this restriction does not apply to single-user, family or assisted-use toilet rooms. In fact, Section 1109.2.1.7 requires that doors to family or assisted-use toilet and bathing rooms be securable from within the room.

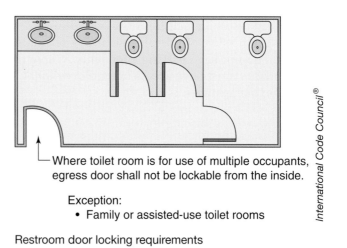

Where toilet room is for use of multiple occupants, egress door shall not be lockable from the inside.

Exception:
• Family or assisted-use toilet rooms

Restroom door locking requirements

CHANGE TYPE: Clarification

CHANGE SUMMARY: Drinking fountains are now allowed to serve multiple tenant spaces provided the fountains are located within the appropriate distances and available for the use of the occupants.

2012 CODE: 2902.5 Drinking Fountain Location. Drinking fountains shall not be required to be located in individual tenant spaces provided that public drinking fountains are located within a travel distance of 500 feet (152 400 mm) of the most remote location in the tenant space and not more than one story above or below the tenant space. Where the tenant space is in a covered or open mall, such distance shall not exceed 300 feet (91 440 mm). Drinking fountains shall be located on an accessible route.

CHANGE SIGNIFICANCE: Similar to the allowances for common-use toilet rooms, in a multi-tenant building it is now permissible to provide separate drinking fountains for each individual tenant space or to provide the drinking fountains in a common area where they may serve multiple tenant spaces. While this option has been fairly well accepted and adequately addressed for the toilet rooms, the allowance was never specifically stated for drinking fountains.

Drinking fountains can either be provided within the individual tenant spaces or in a common area serving multiple tenant spaces and both customers and employees. As long as the number of drinking fountains provided can serve all of the intended occupants that they are designated to serve, drinking fountains in a common area are permitted to serve as the required fixtures for multiple tenants. The restrictions that the code will impose are similar to those found in Section 2902.3 related to toilet facilities. Primarily, they need to be within the specified distances and be available to the occupants they serve by being on an accessible route. The accessible route requirements will coordinate with the provisions for drinking fountains in Section 1109.5 and compare to the requirements for toilet rooms found in Section 2902.3.1.

2902.5
Required Drinking Fountains

Drinking fountains may serve multiple tenant spaces.

3007

Fire Service Access Elevators

CHANGE TYPE: Modification

CHANGE SUMMARY: Many of the provisions addressing fire service access elevators have now been coordinated with those applicable to occupant evacuation elevators to ensure that the fire service access elevators are able to continue to function and serve their intended purpose during an emergency.

2012 CODE:

SECTION 3007
FIRE SERVICE ACCESS ELEVATOR

3007.1 General. Where required by Section 403.6.1, every floor of the building shall be served by fire service access <u>elevators complying with Sections 3007.1 through 3007.10.</u> Except as modified in this section, fire service access elevators shall be installed in accordance with this chapter and ASME A17.1/CSA B44.

<u>**3007.2 Phase I Emergency Recall Operation.**</u> <u>Actuation of any building fire alarm initiating device shall initiate Phase I emergency recall operation on all fire service access elevators in accordance with the</u>

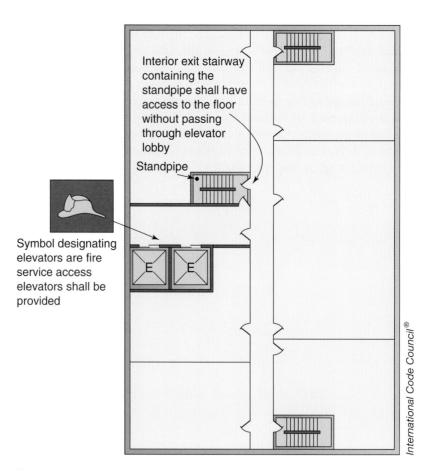

Interior exit stairway containing the standpipe shall have access to the floor without passing through elevator lobby

Standpipe

Symbol designating elevators are fire service access elevators shall be provided

E   E

International Code Council®

Fire service access elevator

requirements in ASME A17.1/CSA B44. All other elevators shall remain in normal service unless Phase I emergency recall operation is manually initiated by a separate, required, three-position, key-operated "Fire Recall" switch or automatically initiated by the associated elevator lobby, hoistway, or elevator machine room smoke detectors. In addition, if the building also contains occupant evacuation elevators in accordance with Section 3008, an independent, three-position, key-operated "Fire Recall" switch conforming to the applicable requirements in ASME A17.1/CSA B44 shall be provided at the designated level for each fire service access elevator.

3007.3 Automatic Sprinkler System. The building shall be equipped throughout by an automatic sprinkler system in accordance with Section 903.3.1.1, except as otherwise permitted by Section 903.3.1.1.1 and as prohibited by Section 3007.3.1.

3007.3.1 Prohibited Locations. Automatic sprinklers shall not be installed in elevator machine rooms, elevator machine spaces, and elevator hoistways of fire service access elevators.

3007.3.2 Sprinkler System Monitoring. The sprinkler system shall have a sprinkler control valve supervisory switch and waterflow-initiating device provided for each floor that is monitored by the building's fire alarm system.

3007.4 Water Protection. An approved method to prevent water from infiltrating into the hoistway enclosure from the operation of the automatic sprinkler system outside the enclosed fire service access elevator lobby shall be provided.

3007.5 Shunt Trip. Means for elevator shutdown in accordance with Section 3006.5 shall not be installed on elevator systems used for fire service access elevators.

~~3007.2~~ 3007.6 Hoistway Enclosures ~~Protection~~. The fire service access elevator hoistway shall be located in a shaft enclosure complying with Section 713.

3007.6.1 Structural Integrity of Hoistway Enclosures. The fire service access elevator hoistway enclosure shall comply with Sections 403.2.3.1 through 403.2.3.4.

~~3007.3~~ 3007.6.2 Hoistway Lighting. When firefighters' emergency operation is active, the entire height of the hoistway shall be illuminated at not less than 1 foot-candle (11 lux) as measured from the top of the car of each fire service access elevator.

~~3007.4~~ 3007.7 Fire Service Access Elevator Lobby. The fire service access elevator shall open into a fire service access elevator lobby in accordance with Sections 3007.7.1 through 3007.7.5.

3007 continues

3007 continued

Exception: Where a fire service access elevator has two entrances onto a floor, the second entrance shall be permitted to open into an elevator lobby in accordance with Section 708.14.1.

~~**3007.4.1**~~ **3007.7.1 Access.**   The fire service access elevator lobby shall have direct access to an ~~exit~~ enclosure <u>for an interior exit stairway</u>.

~~**3007.4.2**~~ **3007.7.2 Lobby Enclosure.**   The fire service access elevator lobby shall be enclosed with a smoke barrier having a ~~minimum 1-hour~~ fire-resistance rating <u>of not less than one hour</u>, except that lobby doorways shall comply with Section 3007.7.3.

Exception: Enclosed fire service access elevator lobbies are not required at the <u>levels of exit discharge</u> ~~street floor~~.

~~**3007.4.3**~~ **3007.7.3 Lobby Doorways.**   <u>Other than the door to the hoist-way,</u> each <u>doorway to a</u> fire service access elevator lobby shall be provided with a ~~doorway that is protected with a~~ ¾-hour fire door assembly complying with Section 716.5. The fire door assembly shall also comply with the smoke and draft control door assembly requirements of Section 716.5.3.1 with the UL 1784 test conducted without the artificial bottom seal.

~~**3007.4.4**~~ **3007.7.4 Lobby Size.**   Each enclosed fire service access elevator lobby shall be ~~a minimum of~~ <u>not less than</u> 150 square feet (14 m²) in area with a minimum dimension of 8 feet (2440 mm).

3007.7.5 Fire Service Access Elevator Symbol.   <u>A pictorial symbol of a standardized design designating which elevators are fire service access elevators shall be installed on each side of the hoistway door frame on the portion of the frame at right angles to the fire service access elevator lobby. The fire service access elevator symbol shall be designed as shown in Figure 3007.7.5 and shall comply with the following:</u>

1. <u>The fire service access elevator symbol shall be not less than 3 inches (76 mm) in height.</u>
2. <u>The vertical centerline of the fire service access elevator symbol shall be centered on the hoistway door frame. Each symbol shall not be less than 78 inches (1981 mm) and not more than 84 (2134 mm) inches above the finished floor at the threshold.</u>

<u>**FIGURE 3007.7.5**</u>
<u>**FIRE SERVICE ACCESS ELEVATOR SYMBOL**</u>

~~**3007.6**~~ **3007.8 Elevator System Monitoring.**   The fire service access elevator shall be continuously monitored at the fire command center by a standard emergency service interface system meeting the requirements of NFPA 72.

~~**3007.7**~~ **3007.9 Electrical Power.**   The following features serving each fire service access elevator shall be supplied by both normal power and Type 60/Class 2/Level 1 standby power:

1. Elevator equipment.
2. Elevator hoistway lighting.

3. Elevator machine room ventilation and cooling equipment.

4. Elevator controller cooling equipment.

3007.7.1 3007.9.1 Protection of Wiring or Cables. Wires or cables that are located outside of the elevator hoistway and machine room and that provide normal or standby power, control signals, communication with the car, lighting, heating, air conditioning, ventilation, and fire-detecting systems to fire service access elevators shall be protected by construction having a ~~minimum 1-hour~~ fire-resistance rating of not less than 2 hours, or shall be circuit integrity cable having a ~~minimum 1-hour~~ fire-resistance rating of not less than 2 hours.

> **Exception:** Wiring and cables to control signals are not required to be protected provided that wiring and cables do not serve Phase II emergency in-car operation.

3007.5 3007.10 Standpipe Hose Connection. A Class I standpipe hose connection in accordance with Section 905 shall be provided in the ~~exit enclosure~~ interior exit stairway and ramp having direct access from the fire service access elevator lobby.

3007.10.1 Access. The exit enclosure containing the standpipe shall have access to the floor without passing through the fire service access elevator lobby.

CHANGE SIGNIFICANCE: In high-rise buildings that are more than 120 feet in height, fire service access elevators must be provided in accordance with Section 3007. In order to ensure that these elevators function as they are intended and are available for the fire department's use in dealing with emergencies, a number of changes have been made to the requirements.

The emergency recall provisions of Section 3007.2 ensure that the elevator can be recalled quickly to the designated level and is therefore available when the responding firefighters arrive on the scene. Inclusion of these provisions in the code helps coordinate with the requirements of the ASME A17.1 elevator standard and provides a standardized method of recalling the elevators for the use of the fire service.

Sprinkler system provisions were added in order to clarify that a sprinkler system is mandated in those high-rise buildings provided with fire service access elevators, however the sprinkler system is prohibited within the associated elevator machine rooms, elevator machine spaces, or hoistways that serve the fire service elevator. Similar to the requirements for occupant evacuation elevators found in Section 3008.3, the provision differs in that the sprinkler system is not to be installed within the hoistway of the fire service access elevator.

The water protection requirements of Section 3007.4 should be viewed as providing performance language that will permit any number of options to prevent water from an operating sprinkler system from finding its way into the elevator hoistway enclosure, including the installation of drains, a sloped floor, or other solutions. Water does cause problems for elevators during a fire, so providing some means of stopping the water from entering the hoistway is important. It should be noted that the

3007 continues

3007 continued requirements are only applicable to sprinklers that are outside of the lobby and are not applicable for sprinklers that are activated within the lobby. The lobby sprinklers are excluded because the elevators will go into fire department recall if there is smoke or a fire within the lobby.

Due to the restrictions of Section 3007.3 regarding the sprinkler system installation and prohibitions, Section 3007.5 will prohibit the installation of the elevator shutdown system (shunt trip) that would typically be required by Section 3006.5. Because Section 3007.3 prohibits the installation of the sprinkler system in the hoistway and machine rooms, the provisions of Section 3006.5 are not necessary.

Section 3007.6.1 dealing with the structural integrity of the hoistway enclosures was added to ensure the shafts protecting the fire service elevator remain intact and are therefore usable by the responding firefighters. This issue of hardening of the hoistway shafts was one of the recommendations that came out of the NIST World Trade Center Report and has been incorporated elsewhere in the IBC.

Section 3007.7.5 provides a means to indicate which elevators in a building are designated as the fire service access elevators. The fire hat symbol that is used is already required inside of the elevator car cab by the requirements of the ASME A17.1 elevator standard and therefore will already be easily recognizable by the fire service.

The wiring protection requirements of Section 3007.9.1 address two separate aspects. The base paragraph has been revised so that any wiring outside of the hoistway or machine room is protected to a level of protection equivalent to the shaft enclosure itself. If the wiring is within the hoistway or machine room, it will inherently have this level of protection. Therefore, the provision only needs to address the supply wiring or feeders that are bringing the power into the shaft to ensure they are protected and can continue to power the elevator for the required time it is needed. The required level of protection for the wires or cables has also been increased from 1 hour to 2 hours. The 2-hour rating was selected because it is consistent with the minimum required fire-resistance rating of the hoistway and the fire pump feeder enclosure rating. This degree of protection helps ensure that the elevator is able to continue to function during the time periods that the fire service needs it. The exception recognizes that elevator landing fixtures that provide control signals such as hall call buttons and hall lanterns do not need to be protected to the same fire-resistance rating, because these signals are not necessary to ensure the viability of the fire service elevator during Phase II operation. It also recognizes that the elevator industry does not generally test the elevator landing fixtures to obtain a fire-resistance rating, therefore, protecting the wiring that serves those fixtures is not necessary.

The standpipe access requirements of Section 3007.10.1 will permit the fire department to connect and advance their attack hose onto the fire floor without opening the door between the elevator lobby and the floor. This additional separation helps to limit the possible spread of smoke from the fire floor into the elevator lobby. By minimizing the potential for smoke to spread into the lobby, as it would if the lobby doors had to be opened to run the hose onto the floor, the fire service elevator will be able to continue its operation and serve as a staging area and supply route for the fire department. If smoke did move into the lobby, it would have the potential to affect the operation of the elevator or cause it to go into recall.

CHANGE TYPE: Modification

CHANGE SUMMARY: The provisions addressing occupant evacuation elevators are now more closely coordinated with those regulating fire service access elevators.

2012 CODE:

3008
Occupant Evacuation Elevators

SECTION 3008
OCCUPANT EVACUATION ELEVATORS

3008.1 General.   Where elevators are to be used for occupant self-evacuation during fires, all passenger elevators for general public use shall comply with ~~this~~ Sections 3008.1 through 3008.11. Where other elevators are used for occupant self-evacuation, they shall also comply with these sections.

~~3008.4~~ 3008.1.1 Additional Exit Stairway.   (no changes to text)

~~3008.2~~ 3008.1.2 Fire Safety and Evacuation Plan.   (no changes to text)

3008.2 Phase I Emergency Recall Operation.   An independent, three-position, key-operated "Fire Recall" switch complying with ASME A17.1/CSA B44 shall be provided at the designated level for each occupant evacuation elevator.

~~3008.3~~ 3008.2.1 Operation.   The occupant evacuation elevators shall be used for occupant self-evacuation only in the normal elevator operating mode prior to Phase I emergency recall operation in accordance with the requirements in ASME A17.1/CSA B44 and the building's fire safety and evacuation plan.

3008 continues

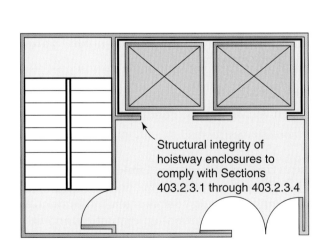

Occupant evacuation elevator

Structural integrity of hoistway enclosures to comply with Sections 403.2.3.1 through 403.2.3.4

Occupant evacuation elevator system activated by any of the following:

- Operation of sprinkler system per Section 3008.3
- Smoke detectors
- Approved manual means

International Code Council®

3008 continued

3008.2.2 Activation. Occupant evacuation elevator systems shall be activated by any of the following:

1. The operation of an automatic sprinkler system complying with Section 3008.3.
2. Smoke detectors required by another provision of the code.
3. Approved manual controls.

3008.10 3008.4 Water Protection. ~~The occupant evacuation elevator hoistway shall be designed utilizing~~ An approved method to prevent water from infiltrating into the hoistway enclosure from the operation of the automatic sprinkler system ~~from infiltrating into the hoistway enclosure.~~ outside the enclosed occupant evacuation elevator lobby shall be provided.

3008.9 3008.6 Hoistway Enclosure Protection. ~~The~~ Occupant evacuation elevators hoistways shall be located in ~~a hoistway~~ shaft enclosure~~(s)~~ complying with Section 713.

3008.6.1 Structural Integrity of Hoistway Enclosures. Occupant evacuation elevator hoistway enclosures shall comply with Section 403.2.3.1 through 403.2.3.4.

3008.11 3008.7 Occupant Evacuation Elevator Lobby. The occupant evacuation elevators shall open into an elevator lobby in accordance with Sections 3008.7.1 through 3008.7.7.

3008.11.3 3008.7.3 Lobby Doorways. Other than the door to the hoistway, each doorway to an occupant evacuation elevator lobby shall be provided with a ~~doorway that is protected with a~~ ¾-hour fire door assembly complying with Section 716.5. The fire door assembly shall also comply with the smoke and draft control assembly requirements of Section 716.5.3.1 with the UL 1784 test conducted without the artificial bottom seal.

3008.15 3008.9 Electrical Power. The following features serving each occupant evacuation elevator shall be supplied by both normal power and Type 60/Class 2/Level 1 standby power:

1. Elevator equipment.
2. Elevator machine room ventilation and cooling equipment.
3. Elevator controller cooling equipment.

3008.15.1 3008.9.1 Protection of Wiring or Cables. Wires or cables that are located outside of the elevator hoistway and machine room and that provide normal or standby power, control signals, communication with the car, lighting, heating, air conditioning, ventilation, and fire-detecting systems to fire service access elevators shall be protected by construction having a ~~minimum 1-hour~~ fire-resistance rating of not less than 2 hours or shall be circuit integrity cable having a ~~minimum 1-hour~~ fire resistance rating of not less than 2 hours.

Exception: Wiring and cables to control signals are not required to be protected provided that wiring and cables do not serve Phase II emergency in-car operation.

~~3008.7~~ 3008.11 ~~High-Hazard Content~~ Hazardous Material Areas. (no changes to text)

CHANGE SIGNIFICANCE: Where elevators are to be used for occupant self-evacuation during a fire, they are required to be constructed and protected in accordance with Section 3008. The revisions that have been made in this section are intended to ensure the elevators are available during emergencies and can safely be used by the occupants on their own.

Section 3008.2 will allow the occupant evacuation elevators (OEEs) to be quickly recalled by the fire department when they enter the building. The required type of switch is specified (independent, three-position, key-operated), as well as compliance with the ASME A17.1 elevator standard, which contains additional information regarding the installation of the control.

The activation requirements of Section 3008.2.2 are intended to provide the means to initiate automatic activation of the OEE system. This type of system is not required but would be permissible to install and is capable of improving the evacuation process. Automatic operation of the OEE permits the elevator system to evaluate which floors to respond to and evacuate first and allows the OEE to function as an express elevator versus stopping at intermediate floors. These provisions need to be considered in conjunction with the operation provisions of Section 3008.2.1, which permit the OEE to be used for occupant self-evacuation "only in the normal elevator operating mode prior to Phase I emergency recall operation." Once the elevator goes into recall, it would only be used for fire department–assisted evacuations, regardless of whether it had been operating in self-evacuation or in an automatic operation mode.

The water protection requirements of Section 3008.4 should be viewed as providing performance language that will permit any number of options to prevent water from an operating sprinkler system from finding its way into the elevator hoistway enclosure, including the installation of drains, a sloped floor, or other solutions. Water causes problems for elevators during a fire, so providing some means of stopping the water from entering the hoistway is important. It is important to note, however, that the requirements are only applicable to sprinklers that are outside of the lobby and are not required for sprinklers that are activated within the lobby. The lobby sprinklers are excluded because the elevators will go into fire department recall if there is smoke or a fire within the lobby.

Section 3008.6.1 dealing with the structural integrity of the hoistway enclosures was added to ensure the shafts protecting the OEEs remain intact and are therefore usable by the occupants. This issue of hardening of the hoistway shafts was one of the recommendations that came out of the NIST World Trade Center Report and has been incorporated into the IBC in other locations. The reference to the provisions in the high-rise section describe how the hoistway enclosure is to be constructed, and this requirement will apply to any OEE hoistway when the high-rise building is more than 420 feet in building height.

3008 continues

3008 continued

The lobby doorway requirements of Section 3008.7.3 correlate with the lobby doorway requirements for fire service access elevators in Section 3007.7.3. The integrity and tenability of elevator lobbies used for occupant evacuation are just as critical as those provided for fire service access. The revision to the first sentence clarifies that the requirement for the rated doors applies to all doors into the lobby, except for the hoistway door.

The wiring protection requirements of Section 3008.9.1 address two separate aspects. The base paragraph has been revised so that any wiring outside of the hoistway or machine room is protected to a level of protection equivalent to that of the shaft enclosure itself. If the wiring is within the hoistway or machine room, it will inherently have this level of protection. Therefore, the provision only needs to address the supply wiring or feeders that are bringing the power into the shaft to ensure they are protected and can continue to power the elevator for the required time it is needed. The required level of protection for the wires or cables has also been increased from 1 hour up to 2 hours. The 2-hour rating was selected because it is consistent with the minimum required fire-resistance rating of the hoistway and the fire pump feeder enclosure rating. This should ensure that the elevator is able to continue to function during the time period needed to evacuate the building. The exception recognizes that elevator landing fixtures that provide control signals such as hall call buttons and hall lanterns do not need to be protected to the same fire-resistance rating because these signals are not necessary to ensure the viability of the OEE.

CHANGE TYPE: Modification

CHANGE SUMMARY: The referenced standard for structural design of antenna supporting towers, TIA 222-G, contains exemptions for seismic design that are not consistent with the requirements of IBC Chapter 16 and ASCE 7, therefore Section 3108.1 has been modified so that the exemptions are not applicable.

2012 CODE: 3108.1 General. Towers shall be designed and constructed in accordance with the provisions of TIA-222. <u>Towers shall be designed for seismic loads. Exceptions related to seismic design listed in Section 2.7.3 of TIA-222 shall not apply. In Section 2.6.6.2 of TIA 222, the horizontal extent of Topographic Category 2, escarpments, shall be 16 times the height of the escarpment.</u>

> **Exception:** Single, freestanding poles used to support antennas not greater than 75 feet (22 860 mm), measured from the top of the pole to grade, shall not be required to be noncombustible.

CHANGE SIGNIFICANCE: The referenced standard for structural design of antenna supporting towers is TIA 222-G, *Structural Standard for Steel Antenna Towers and Antenna Supporting Structures*. The TIA standard contains exemptions from seismic design that are not consistent with the requirement that referenced standards provide the same basic level of public safety as the IBC. Section 11.1.2 of ASCE 7 requires every structure, and portion thereof—including nonstructural components—to be designed and constructed to resist the effects of earthquake ground motions as prescribed by the seismic requirements of the standard. Certain nonbuilding structures, such as communications towers, are also within the scope and therefore must be designed and constructed in accordance with the seismic requirements of the IBC and ASCE 7. Chapter 15 of ASCE 7 contains specific requirements for design of communications towers, and no exemptions are included. Further, because TIA 222-G is not a referenced standard for seismic design in Section 1613, the seismic design requirements of these structures are governed by ASCE 7. Thus, Section 3108.1 in the 2012 IBC precludes the exceptions related to seismic design contained in Section 2.7.3 of TIA 222-G.

3108

Telecommunication and Broadcast Towers

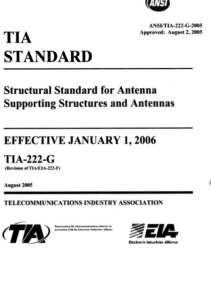

TIA standard for telecommunication and broadcast towers

International Code Council®

3302.3, 3303.7, 3313

Fire Safety during Construction

CHANGE TYPE: Addition

CHANGE SUMMARY: Construction protection requirements of the IFC have been incorporated into the IBC to ensure they are not overlooked.

2012 CODE: 3302.3 Fire Safety during Construction. Fire safety during construction shall comply with the applicable requirements of this code and the applicable provisions of Chapter 33 of the *International Fire Code.*

3303.7 Fire Safety during Demolition. Fire safety during demolition shall comply with the applicable requirements of this code and the applicable provisions of Chapter 33 of the *International Fire Code.*

SECTION 3311
STANDPIPES

~~**3311.4 Water Supply.** Water supply for fire protection, either temporary or permanent, shall be made available as soon as combustible material accumulates.~~

SECTION 3313
WATER SUPPLY FOR FIRE PROTECTION

3313.1 Where Required. An approved water supply for fire protection, either temporary or permanent, shall be made available as soon as combustible material arrives on the site.

CHANGE SIGNIFICANCE: The IBC construction safeguard requirements have now been coordinated with those of the IFC. IFC Chapter 33 has all of the fire safety requirements for fire-safe operations during construction

The IBC includes provisions for fire safety during construction.

International Code Council®

and demolition. Providing a reference to IFC Chapter 33 within the IBC ensures that users are aware of the IFC requirements and do not overlook those important provisions.

The water supply provisions were revised in an effort to correlate Chapter 14 of the *International Existing Building Code* (IEBC), Chapter 33 of the IBC, and the requirements of IFC Chapter 33.

3401.3
Compliance for Existing Buildings

CHANGE TYPE: Modification

CHANGE SUMMARY: It has now been specifically established that the existing building provisions of Chapter 34 are viewed as specific provisions and therefore take precedence over requirements in other codes.

2012 CODE: 3401.3 Compliance.   Alterations, repairs, additions, and changes of occupancy to, or relocation of, existing buildings and structures shall comply with the provisions for alterations, repairs, additions, and changes of occupancy or relocation, respectively, in the *International Fire Code, International Fuel Gas Code, International Mechanical Code, International Plumbing Code, International Property Maintenance Code, International Private Sewage Disposal Code, International Residential Code,* and NFPA 70. Where provisions of the other codes conflict with provisions of this chapter, the provisions of this chapter shall take precedence.

CHANGE SIGNIFICANCE: Although the requirements of Section 3401.3 and Chapter 1 (Sections 101.4 and 102.4) will generally require compliance with the various other codes that are listed, it has been specifically established that the provisions of Chapter 34 take precedence over those other codes and their provisions where there is a conflict. Conceptually, the provision is the equivalent of Section 102.1, which states that "where there is a conflict between a general requirement and a specific requirement, the specific requirement shall be applicable." Because Chapter 34 provides specific requirements related to code

Chapter 34 takes precedence if there are conflicts with other codes.

International Code Council®

requirements for existing buildings, these requirements should be viewed as taking precedence over the general provisions of the other documents.

The addition of the phrase "or relocation" at two locations helps support the fact the IBC is applicable to moved structures. The additional text coordinates the provisions of this section with other provisions within Chapter 34. Section 3410 of the code requires that structures that are moved into or within the jurisdiction are regulated and are required to comply with the requirements for new structures.

3411

Type B Units in Existing Buildings

CHANGE TYPE: Modification

CHANGE SUMMARY: Type B units are now required in existing buildings when there is a change of occupancy or an alteration and more than 50 percent of the building is affected.

2012 CODE: 3411.1 Scope. The provisions of Sections 3411.1 through 3411.9 apply to maintenance, change of occupancy, additions, and alterations to existing buildings, including those identified as historic buildings.

> **Exception:** ~~Type B dwelling or sleeping units required by Section 1107 of this code are not required to be provided in existing buildings and facilities being altered or undergoing a change of occupancy.~~

3411.4 Change of Occupancy. Existing buildings that undergo a change of group or occupancy shall comply with this section.

> **Exception:** <u>Type B dwelling or sleeping units required by Section 1107 of this code are not required to be provided in existing buildings and facilities undergoing a change of occupancy in conjunction with alterations where the work area is 50 percent or less of the aggregate area of the building.</u>

3411.4.1 Partial Change in Occupancy. Where a portion of the building is changed to a new occupancy classification, any alterations shall comply with Sections 3411.6, 3411.7, and 3411.8.

Type B dwelling units may be required in alterations of existing buildings.

International Code Council®

3411.4.2 Complete Change of Occupancy. Where an entire building undergoes a change of occupancy, it shall comply with Section 3411.4.1 and shall have all of the following accessible features:

1. At least one accessible building entrance.
2. At least one accessible route from an accessible building entrance to primary function areas.
3. Signage complying with Section 1110.
4. Accessible parking, where parking is being provided.
5. At least one accessible passenger loading zone, when loading zones are provided.
6. At least one accessible route connecting accessible parking and accessible passenger loading zones to an accessible entrance.

Where it is technically infeasible to comply with the new construction standards for any of these requirements for a change of group or occupancy, the above items shall conform to the requirements to the maximum extent technically feasible.

> **Exception:** The accessible features listed in Items 1 through 6 are not required for an accessible route to Type B units.

3411.6 Alterations. A ~~building~~ facility ~~or element~~ that is altered shall comply with the applicable provisions in Chapter 11 of this code ~~and ICC A117.1~~, unless technically infeasible. Where compliance with this section is technically infeasible, the alteration shall provide access to the maximum extent technically feasible.

Exceptions:
1. The altered element or space is not required to be on an accessible route, unless required by Section 3411.7.
2. Accessible means of egress required by Chapter 10 are not required to be provided in existing facilities.
3. The alteration to Type A individually owned dwelling units within a Group R-2 occupancy shall be permitted to meet the provision for a Type B dwelling unit ~~and shall comply with the applicable provisions in Chapter 11 and ICC A117.1~~.
4. Type B dwelling or sleeping units required by Section 1107 of this code are not required to be provided in existing buildings and facilities undergoing a change of occupancy in conjunction with alterations where the work area is 50 percent or less of the aggregate area of the building.

3411.7 Alterations Affecting an Area Containing a Primary Function. Where an alteration affects the accessibility to, or contains an area of primary function, the route to the primary function area shall be accessible. The accessible route to the primary function area shall include toilet facilities or drinking fountains serving the area of primary function.

3411 continues

3411 continued

Exceptions:

1. The costs of providing the accessible route are not required to exceed 20 percent of the costs of the alterations affecting the area of primary function.

2. This provision does not apply to alterations limited solely to windows, hardware, operating controls, electrical outlets, and signs.

3. This provision does not apply to alterations limited solely to mechanical systems, electrical systems, installation or alteration of fire protection systems, and abatement of hazardous materials.

4. This provision does not apply to alterations undertaken for the primary purpose of increasing the accessibility of ~~an existing building,~~ a facility ~~or element~~.

5. This provision does not apply to altered areas limited to Type B dwelling and sleeping units.

3411.8 Scoping for Alterations. The provisions of Sections 3411.8.1 through 3411.8.14 shall apply to alterations to existing buildings and facilities.

3411.8.7 Accessible Dwelling or Sleeping Units. Where Group I-1, I-2, I-3, R-1, R-2, or R-4 dwelling or sleeping units are being altered or added, the requirements of Section 1107 for accessible units apply only to the quantity of spaces being altered or added.

3411.8.8 Type A Dwelling or Sleeping Units. Where more than 20 Group R-2 dwelling or sleeping units are being altered or added, the requirements of Section 1107 for Type A units apply only to the quantity of the spaces being altered or added.

3411.8.9 Type B Dwelling or Sleeping Units. Where four or more Group I-1, I-2, R-1, R-2, R-3, or R-4 dwelling or sleeping units are being added, the requirements of Section 1107 for Type B units apply only to the quantity of the spaces being added. Where Group I-1, I-2, R-1, R-2, R-3, or R-4 dwelling or sleeping units are being altered and where the work area is greater than 50 percent of the aggregate area of the building, the requirements of Section 1107 for Type B units apply only to the quantity of the spaces being altered.

3411.9 Historic Buildings. These provisions shall apply to ~~buildings or~~ facilities designated as historic structures that undergo alterations or a change of occupancy, unless technically infeasible. Where compliance with the requirements for accessible routes, entrances or toilet ~~facilities~~ rooms would threaten or destroy the historic significance of the ~~building or~~ facility, as determined by the applicable governing authority, the alternative requirements of Sections 3411.9.1 through 3411.9.4 for that element shall be permitted.

Exception: Type B dwelling or sleeping units required by Section 1107 of the *International Building Code* are not required to be provided in historical buildings.

CHANGE SIGNIFICANCE: Type B dwelling and sleeping units must now be provided when an existing building undergoes a change of occupancy or is altered and more than 50 percent of the building is affected. Currently, the federal Fair Housing Act (FHAct) does not require that fair-housing-type units be provided in existing buildings. The inclusion of provisions addressing Type B units in the IBC and ICC A117.1 standard has resulted in requirements that meet or exceed the fair-housing unit requirements.

It is important to understand the requirements of the Fair Housing Act in order to fully understand the significance of these changes. The type of uses covered under the FHAct include apartments, condominiums, dormitories, fraternities, sororities, convents, monasteries, assisted living facilities, nursing homes, group homes, etc. The FHAct is applicable to these residential-type buildings if they were first occupied after March 13, 1991. However, if a building was constructed for some use other than housing, then a change of occupancy to one of the covered housing uses occurs, the fair-housing provisions do not require construction of the fair-housing units even if the change of use or alteration occurred after the March 1991 date. On the other hand, if the building was not properly constructed when it was built, then the building would require alterations to bring it into compliance. Such a noncomplying building would not have been able to use the existing building exception that previously was in the IBC.

When a building now undergoes what the IEBC would consider a Level 3 alteration, or a change of occupancy that includes a Level 3 alteration, whatever elements are altered must be brought up to comply with the Type B unit requirements. If the element is not part of the alteration, it is not required to be changed or brought into compliance with the new requirement. Because the IBC does not include a definition for "Level 3 alterations," the IBC provisions are based on the terminology "work area exceeds 50 percent of the aggregate area of the building," which matches the definition found in IEBC Section 405.1.

When a change of use or major alteration is being done, a prime opportunity becomes available to have an existing building approach compliance. The improvements will not only be a benefit for people that need that housing to live in, but will also help the building owners lessen or avoid complaints filed under the FHAct. According to the latest U.S. Census information, 41 percent of people 65 years of age and older have some level of disability. With the fastest-growing group in the United States being people over 65 years of age, there will be an increasing need for housing that meets these very basic levels of accessibility.

The following is a brief review of the changes on a section-by-section basis.

The change of occupancy and alteration requirements of Section 3411 now generally require Type B units to be provided within existing buildings that undergo any type of a change. Previously, the exception to Section 3411.1 provided a blanket exemption for any requirement related to Type B units within an existing building. However, a new exception to Section 3411.4 is applicable to buildings that undergo a change of occupancy. In this situation, if less than 50 percent of the building is altered, then Type B units are not required. As mentioned previously, this 50 percent requirement is based on the IEBC's "Level 3" alteration requirements.

3411 continues

3411 continued

When the building undergoes a full change of occupancy, a new exception in Section 3411.4.2 eliminates the need to provide the listed accessible features for the Type B units. In general, these six listed items would typically be required when any building underwent a complete change of occupancy. This exception recognizes that some of these elements may be difficult to make accessible, and it is important to remember that this will still meet or exceed the federal law. Therefore, granting the use of this exception does not conflict with any provision of the Fair Housing Act. In addition, this exception helps to recognize that the site impracticality provisions of Section 1107.7.4 may be applicable.

The new exception in Section 3411.6 for alterations is very similar to the exception that was added into Section 3411.4 and discussed earlier. The difference is simply that this section applies to building alterations, while the previous section applied to alterations that are undertaken as a part of a change of occupancy. This exception will exempt buildings that have less than 50 percent of the building area altered from needing to be provided with Type B units.

The primary function area requirements of Section 3411.7 also contain a new exception that exempts the requirements from application to Type B units. Philosophically, because the other alteration requirements already address the Type B units in a variety of ways, this requirement for the accessible route to be made accessible was deemed unnecessary. Where small projects are undertaken, it is necessary that certain elements are provided on an accessible route serving the space. Here it is important to note that the FHAct does not impose this requirement and that the IBC will now address buildings that undergo a substantial alteration (greater than 50 percent of the building area)

Previously, Sections 3411.8.8 and 3411.8.9 only applied when a certain number of units were added to an existing building. Now when 20 or more units are "*altered or* added," then 2 percent, but not less than one, of the units being altered or added are required to be Type A units. The added text within Section 3411.8.9 dealing with the Type B units will also address altered units, but it does come with the 50 percent of the building area limitation that was discussed previously. Historically, under both the Fair Housing Act and the IBC, Type B units were only required in existing buildings when four or more units were added.

It is important that code users are aware of other sections and how they may affect the changes that were discussed earlier. First, the requirement of Section 3411.3 should be noted. Section 1107.7 and other provisions of the code can reduce the number of Type A and Type B units that may be required. For example, in a building without an elevator, Section 1107.7.1 would typically only require the accessible ground level to have accessible units. Other levels would not require the Type A or Type B units unless there were no units on the ground level, the building had an elevator, or if there was a sloping site and multiple levels were accessible from grade. It is intended that the exceptions for nonelevator buildings, site limitations, and flood zones currently indicated in Section 1107.7 are still applicable and will possibly limit the application of Section 3411. In addition, the provisions of Section 1107.7 and some of the exceptions in Section 3411 help reinforce the intent that these new provisions for existing buildings are not meant to require elevators when alterations are performed on upper floors in nonelevator buildings.

Appendix L
Earthquake-Recording Instruments

CHANGE TYPE: Addition

CHANGE SUMMARY: A new appendix requires earthquake-recording instruments to be installed in certain buildings located where the 1-second spectral response acceleration, S_1, is greater than 0.40.

2012 CODE: **1613.8 Earthquake-Recording Instrumentations.** For earthquake-recording instrumentations, see Appendix L.

APPENDIX L
EARTHQUAKE-RECORDING INSTRUMENTATION

SECTION L101
GENERAL

L101.1 General. Every structure located where the 1-second spectral response acceleration, S_1, in accordance with Section 1613.5 is greater than 0.40 that either (1) exceeds six stories in height with an aggregate floor area of 60,000 square feet (5574 m^2) or more or (2) exceeds ten stories in height regardless of floor area shall be equipped with not less than three approved recording accelerographs. The accelerographs shall be interconnected for common start and common timing.

L101.2 Location. As a minimum, instruments shall be located at the lowest level, midheight, and near the top of the structure. Each instrument shall be located so that access is maintained at all times and is unobstructed by room contents. A sign stating "MAINTAIN CLEAR ACCESS TO THIS INSTRUMENT" in one-inch block letters shall be posted in a conspicuous location.

Appendix L continues

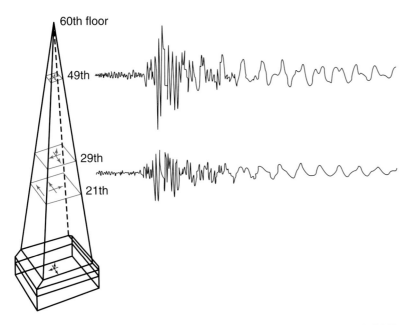

San Francisco transamerica tower *Courtesy of U.S. Geological Survey (USGS)*

Appendix L continued

L101.3 Maintenance.   Maintenance and service of the instrumentation shall be provided by the owner of the structure. Data produced by the instrument shall be made available to the building official on request.

Maintenance and service of the instruments shall be performed annually by an approved testing agency. The owner shall file with the building official a written report from an approved testing agency, certifying that each instrument has been serviced and is in proper working condition. This report shall be submitted when the instruments are installed and annually thereafter. Each instrument shall have affixed to it an externally visible tag specifying the date of the last maintenance or service and the printed name and address of the testing agency.

CHANGE SIGNIFICANCE: Earthquake-recording instrumentation measurements provide fundamental information needed to cost effectively improve understanding of the seismic response and performance of buildings subjected to earthquake ground motions. The language of the new provision in the IBC requiring earthquake-recording instrumentation originated with the 1997 Uniform Building Code (UBC). When the 2000 IBC was initially developed, this provision was inadvertently left out. The requirement only applies to newly constructed buildings of a specified size and located where the 1-second spectral response acceleration, S_1, is greater than 0.40. Because the provision is in an appendix chapter, it is not mandatory unless specifically adopted by the jurisdiction.

Appendix M
Tsunami-Generated Flood Hazards

CHANGE TYPE: Addition

CHANGE SUMMARY: A new appendix provides requirements for coastal communities that have a potential for being inundated by the effects of tsunami waves.

2012 CODE:

APPENDIX M
TSUNAMI-GENERATED FLOOD HAZARD

SECTION M101
TSUNAMI-GENERATED FLOOD HAZARD

M101.1 General. The purpose of this appendix is to provide tsunami regulatory criteria for those communities that have a tsunami hazard and have elected to develop and adopt a map of their tsunami hazard inundation zone.

M101.2 Definitions. The following words and terms shall, for the purposes of this appendix, have the meanings shown herein.

TSUNAMI HAZARD ZONE MAP. A map adopted by the community that designates the extent of inundation by a design event tsunami. This map shall be based on the tsunami inundation map, which is developed and provided to a community by either the applicable State agency or the

Appendix M continues

Tsunami hazard zone sign *(Courtesy of UNESCO)*

Appendix M continued   National Atmospheric and Oceanic Administration (NOAA) under the National Tsunami Hazard Mitigation Program, but the map shall be permitted to utilize a different probability or hazard level.

TSUNAMI HAZARD ZONE.   The area vulnerable to being flooded or inundated by a design event tsunami as identified on a community's Tsunami Hazard Zone Map.

M101.3 Establishment of Tsunami Hazard Zone.   Where a community has adopted a Tsunami Hazard Inundation Map, that map shall be used to establish a community's Tsunami Hazard Zone.

M101.4 Construction within the Tsunami Hazard Zone.   Buildings and structures designated Occupancy Category III or IV in accordance with Section 1604.5 shall be prohibited within a Tsunami Hazard Zone.

Exceptions:

1. A vertical evacuation tsunami refuge shall be permitted to be located in a Tsunami Hazard Zone provided it is constructed in accordance with FEMA P646.

2. Community critical facilities shall be permitted to be located within the Tsunami Hazard Zone when such a location is necessary to fulfill their function, providing suitable structural and emergency evacuation measures have been incorporated.

SECTION M102
REFERENCED STANDARDS

FEMA P646—08.   *Guidelines for Design of Structures for Vertical Evacuation from Tsunamis M101.4*

CHANGE SIGNIFICANCE: The areas designated on State or National Oceanic and Atmospheric Administration (NOAA) Tsunami Hazard Inundation Maps are most likely to suffer significant damage during a design tsunami event. Given the potentially serious life-safety risk presented to structures within these areas, the intent of the new requirement is to limit the presence of high-hazard and high-occupancy structures (Risk Categories III and IV) within the designated Tsunami Hazard Zone. Buildings within the designated hazard zone are only permitted under certain conditions. A vertical evacuation tsunami refuge is permitted when constructed in accordance with FEMA P646 or where critical facilities are located within the hazard zone to fulfill their function and they incorporate adequate structural and emergency evacuation features. Vertical evacuation is a central part of the National Tsunami Hazard Mitigation Program, driven by the fact that there are coastal communities along the West Coast of the United States that are vulnerable to tsunamis that could be generated within minutes of an earthquake on the Cascadia Subduction Zone. Vertical evacuation structures provide a means to create areas of refuge for communities in which evacuation out of the inundation zone is not feasible. The referenced FEMA guide includes information to assist in the planning and design of tsunami vertical evacuation structures. Because the provision is in an appendix chapter, it is not mandatory unless specifically adopted by the jurisdiction.

Index

Don't Miss Out On Valuable ICC Membership Benefits. Join ICC Today!

Join the largest and most respected building code and safety organization. As an official member of the International Code Council®, these great ICC® benefits are at your fingertips.

EXCLUSIVE MEMBER DISCOUNTS

ICC members enjoy exclusive discounts on codes, technical publications, seminars, plan reviews, educational materials, videos, and other products and services.

TECHNICAL SUPPORT

ICC members get expert code support services, opinions, and technical assistance from experienced engineers and architects, backed by the world's leading repository of code publications.

FREE CODE—LATEST EDITION

Most new individual members receive a free code from the latest edition of the International Codes®. New corporate and governmental members receive one set of major International Codes (Building, Residential, Fire, Fuel Gas, Mechanical, Plumbing, Private Sewage Disposal).

FREE CODE MONOGRAPHS

Code monographs and other materials on proposed International Code revisions are provided free to ICC members upon request.

PROFESSIONAL DEVELOPMENT

Receive Member Discounts for on-site training, institutes, symposiums, audio virtual seminars, and on-line training! ICC delivers educational programs that enable members to transition to the I-Codes®, interpret and enforce codes, perform plan reviews, design and build safe structures, and perform administrative functions more effectively and with greater efficiency. Members also enjoy special educational offerings that provide a forum to learn about and discuss current and emerging issues that affect the building industry.

ENHANCE YOUR CAREER

ICC keeps you current on the latest building codes, methods, and materials. Our conferences, job postings, and educational programs can also help you advance your career.

CODE NEWS

ICC members have the inside track for code news and industry updates via e-mails, newsletters, conferences, chapter meetings, networking, and the ICC website (www.iccsafe.org). Obtain code opinions, reports, adoption updates, and more. Without exception, ICC is your number one source for the very latest code and safety standards information.

MEMBER RECOGNITION

Improve your standing and prestige among your peers. ICC member cards, wall certificates, and logo decals identify your commitment to the community and to the safety of people worldwide.

ICC NETWORKING

Take advantage of exciting new opportunities to network with colleagues, future employers, potential business partners, industry experts, and more than 50,000 ICC members. ICC also has over 300 chapters across North America and around the globe to help you stay informed on local events, to consult with other professionals, and to enhance your reputation in the local community.

JOIN NOW! 1-888-422-7233, x33804 | www.iccsafe.org/membership

INTERNATIONAL CODE COUNCIL®

People Helping People Build a Safer World™

09-01530

Most Widely Accepted and Trusted

ICC EVALUATION SERVICE

The leader in evaluating building products for code compliance & sustainable attributes

- Close to a century of experience in building product evaluation
- Most trusted and most widely accepted evaluation body
- In-house staff of licensed engineers
- Leader in innovative product evaluations
- Direct links to reports within code documents for easy access

Three signature programs:

1. ICC-ES Evaluation Report Program
2. Plumbing, Mechanical and Fuel Gas (PMG) Listing Program
3. Sustainable Attributes Verification and Evaluation (SAVE) Program

ICC-ES is a subsidiary of the International Code Council (ICC), the developer of the International Codes. Together, ICC-ES and ICC incorporate proven techniques and practices into codes and related products and services that foster safe and sustainable design and construction.

For more information, please contact us at es@icc-es.org or call 1.800.423.6587 (x42237).

www.icc-es.org

10-03908

INTERNATIONAL CODE COUNCIL®

People Helping People Build a Safer World™

When it comes to code education, ICC has you covered.

ICC publishes building safety, fire prevention and energy efficiency codes that are used in the construction of residential and commercial buildings. Most U.S. cities, counties, and states choose the I-Codes based on their outstanding quality.

ICC also offers the highest quality training resources and tools to properly apply the codes.

TRAINING RESOURCES

- **Customized Training:** Training programs tailored to your specific needs.
- **Institutes:** Explore current and emerging issues with like-minded professionals.
- **ICC Campus Online:** Online courses designed to provide convenience in the learning process.
- **Webinars:** Training delivered online by code experts.
- **ICC Training Courses:** On-site courses taught by experts in their field at select locations and times.

TRAINING TOOLS

- **Online Certification Renewal Update Courses:** Need to maintain your ICC certification? We've got you covered.
- **Training Materials:** ICC has the highest quality publications, videos and other materials.

For more information on ICC training, visit http://www.iccsafe.org/Education or call 1-888-422-7233, ext. 33818.

10-04077

INTERNATIONAL CODE COUNCIL

People Helping People Build a Safer World™

Dedicated to the Support of Building Safety and Sustainability Professionals

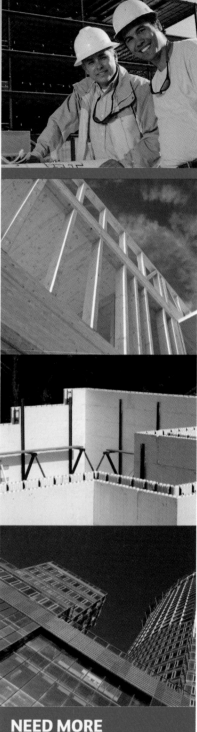

An Overview of the International Code Council

The International Code Council (ICC) is a membership association dedicated to building safety, fire prevention and sustainability in the design and construction of residential and commercial buildings, including homes and schools. Most U.S. cities, counties, states and U.S. territories, and a growing list of international bodies, that adopt building safety codes use ones developed by the International Code Council.

Services of the ICC

The organizations that comprise the International Code Council offer unmatched technical, educational and informational products and services in support of the International Codes, with more than 250 highly qualified staff members at 16 offices throughout the United States, Latin America and the Middle East. Some of the products and services readily available to code users include:

- **CODE APPLICATION ASSISTANCE**
- **EDUCATIONAL PROGRAMS**
- **CERTIFICATION PROGRAMS**
- **TECHNICAL HANDBOOKS AND WORKBOOKS**
- **PLAN REVIEW SERVICES**
- **CODE COMPLIANCE EVALUATION SERVICES**
- **ELECTRONIC PRODUCTS**
- **MONTHLY ONLINE MAGAZINES AND NEWSLETTERS**

- **PUBLICATION OF PROPOSED CODE CHANGES**
- **TRAINING AND INFORMATIONAL VIDEOS**
- **BUILDING DEPARTMENT ACCREDITATION PROGRAMS**
- **GREEN BUILDING PRODUCTS AND SERVICES INCLUDING PRODUCT SUSTAINABILITY TESTING**

The ICC family of non-profit organizations include:

ICC EVALUATION SERVICE (ICC-ES)

ICC-ES is the United States' leader in evaluating building products for compliance with code. A nonprofit, public-benefit corporation, ICC-ES does technical evaluations of building products, components, methods, and materials.

ICC FOUNDATION (ICCF)

ICCF is dedicated to consumer education initiatives, professional development programs to support code officials and community service projects that result in safer, more sustainable buildings and homes.

INTERNATIONAL ACCREDITATION SERVICE (IAS)

IAS accredits testing and calibration laboratories, inspection agencies, building departments, fabricator inspection programs and IBC special inspection agencies.

NEED MORE INFORMATION? CONTACT ICC TODAY!
1-888-ICC-SAFE
(422-7233)
www.iccsafe.org

10-03430